Proton and Carbon Ion Therapy

IMAGING IN MEDICAL DIAGNOSIS AND THERAPY

William R. Hendee, Series Editor

Proton and Carbon Ion Therapy

Edited by

C.-M. Charlie Ma
Tony Lomax

CRC Press
Taylor & Francis Group
Boca Raton London New York

CRC Press is an imprint of the
Taylor & Francis Group, an **informa** business

A TAYLOR & FRANCIS BOOK

CRC Press
Taylor & Francis Group
6000 Broken Sound Parkway NW, Suite 300
Boca Raton, FL 33487-2742

First issued in paperback 2020

© 2013 by Taylor & Francis Group, LLC
CRC Press is an imprint of Taylor & Francis Group, an Informa business

No claim to original U.S. Government works

Version Date: 20120628

ISBN 13 : 978-0-367-57666-0 (pbk)
ISBN 13 : 978-1-4398-1607-3 (hbk)

Library of Congress Cataloging-in-Publication Data

Proton and carbon ion therapy / editors, C-M Charlie Ma, Tony Lomax.
    p. ; cm. -- (Imaging in medical diagnosis and therapy)
  Includes bibliographical references and index.
  ISBN 978-1-4398-1607-3 (hardcover : alk. paper)
  I. Ma, Chang-Ming Charlie. II. Lomax, Tony. III. Series: Imaging in medical diagnosis and therapy.
  [DNLM: 1. Radiotherapy--methods. 2. Carbon Isotopes--therapeutic use. 3. Protons--therapeutic use. 4. Radiometry--methods. 5. Radiotherapy Dosage. WN 250]

615.8'42--dc23

2012025719

Visit the Taylor & Francis Web site at
http://www.taylorandfrancis.com

and the CRC Press Web site at
http://www.crcpress.com

Contents

SECTION I

SECTION II

SECTION III

Series Preface

Advances in the science and technology of medical imaging and radiation therapy have been more profound and rapid than ever since their inception over a century ago. Furthermore, the disciplines are increasingly cross-linked as imaging methods become more widely used to plan, guide, monitor, and assess treatments in radiation therapy. Today, the technologies of medical imaging and radiation therapy are so complex and computer-driven that it is difficult for the people (physicians and technologists) responsible for their clinical use to know exactly what is happening at the point of care, when a patient is being examined or treated. The people best equipped to understand the technologies and their applications are medical physicists, and these individuals are assuming greater responsibilities in the clinical arena to ensure that what is intended for the patient is actually delivered in a safe and effective manner.

The growing responsibilities of medical physicists in the clinical arenas of medical imaging and radiation therapy, however, are not without their challenges. Most medical physicists are knowledgeable in either radiation therapy or medical imaging and expert in one or a small number of areas within their discipline. They sustain their expertise in these areas by reading scientific articles and attending scientific talks at meetings. In contrast, their responsibilities increasingly extend beyond their specific areas of expertise. To meet these responsibilities, medical physicists periodically must refresh their knowledge of advances in medical imaging or radiation therapy, and they must be prepared to function at the intersection of these two fields. How to accomplish these objectives is a challenge.

At the 2007 Annual Meeting of the American Association of Physicists in Medicine in Minneapolis, this challenge was the topic of conversation during a lunch that was hosted by Taylor & Francis Publishers and involved a group of senior medical physicists (Arthur L. Boyer, Joseph O. Deasy, C.-M. Charlie Ma, Todd A. Pawlicki, Ervin B. Podgorsak, Elke Reitzel, Anthony B. Wolbarst, and Ellen D. Yorke). The conclusion of this discussion was that a book series should be launched under the Taylor & Francis banner, with each volume in the series addressing a rapidly advancing area of medical imaging or radiation therapy of importance to medical physicists. The aim would be for each volume to provide medical physicists with the information needed to understand technologies driving a rapid advance and their applications to safe and effective delivery of patient care.

Each volume in the series is edited by one or more individuals with recognized expertise in the technological area encompassed by the book. The editors are responsible for selecting the authors of individual chapters and ensuring that the chapters are comprehensive and intelligible to someone without such expertise. The enthusiasm of volume editors and chapter authors has been gratifying and reinforces the conclusion of the Minneapolis luncheon that this series of books addresses a major need of medical physicists.

Imaging in Medical Diagnosis and Therapy would not have been possible without the encouragement and support of the series manager, Luna Han of Taylor & Francis Publishers. The editors and authors are indebted to her steady guidance of the entire project.

William Hendee
Series Editor
Rochester, MN

Preface

This book aims at providing a comprehensive and practical guide to proton and carbon ion therapy for a broad range of personnel in medical physics and radiation therapy, including medical physicists, radiation oncologists, dosimetrists, therapists, medical engineers, and trainees in these disciplines. The scope of this book includes physics and radiobiology basics of proton and ion beams, dosimetry methods and radiation measurements, treatment delivery systems, and practical guidance on patient setup, target localization, and treatment planning for clinical proton and carbon ion therapy, as well as detailed reports on the treatment of pediatric cancers, lymphomas, and various other cancers. The aim is to offer a balanced and critical assessment of state-of-the-art proton and carbon ion therapy along with a discussion of major challenges and future outlook.

This book contains 11 chapters. The first chapter provides an overview, and the final chapter describes future challenges. The remaining content is divided into three sections. The first section provides the basic aspects of proton and carbon ion therapy equipment covering accelerators, gantries, and delivery systems. Chapter 2 describes accelerators for proton and carbon ion therapy, including cyclotron, synchrotron, and other acceleration developments (such as dielectric wall accelerators, laser-accelerated particle beams, and new concepts, including cyclinac and compact cyclotron solutions). Chapter 3 describes proton and carbon ion therapy systems, with a focus on their major components, including energy selection systems, beam transport systems, control systems, gantries, and beam delivery systems. Chapter 4 summarizes the requirements for proton and carbon ion therapy facilities covering performance and dosimetric specifications, room and building designs, shielding calculations, and economic considerations. The second section provides useful material on dosimetry, biology, imaging, and treatment planning basics for proton and carbon ion therapy. Chapter 5 provides a detailed review of radiobiology for proton and carbon ion therapy, including Bragg peak and linear energy transfer (LET), relative radiobiological effectiveness (RBE), oxygen enhancement ratio (OER), and fractionation effect and radiobiological modeling. Chapter 6 gives a thorough review of dosimetry for proton and carbon ion therapy covering detectors, measurement techniques, dosimetry protocols, acceptance testing, beam commissioning, clinical quality assurance procedures, and analysis of dosimetry uncertainty. Chapter 7 introduces image-guided proton and carbon ion therapy. It begins with the complex radiation therapy process and then discusses the application of various image methods for patient immobilization, target definition and structure delineation, patient setup and target localization, dosimetry verification and treatment assessment, and analysis of treatment uncertainty. Chapter 8 focuses on treatment planning for proton and carbon ion therapy with detailed descriptions of treatment-planning systems, dose calculation algorithms, basic planning techniques, and advanced techniques for beam scanning, intensity modulation, and moving targets. The third section provides clinical guidelines for proton and carbon ion therapy for the treatment of specific cancers (sites). Chapters 9 and 10 give a detailed review of the current techniques and treatment outcomes of proton and carbon ion therapy for various treatment sites with four major components: 1) treatment protocols; 2) planning techniques; 3) treatment procedure; and 4) clinical results with a comparison to high-energy electron and photon radiotherapy.

Proton and carbon ion therapy has experienced a revolutionary transition over the last 10 years with remarkable advances in accelerator design, treatment delivery, treatment planning, and quality assurance systems. For the next decade, further improvements are expected in accuracy, efficiency, and robustness. More hospital-based, state-of-the-art proton and carbon ion therapy facilities will be built, and increasing numbers of patients will be treated. We attempt to provide a comprehensive monograph on the important aspects of proton and carbon ion therapy that we hope will serve as an introduction for those who are new to this field and a useful resource for those who are already using proton and carbon ion therapy clinically. Contributions dealing with topics that are still in rapid development present our best current knowledge of and guidance for proton and carbon ion therapy. Technologies and clinical practice may change with time, but we expect that the many basic components presented in this book will remain relevant and useful to the radiotherapy community in the future.

C.-M. Charlie Ma

Acknowledgments

First, we thank our patients for their continuous support and the trust they have placed in us. Their faith and courage in fighting their cancers have always inspired us in the search for better beam modalities and treatment techniques to reduce the burden of this dreadful disease on our society.

We acknowledge our dear family members and friends for their unconditional support and understanding during the last two years, especially on many evenings, weekends, and holidays in the preparation of this book.

We are indebted to all of the authors for their tremendous efforts and precious time in contributing to the chapters of this book, especially to those who stepped in to complete the chapters that had failed the initial deadlines on short notice, notably, A. Carabe, A. Nahum, L. Dong, R. Zhu, A. Jensen, M. Münter, A. Nikoghosyan, and J. Debus.

We are also thankful to our colleagues, past and present, for teaching us the basics and practice of proton and carbon ion therapy and for their continuous support and discussions that have inspired us throughout our careers. C.-M. Charlie Ma is grateful to Lorraine Medoro for her proofreading and for reformatting the references.

Finally, we would like to express our gratitude to Bill Hendee, the Series Editor, and Luna Han of Taylor & Francis Publishers. It would be impossible to complete this ambitious project without their consistent support and guidance.

Editors

C.-M. Charlie Ma is affiliated with Fox Chase Cancer Center, Philadelphia, Pennsylvania, where he is a professor, the director of Radiation Physics, and vice chairman of the Department of Radiation Oncology. Dr. Ma earned his Ph.D. and completed his postdoctorate fellowship in medical physics at The Institute of Cancer Research and Royal Marsden Hospital, University of London, London, U.K. He received his B.S. degree in laser instruments from Zhejiang University, Hangzhou, China, and his M.S. degree in precision instruments from the University of Shanghai for Science and Technology, Shanghai, China. He is an internationally recognized expert in the physics of intensity modulated radiation therapy (IMRT) and image-guided radiation therapy (IGRT) and has lead a research team focused on laser-accelerated proton beams for radiation therapy, along with clinical research and developments for advanced radiotherapy treatments. Dr. Ma is an active member of the American Association of Physicists in Medicine, the American College of Medical Physics, the American Society for Therapeutic Radiology and Oncology, the European Society for Therapeutic Radiology and Oncology, and the Canadian Organization of Medical Physicists.

Tony Lomax is the head of medical physics in the Center for Proton Therapy, Paul Scherrer Institute (PSI), Zurich, Switzerland. He obtained his B.Sc. degree in physics and physical electronics from Brighton Polytechnic in the United Kingdom, and his M.Sc. and Ph.D. degrees in medical physics from the University of Aberdeen, Aberdeen, Scotland, U.K. He served as a medical physicist at the Dryburn Hospital, Durham, England, U.K. until 1992, when he joined PSI. His research interests include the medical physics aspects of proton therapy, treatment planning, optimization and robustness analysis for treatment plans, image-guided therapy, and motion management. He serves on the editorial board of the *European Journal of Medical Physics* and is an active member of the American Association of Physicists in Medicine and the European Society for Radiotherapy and Oncology.

Contributors

David A. Bush
Department of Radiation Medicine
Loma Linda University Medical Center
Loma Linda, California

Alejandro Carabe
Department of Radiation Oncology
University of Pennsylvania
Philadelphia, Pennsylvania

Joey P. Cheung
Department of Radiation Physics
The University of Texas MD Anderson Cancer
    Center
Houston, Texas

Jürgen Debus
Department of Radiation Oncology
University Hospital of Heidelberg
Heidelberg, Germany

Lei Dong
Scripps Proton Therapy Center
San Diego, California

Jonathan B. Farr
Division of Radiological Sciences
St. Jude Children's Research Hospital
Memphis, Tennessee

Jacob Flanz
Northeast Proton Therapy Center
Massachusetts General Hospital
Boston, Massachusetts

Oliver Jäkel
Department of Medical Physics
University of Heidelberg
and
Heidelberg Ion Beam Therapy Center and German
    Cancer Research Center (DKFZ)
Heidelberg, Germany

Alexandra D. Jensen
Department of Radiation Oncology
University Hospital of Heidelberg
Heidelberg, Germany

Tony Lomax
Centre for Proton Radiotherapy
Paul Scherrer Institute (PSI)
Villigen, Switzerland

C.-M. Charlie Ma
Department of Radiation Oncology
Fox Chase Cancer Center
Philadelphia, Pennsylvania

Richard L. Maughan
Division of Medical Physics
University of Pennsylvania
Philadelphia, Pennsylvania

M. W. Münter
Department of Radiation Oncology
University Hospital of Heidelberg
Heidelberg, Germany

Anna V. Nikoghosyan
Department of Radiation Oncology
University Hospital of Heidelberg
Heidelberg, Germany

Marco Schippers
Paul Scherrer Institute (PSI)
Villigen, Switzerland

Jerry D. Slater
Department of Radiation Medicine
Loma Linda University Medical Center
Loma Linda, California

Alfred R. Smith
Houston, Texas

X. Ronald Zhu
Department of Radiation Physics
The University of Texas MD Anderson Cancer
  Center
Houston, Texas

1. Introduction to Proton and Carbon Ion Therapy

C.-M. Charlie Ma

1.1 Discovery of Radiation Particles

1.1.1 Discovery of Electrons

In 1869, the German physicist Johann Wilhelm Hittorf saw a glow that was emitted from a cathode when he was investigating electrical conductivity in rarefied gases. In 1876, another German physicist, Eugen Goldstein, found that the rays from this glow cast a shadow, and he called them cathode rays. During the same period, the English chemist and physicist Sir William Crookes showed that the luminescence rays appearing within his cathode ray tube with a high vacuum (Dekosky 1983) carried energy and moved from the cathode to the anode. With a magnetic field, he also demonstrated that the beam behaved as though it were negatively charged (Leicester 1971). He explained these properties by what he called "radiant matter," which was a fourth state of matter, consisting of negatively charged molecules that were being projected with high velocity from the cathode (Zeeman 1907).

In 1890, the German-born British physicist Arthur Schuster tried to determine the charge-to-mass ratio of the ray components by measuring the amount of deflection for a given level of current. His result was not well received, which indicated a value that was more than a thousand times greater than what was expected (Leicester 1971). Six years later, the British physicist J. J. Thomson and his colleagues John S. Townsend and H. A. Wilson (Peters 1995) demonstrated that cathode rays were unique particles rather than waves, atoms, or molecules, as was believed earlier (Thomson 1897). Thomson called these particles "corpuscles," which had perhaps one thousandth of the mass of the least massive ion known, hydrogen (Wilson 1997; Thomson 1897). He showed that their charge-to-mass ratio e/m was independent of the cathode material and the negatively charged particles produced by radioactive materials, heated materials, and illuminated materials were universal (Thomson 1897; Thomson 1906). The Irish physicist George F. Fitzgerald proposed to call these particles *electrons*, which quickly gained universal acceptance (Dekosky 1983).

In 1896, the French physicist Henri Becquerel discovered that some naturally fluorescing minerals emitted radiation without any exposure to an external energy source. Later on, the New Zealand physicist Ernest

Proton and Carbon Ion Therapy Edited by C.-M. Charlie Ma and Tony Lomax © 2013 Taylor & Francis Group, LLC. ISBN: 978-1-4398-1607-3.

Chapter 1

Rutherford discovered that these materials emitted particles. He designated these particles alpha and beta, on the basis of their ability to penetrate matter (Trenn 1976). In 1900, Becquerel proved that the beta rays emitted by radium could be deflected by an electric field and their mass-to-charge ratio was the same as for cathode rays (Becquerel 1900). This evidence strengthened the view that electrons existed as components of atoms (Myers 1976).

1.1.2 Discovery of X-Rays

The discovery of X-rays has been credited to the German physicist Wilhelm Röntgen, because he was the first to systematically study them in 1895 (Filler 2009), although he was not the first to observe their effects. He also gave them the name X-rays, although these are still referred to as Röntgen rays in some languages, e.g., in German, Russian, Japanese, and Dutch.

On November 8, 1895, Wilhelm Röntgen stumbled upon X-rays while experimenting with Lenard and Crookes tubes (Peters 1995). His experimental devices included a fluorescent screen painted with barium platinocyanide and a Crookes tube that he had wrapped in black cardboard so that the visible light from the tube would not interfere. He saw a faint green glow from the screen, about 1 m away. He realized that some invisible rays coming from the tube were passing through the cardboard to make the screen glow. These rays could also pass through books and papers on his desk. He started investigating these unknown rays systematically. He wrote an initial report "On a new kind of ray: A preliminary communication" and, on December 28, 1895, submitted it to the Würzburg's Physical-Medical Society Journal (Stanton 1896). This was the first paper written on X-rays. Röntgen referred to the radiation as X, to indicate that it was an unknown type of radiation. Röntgen discovered its medical use when he made a picture of his wife's hand on a photographic plate formed due to X-rays (see Figure 1.1; Crane 2010). Röntgen received the first Nobel Prize in physics for his discovery.

X-rays were found emanating from Crookes tubes—experimental discharge tubes invented around 1875—by scientists investigating the cathode rays, which were energetic electron beams. Crookes tubes created free electrons by the ionization of the residual air in the tube by a high-DC voltage of anywhere between a few kilovolts and 100 kV. This voltage accelerated the electrons coming from the cathode to a high-enough velocity such that they created X-rays when they struck the

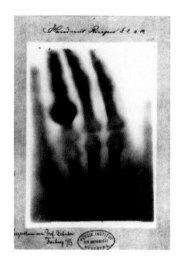

FIGURE 1.1 Hand mit Ringen (hand with rings): Print of Wilhelm Röntgen's first "medical" X-ray of his wife's hand, taken on December 22, 1895, and presented to Ludwig Zehnder of the Physik Institut, University of Freiburg, on January 1, 1896. (Kevles, B. 1996. *Naked to the Bone, Medical Imaging in the Twentieth Century*, Camden, NJ: Rutgers University Press: 19–22.)

anode or the glass wall of the tube. Many of the early Crookes tubes undoubtedly radiated X-rays, because early researchers noticed effects that were attributable to them.

The German physicist Johann Hittorf (1824–1914) found that, when he placed unexposed photographic plates near the tube, some were flawed by shadows, although he did not investigate this event further. In 1877, the Ukrainian-born experimental physicist Pulyui constructed various designs of vacuum discharge tubes to investigate their properties. In 1886, he found that sealed photographic plates became dark when exposed to the emanations from the tubes. Early in 1896, just a few weeks after Röntgen published his first X-ray photograph, Pulyui published high-quality X-ray images in journals in Paris and London (Gaida 1997).

In 1887, Nikola Tesla investigated X-rays using high voltages and tubes of his own design. He developed a special single-electrode X-ray tube based on the Bremsstrahlung process, in which a high-energy secondary X-ray emission is produced when charged particles (such as electrons) pass through matter. By 1892, he performed several such experiments, but he did not categorize the emissions as what were later called X-rays. Tesla generalized the phenomenon as radiant energy of "invisible" kinds (Cheney 2001). His X-ray experimentation by vacuum high-field emissions also led him to alert the scientific community to the biological hazards associated with X-ray exposure (Cheney, Uth, and Glenn 1999).

In 1891, the Stanford University physics professor Fernando Sanford (1854–1948) experimented on X-rays. From 1886 to 1888, he studied in the Hermann Helmholtz Laboratory, Berlin, Germany, where he became familiar with the cathode rays generated in vacuum tubes when a voltage was applied across separate electrodes. His letter on January 6, 1893 (describing his discovery as "electric photography") was published in *The Physical Review* and an article entitled "Without Lens or Light, Photographs Taken with Plate and Object in Darkness" appeared in the San Francisco Examiner (Wyman 2005).

Meanwhile, Philipp Lenard, a student of Heinrich Hertz, built a Crookes tube (later called a "Lenard tube") with a "window" in the end made of thin aluminum, facing the cathode so that the cathode rays would strike it (Thomson 1903). He found that something came through, which would expose photographic plates and cause fluorescence. He measured the penetrating power of these rays through various materials. It was believed that at least some of these "Lenard rays" were actually X-rays.

1.1.3 Discovery of Protons

In 1815, William Prout proposed that all atoms are composed of hydrogen atoms based on a simplistic interpretation of early values of atomic weights (see Prout's hypothesis), which was disproved when more accurate values were measured. The concept of a hydrogen-like particle as a constituent of other atoms was developed over a long period.

In 1886, Eugen Goldstein discovered anode rays and showed that they were positively charged particles (ions) produced from gases. Unlike the negative electrons discovered by J. J. Thomson, however, they could not be identified with a single particle, because particles from different gases had different values of charge-to-mass ratio (e/m).

In 1899, Ernest Rutherford studied the absorption of radioactivity using thin sheets of metal foil and found two forms of radiation: 1) *alpha* (α), which is absorbed by a few thousandths of a centimeter of metal foil, and 2) *beta* (β), which can pass through one hundred times as much foil before it is absorbed. Later, a third form of radiation, called *gamma* (γ), was discovered, which can penetrate as much as several centimeters of lead. These three radiation particles also differ in the way they are affected by electric and magnetic fields.

In 1911, Ernest Rutherford discovered the atomic nucleus. Soon after, Antonius van den Broek suggested that the place of each element in the periodic table (its atomic number) is equal to its nuclear charge. This was confirmed experimentally by Henry Moseley using X-ray spectra in 1913. In 1919, Rutherford reported his experimental results, showing that the hydrogen nucleus is present in other nuclei—a result usually described as the discovery of the proton (Petrucci, Harwood, and Herring 2002). He described that, when he shot alpha particles into air, especially into pure nitrogen gas, his scintillation detectors showed the signatures of hydrogen nuclei. Rutherford determined that this hydrogen could have come only from the nitrogen, and therefore, nitrogen must contain hydrogen nuclei. One hydrogen nucleus was being knocked off by the impact of the alpha particle, producing oxygen-17 in the process. This was the first reported nuclear reaction, $^{14}N + \alpha \rightarrow ^{17}O + p$. Rutherford named the hydrogen nucleus the proton, after the neuter singular of the Greek word for "first" ($\pi\rho\tilde{\omega}\tau o\nu$).

1.1.4 Discovery of Neutrons

Ernest Rutherford was the first to conceptualize the possible existence of the neutron. In 1920, Rutherford proposed that the disparity found between the atomic number of an atom and its atomic mass could be explained by the existence of a neutrally charged particle within the atomic nucleus. He actually believed that the neutron is a neutral double consisting of an electron orbiting a proton. In 1930, Viktor Ambartsumian and Dmitri Ivanenko in the USSR proved that the nucleus cannot consist of protons and electrons; some neutral particles must exist besides the protons (Astrofizika 2008).

In 1931, Walther Bothe and Herbert Becker in Germany discovered that an unusually penetrating radiation was produced when very energetic alpha particles emitted from polonium fell on certain light elements, specifically beryllium, boron, or lithium. This radiation was more penetrating than any gamma rays known, and the details of the experimental results were very difficult to interpret on this basis. A year later, Irène Joliot-Curie and Frédéric Joliot in Paris showed that, if this unknown radiation fell on paraffin or any other hydrogen-containing compound, it ejected protons of very high energy. In 1932, James Chadwick performed a series of experiments at the University of Cambridge, showing that the gamma ray hypothesis was untenable. He suggested that the new radiation consisted of uncharged particles of approximately the mass of the proton, and he performed a series of experiments verifying his suggestion (Chadwick 1932). These uncharged particles were called *neutrons*, apparently from the Latin root for *neutral* and the Greek ending *-on* (by the imitation of *electron* and *proton*). He published his findings with characteristic modesty in a first paper entitled, "Possible Existence of Neutron." In 1935, he received the Nobel Prize for his discovery.

1.2 Early Development of Particle Therapy

1.2.1 First Proposal

In 1930, E. O. Lawrence led the development of the cyclotron at the Lawrence Berkeley National Laboratory (LBL), University of California, Berkeley, and won the Nobel Prize for this work in 1939. Robert Wilson was the first to point out the advantageous dose distributions of protons and their potential for cancer therapy (Wilson 1946). Proton depth dose distributions are characterized by a relatively low dose at the shallow depths, a peak near the end of the proton range, and then a rapid fall-off (Figure 1.2). For protons, α-rays, and other ion rays, the peak occurs immediately before the particles come to rest. This is called the Bragg peak, for William Henry Bragg, who discovered it in 1903. The idea is to deliver a high dose of ionizing radiation to a deep-seated tumor while not exceeding the tolerance dose of the intervening normal tissues, and no dose will be given to normal tissues beyond the tumor. Wilson also proposed several innovative concepts that were subsequently used in the delivery of proton beams in cancer therapy, including the use of range modulation wheels for producing spread-out Bragg peaks (SOBPs) that cover larger targets that can be treated with pristine Bragg peaks (Figure 1.3).

1.2.2 Initial Treatments

The first proton treatment of human patients was carried out by C. A. Tobias, J. H. Lawrence, and others on the LBL cyclotron in the late 1950s. The treatment target was the pituitary gland. They utilized Bragg peak techniques that stopped the Bragg peak in the pituitary target with beams that passed entirely through the brain in a path that intersected the pituitary gland (Lawrence et al. 1958). In 1958, B. Larsson and L. Leksel reported the first use of range modulation to form an SOBP and beam scanning to produce large treatment fields in the lateral dimension. They developed radiosurgical techniques for the treatment of brain tumors using proton beams from the Uppsala cyclotron (Larsson et al. 1958). In 1961, R. Kjellberg, a neurosurgeon at the Massachusetts General Hospital (MGH), and his colleagues treated small intracranial

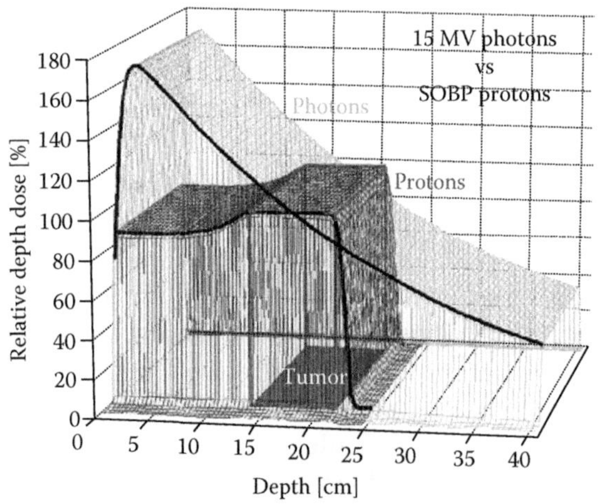

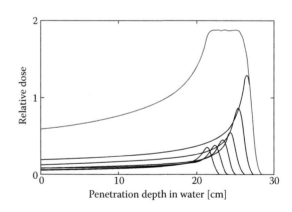

FIGURE 1.2 A comparison of depth doses for 15-MV photons and range/intensity modulated protons of variable energy. The proton SOBP provides a region of high, uniform dose at the tumor target. The dark lines indicate an "ideal" dose distribution that is uniform within the tumor region and zero elsewhere. (Smith, A. 2006. Proton therapy. *Phys. Med. Biol.* 51: R491–R504.)

FIGURE 1.3 Range and intensity modulation of Bragg peaks to achieve an SOBP. SOBPs can be produced by the use of a physical device (ridge filter or modulation wheel) or by energy selection from the accelerator in conjunction with the variable weighting of each individual Bragg peak. SOBPs can be produced for variable widths. (Smith, A. 2006. Proton therapy. *Phys. Med. Biol.* 51: R491–R504.)

targets with radiosurgical techniques at the Harvard Cyclotron Laboratory (HCL), Cambridge, MA (Kjellberg et al. 1962).

1.2.3   Further Particle Therapy Developments

From the late 1960s until 1980, there were significant efforts in the development of proton therapy at several physics research facilities around the world. Proton therapy programs were initiated in Russia at the Joint Institute for Nuclear Research, Dubna, the Moscow Institute for Theoretical and Experimental Physics in 1969, and St. Petersburg in 1975. In 1979, proton treatments started at the National Institute for Radiological Sciences, Chiba, Japan, where a spot scanning system for proton treatment delivery was developed in 1980. Between 1980 and 2000, there was a flurry of activity in proton therapy around the world, with patient treatment starting in Clatterbridge, England, in 1989; Nice and Orsay, France, in 1991; iThemba Labs, Cape Town, South Africa, in 1993, Paul Scherrer Institute (PSI), Villigen, Switzerland,

in 1996; Hahn-Meitner Institute (HMI), Berlin, Germany, in 1998; National Cancer Center (NCC), Kashiwa, Japan, in 1998; and Dubna, Russia in 1999.

In 1975, a heavy ion therapy program began at the LBL, where physicians and medical physicists from the University of California, San Francisco, provided the medical expertise, and high-energy and accelerator physicists from LBL provided the beam delivery technology and software (Lyman et al. 1979). Pioneering work in heavy ion therapy was performed at LBL, where 2054 patients were treated with helium ions between 1957 and 1992, and then, the program was terminated (Castro 1995). Between 1975 and 1992, 433 patients were treated with heavier ions (Ne, N, O, C, Si, and Ar). Currently, heavy ion therapy is performed at three centers: 1) Heavy Ion Medical Accelerator in Chiba (HIMAC), Chiba, Japan; 2) Hyogo Ion Beam Medical Center (HIBMC), Hyogo, Japan; and 3) Gesellschaft für Schwerionenforschung (GSI; now GSI Helmholtz Centre for Heavy Ion Research GmbH), Darmstadt, Germany (Tsujii et al. 2007; Torikoshi et al. 2007; Hishikawa et al. 2004). These centers all

Table 1.1   Particle Therapy Facilities Out of Operation (Including Patient Statistics)

Institution/City	Country	Particle	First Treatment	Last Treatment	Patient Total	Comments
Louvain-la-Neuve	Belgium	p	1991	1993	21	Ocular tumors only
Vancouver (TRIUMF)	Canada	π−	1979	1994	367	Ocular tumors only
Darmstadt (GSI)	Germany	Ion	1997	2009	440	
Tsukuba (PMRC)	Japan	p	1983	2000	700	
Chiba	Japan	p	1979	2002	145	Ocular tumors only
WERC	Japan	p	2002	2009	62	
Dubna (1)	Russia	p	1967	1996	124	
Uppsala (1)	Sweden	p	1957	1976	73	
Villigen PSI (SIN-Piotron)	Switzerland	π−	1980	1993	503	
Villigen PSI (OPTIS 1)	Switzerland	p	1984	2010	5458	Ocular tumors only
Berkeley 184	USA	p	1954	1957	30	
Berkeley	USA	He	1957	1992	2054	
Berkeley	USA	Ion	1975	1992	433	
Bloomington (MPRI)	USA	p	1993	1999	34	Ocular tumors only
Harvard	USA	p	1961	2002	9116	
Los Alamos	USA	π−	1974	1982	230	

Table 1.2 Particle Therapy Facilities in Operation (Including Patient Statistics)

Institution/City	Country	Particle	S/Cᵃ. Max. Energy (MeV)	Beam Direction	First Treatment	Patient Total	Date of Total
ITEP, Moscow	Russia	p	S 250	1 horizontal	1969	4246	December 2010
St. Petersburg	Russia	p	S 1000	1 horizontal	1975	1362	December 2010
PSI, Villigen	Switzerland	pᵇ	C 250	1 gantry, 1 horizontal	1996	772	December 2010
Dubna	Russia	p	C 200ᵈ	horizontal	1999	720	December 2010
Uppsala	Sweden	p	C 200	1 horizontal	1989	1000	December 2010
Clatterbridge	U.K.	p	C 62	1 horizontal	1989	2021	December 2010
Loma Linda	CA, U.S.	p	S 250	3 gantry, 1 horizontal	1990	15,000	January 2011
Nice	France	p	C 65	1 horizontal	1991	4209	December 2010
Orsay	France	pᵉ	C 230	1 gantry, 1 horizontal	1991	5216	December 2010
iThemba Labs	South Africa	p	C 200	1 horizontal	1993	511	December 2009
IU Health PTC, Bloomington	IN, U.S.	p	C 200	2 gantry, 1 horizontal	2004	1145	December 2010
UCSF	CA, U.S.	p	C 60	1 horizontal	1994	1285	December 2010
HIMAC, Chiba	Japan	C-ion	S 800/u	horizontal, vertical	1994	5497	August 2010
TRIUMF, Vancouver	Canada	p	C 72	1 horizontal	1995	152	December 2010
HZB (HMI), Berlin	Germany	p	C 72	1 horizontal	1998	1660	December 2010
NCC, Kashiwa	Japan	p	C 235	2 gantry	1998	772	December 2010
HIBMC, Hyogo	Japan	P C-ion	S 230 S 320/u	1 gantry P horizontal, vertical C-ion			
PMRC(2), Tsukuba	Japan	P	S 250	gantry	2001	1849	December 2010
NPTC, MGH Boston	MA, U.S.	p	C 235	2 gantry, 1 horizontal	2001	4967	December 2010
INFN-LNS, Catania	Italy	p	C 60	1 horizontal	2002	174	March 2009
Shizuoka Cancer Center	Japan	p	S 235	3 gantry, 1 horizontal	2003	986	December 2010

(continued)

Table 1.2 Particle Therapy Facilities in Operation (Including Patient Statistics) (Continued)

Institution/City	Country	Particle	S/C[a]. Max. Energy (MeV)	Beam Direction	First Treatment	Patient Total	Date of Total
Southern Tohoku PTC, Fukushima	Japan	p	C 230	2 gantry, 1 horizontal	2008	No data	December 2010
WPTC, Zibo	China	p	C 230	2 gantry, 1 horizontal	2004	1078	December 2010
MD Anderson Cancer Center, Houston	TX, U.S.	p[c]	S 250	3 gantry, 1 horizontal	2006	2700	April 2011
UFPTI, Jacksonville	FL, U.S.	p	C 230	3 gantry, 1 horizontal	2006	2679	December 2010
NCC, Ilsan	South Korea	p	C 230	2 gantry, 1 horizontal	2007	648	December 2010
RPTC, Munich	Germany	p[b]	C 250	4 gantry, 1 horizontal	2009	446	December 2010
ProCure PTC, Oklahoma City	OK, U.S.	p	C 230	1 gantry, 1 horizontal, 2 horizontal/60°	2009	21	December 2009
HIT, Heidelberg	Germany	p[b] C-ion[b]	S 250 S 430/u	2 horizontal P[b] 2 horizontal C-ion[b]	2009	Treatment started	November 2009
UPenn, Philadelphia	PA, U.S.	p	C 230	4 gantry, 1 horizontal	2010	150	December 2010
GHMC, Gunma	Japan	C-ion	S 400/u	3 horizontal, vertical	2010	Treatment started	March 2010
IMPCAS, Lanzhou	China	C-ion	S 400/u	1 horizontal	2006	126	December 2010
CDH Proton Center, Warrenville	IL, U.S.	p	C 230	1 gantry, 1 horizontal 2 horizontal/60°	2010	Treatment started	October 2010
HUPTI, Hampton	VA, U.S.	p	C 230	4 gantry, 1 horizontal	2010	Treatment started	August 2010
IFJ PAN, Krakow	Poland	p	C 60	1 horizontal	2011	9	April 2011
Medipolis MedResInst, Ibusuki	Japan	p	S 250	3 gantry	2011	Treatment started	January 2011

[a] S/C = synchrotron (S), cyclotron (C).

[b] With beam-scanning at gantry and spread beam at OPTIS2 (since October 2010).

[c] With spread beam and beam-scanning (since 2008).

[d] Degraded beam.

[e] New cyclotron with fixed beam operational since July 2010; the gantry operational since October 2010.

use carbon ions and had treated approximately 3500 patients (3000 patients at HIMAC until December 2006, 430 at GSI until August 2008, and 93 at HIBMC until February 2007). New heavy ion facilities, which will be brought into clinical operation, include the Heidelberg Ion Therapy Center (HIT), Heidelberg, Germany (Heeg et al. 2004; Haberer et al. 2004), the National Center for Oncological Hadrontherapy (CNAO), Pavia, Italy (Amaldi 2004; Amaldi and Kraft 2007), and Gunma University Heavy Ion Medical Center (GHMC), Gunma, Japan. An experimental treatment facility at a research laboratory was also set up in Lanzhou, China, and started treating patients with superficially seated tumors in 2006 (a maximum energy of 100 MeV/u was used; Li et al. 2007).

Two review papers published in the journal *Medical Physics* in the late 2000s are excellent sources of information on the history and future developments of proton (Smith 2009) and heavy ion therapy (Jäkel et al. 2008).

1.2.4 Hospital-Based Proton Therapy Facilities

Before 1990, all particle therapy patients were treated at research facilities with protons and other charged particles. Table 1.1 shows research facilities in different countries that offered particle therapy treatments and are no longer in operation. The total number of patients treated at these facilities are also included together with the year when the facility was closed (PTCOG 2011b).

The first clinical proton therapy facility in the United States was built at Loma Linda University Medical Center (LLUMC), Loma Linda, California, in the late 1980s, which was the result of the vision and work of Dr. James Slater, who was the chairman of the Department of Radiation Medicine. The facility has a 250-MeV synchrotron, which was designed and built at the Fermi National Laboratory, and three isocentric gantries, which were designed by Andy Koehler at the HCL (Koehler 1987). In 1990, the facility was officially opened for patient treatments (Slater et al. 1991). Later, the Northeast Proton Therapy Center (NPTC), MGH, was brought online, and the HCL treatment program was transferred to it during 2001 and 2002. The NPTC was designed with a cyclotron accelerator, two isocentric gantries, and a fixed beam room with horizontal beams for ocular melanoma treatments and for radiosurgery. There were several novel aspects of the NPTC beam delivery system: The nozzles had the ability to use both passive scattering and beam-wobbling techniques to spread the beam laterally, and the range modulation wheels were contained completely within the nozzles. The beam intensity incident upon the modulation wheels was modulated so as to extend the energy range over which the wheels could achieve uniform SOBPs (Flanz et al. 1997). By 2010, an additional seven regional hospital-based proton therapy centers were established in the United States alone, and many more worldwide. By the end of 2011, there were a total of 36 particle therapy centers operating in Canada, China, France, Germany, Italy, Japan, Korea, Poland, Russia, South Africa, Sweden, Switzerland, the United Kingdom, and the United States (see Table 1.2; PTCOG 2011a). The total number of patients treated by particle therapy was 84,492 (73,804 treated by proton therapy) at facilities either in or out of operation by the end of 2010 (PTCOG 2011a).

1.3 Development of Particle Therapy Techniques

1.3.1 Three-Dimensional Conformal Particle Therapy and Range Compensation

Because charged particle beams stop in tissue at depths determined by their incident energy and interactions in tissues traversed in their path, it is important to know the three-dimensional (3-D) anatomy in the treatment volume, including the location and density of each tissue element. The three charged particle programs—MGH/HCL, UCSF/LBL, and UNM/LAMPF—were the first radiation therapy programs to install dedicated computed tomography (CT) scanners for particle therapy treatment planning. Because both HCL and LBL treated patients in a sitting position with horizontal beams, these facilities purchased and modified CT scanners for vertical scanning capabilities so that patients could be scanned in the sitting position. The pion transport channel in Los Alamos was vertical; therefore, it was possible to use a conventional CT scanner and scan patients in supine or prone positions. With the availability of 3-D patient CT data, it became possible to perform volumetric treatment planning for particle therapy. At MGH, Michael Goitein developed a proton treatment planning system that incorporated several important treatment planning tools, including beam's eye view, dose volume histograms, and error analysis. These tools have been utilized in virtually every modern treatment planning system (Smith 2009).

Another 3-D treatment planning aspect essential for passive scattered charged particle therapy is range compensation. Range compensation allows the shaping of the dose to match the distal surface of the target volume. It corrects for the shape of the patient surface, the tissue heterogeneities (e.g., air and bone) in the beam path, and the shape of the distal target volume surface. A range compensator was used in Los Alamos for every patient treatment field. In the beginning, the compensators were calculated by hand, and a 3-D model was constructed by the use of Styrofoam. This model was then used to construct a mold that was then poured with wax to form a 3-D range compensator. Practical techniques were also developed to shape the dose to the proximal surface of the target volume in cases where there were critical structures proximal to the tumor. This was probably the first use of both the double-sided range compensator and shaping to the proximal target volume surface. At both HCL and LBL, range compensators were made on milling machines using acrylics such as Lucite. At MGH, a method for incorporating patient setup and calculation errors into the range compensator to compensate for small misalignments between the compensator and target volume was developed (Smith 2009).

1.3.2 Intensity-Modulated Particle Therapy

Because the pristine Bragg peaks of charged particle beams are too narrow to treat any but the smallest of targets such as the pituitary gland, it is necessary to spread out the dose distribution in depth using range modulation. Both spinning propellers and ridge filters that placed multiple layers of absorbers in the beam for various time periods were used to modulate the intensity (weight) of individual Bragg peaks to achieve a uniform SOBP tailored to match the greatest extent of the target volume in depth (Figure 1.2). This method

of achieving SOBPs is simple intensity modulation, as summarized by A. Smith (2009), who also pointed out another aspect of simple intensity modulation (e.g., the use of several nonuniform fields to achieve an overall uniform dose distribution in a target volume to treat concave target volumes associated with tumors such as skull base chordomas and chondrosarcomas). Modern-day proton therapy utilizes scanning spot beams to achieve full-blown intensity modulation. The first intensity-modulated proton therapy (IMPT) treatment was performed at the PSI in Switzerland.

1.3.3 Image-Guided Particle Therapy

The use of imaging to guide the beam delivery process has been a topic of great interest for proton and ion therapy since its early days. Due to the precision required for high-dose particle therapy, the desired alignment of the target to the beam could only be achieved with imaging techniques. The imaging systems developed at the early charged particle facilities were precursors to the onboard-imaging systems that are currently implemented in modern radiation therapy facilities using high-energy photon and electron beams.

At both HCL and LBL, stereotactic imaging systems were installed in the treatment room to position the patient accurately for each treatment field. In Los Alamos, an imaging suite was established outside the treatment room to perform pretreatment patient setup. The patients were then aligned with lasers and transported to the treatment room on a couch that efficiently and accurately docked with a ram that rose out of the treatment room floor. The treatment aperture and range compensator were fixed to the couch and also aligned with lasers outside the treatment room. This procedure ensures efficient patient exchange and accurate treatment setup for precision particle therapy. Because the dose rates of the pion beam were very low, this transfer mechanism was necessary to minimize the patient exchange and setup time in the treatment room.

1.4 Interests and Trends in Particle Therapy

1.4.1 Advances in Technology

In the last decade, commercial manufacturers have replaced the role of research laboratories in the development of isocentric gantries, compact cyclotrons, medical synchrotrons, and advanced treatment delivery systems (Smith 2008). Superconducting technology has led to the development of smaller and higher

energy cyclotrons (Klein et al. 2005), and the efficiency of cyclotrons in terms of beam extraction and power consumption has been substantially improved. A noteworthy technological achievement was made at the PSI in Switzerland, where techniques for proton pencil-beam scanning were developed and implemented. This technology was first used for single-field, uniform-dose treatments and then extended to IMPT treatments

(Pedroni et al. 1995; Lomax et al. 2004). Several vendors now offer advanced proton therapy systems that have U.S. Food and Drug Administration (FDA) 510k clearance. These achievements have matured proton therapy into fully integrated, modern radiation therapy systems that provide advanced cancer treatments in state-of-the-art hospital settings.

Because of the significant capital cost of a proton and/or ion therapy facility, it is essential to improve the treatment efficiency and efficacy of particle therapy in order to improve its cost-effectiveness. In recent years, commercial treatment planning and data management systems have been developed to increase clinical efficiency and capabilities. Robotics has been introduced into particle therapy systems, resulting in improved efficiency in the management of patient transport, treatment setup, and treatment appliances. Image guidance is playing an increasing role in the precise delivery of 3-D conformal treatment and IMPT (Mazal et al. 1997; Allgower et al. 2007). Vendors have made major improvements in treatment delivery techniques aiming at better clinical outcome. New technologies include the following: 1) beam-scanning techniques; 2) "universal" nozzles that have the ability to deliver both passive scattering and spot-scanning treatments; 3) range modulation wheels installed permanently in the nozzle that produce variable SOBPs by a combination of beam gating and beam intensity variation; 4) advanced treatment control and safety systems; 5) improved imaging systems; and 6 multileaf collimators.

1.4.2   Clinical Results

Several publications in the literature have confirmed improvements in both local control and normal tissue complications with particle therapy over those achieved with photon therapy; however, these comparisons were predominantly made against historical control results with conventional photon treatments (Schulz-Ertner and Tsujii 2007; Tsujii et al. 1993; Suit et al. 2003). It could also be argued that many of the early proton therapy results were obtained from research-based facilities with horizontal, passively scattered beams, limited energies, and restricted flexibility in the complexity and types of treatments that could be carried out (e.g., no patients were treated with IMPT techniques). It is reasonable to assume that, should patients be treated with proton beams in modern, hospital-based facilities with improved treatment technology and improved treatment techniques, additional clinical gains could be achieved. Recent debates in the journal *Medical Physics* have tried to answer the question of whether further improvements in dose distributions with protons over those already obtainable with IMRT will result in outcomes that will be improved significantly enough to justify the increased cost (Schulz et al. 2007) or simply whether proton therapy should be used for prostate treatment, considering the success with intensity-modulated photon therapy and brachytherapy (Moyers et al. 2007). A follow-up debate also focused on the topic of whether, within the next 10–15 years, protons will likely replace photons as the most common type of radiation for curative radiotherapy, which raised many issues on the further development of proton and ion therapy (Maughan et al. 2008). It can be concluded that proton and ion therapy as a new and rapidly growing treatment modality has great potential for some disease sites that can bring substantial benefits to cancer patients. Both significant technical innovations and clinical advances are needed to reduce the cost and quantify the efficacy of proton and ion therapy in order to justify the further development of large proton and ion therapy facilities. The latter can only be achieved by large-scale, randomized clinical trials using the most advanced technologies and treatment techniques available in brachytherapy, external beam photon and electron therapy, and proton and ion therapy.

1.4.3   Reimbursement

In the United States, proton treatments were not assigned procedure codes by the American Medical Association (AMA) until the late 1990s. Reimbursements were evaluated by Medicare and third-party insurance carriers using special procedure codes on a case-by-case basis. Both LLUMC and MGH proton treatment facilities relied heavily on institutional and government funding. This created a stumbling block in the commercial development of proton and ion therapy facilities due to the lack of financial incentives. In the late 1990s, MGH and LLUMC successfully applied to the AMA for proton-specific treatment delivery procedure codes. Medicare and private insurance carriers then assigned payment rates to the new procedure codes. Moreover, when Medicare set rates for proton treatment delivery, they stated that proton treatments were *not considered to be investigational*. The achievement of reimbursement provided financial incentives for the private sector to finance, build, and operate proton therapy facilities, with the expectation that they could realize a reasonable return on investments. Current reimbursement rates for proton therapy delivery remain attractive; however, considering what has happened to IMRT with photons over the last decade, one should not expect that these rates will persist in the long term.

1.4.4 Commercial Developments

Early developments of particle therapy were almost entirely based on government funding, and therefore, all proton and ion therapy treatments were carried out in research-based facilities with equipment developed by the clinical researchers and engineers who worked in those facilities. The lack of a sizable market was responsible for the unavailability of private investors and commercial vendors to spend large sums of money to develop particle therapy equipment and to establish hospital-based particle therapy facilities. During the 1990s and early 2000s, especially with the achievement of reimbursement from Medicare and private insurance companies for proton therapy, several commercial vendors started marketing particle therapy systems and offering systems that provided both protons and carbon ions for cancer therapy. New accelerators based on superconducting technology have been developed, which will make it possible to build more compact isocentric gantries and single-room therapy systems. Private investors and new companies have teamed up to develop and operate proton therapy facilities and others offer various levels of consulting e.g., feasibility studies, marketing studies, business plans, technology evaluation, and staffing and operating plans. The success of new hospital-based proton therapy projects, such as at the Shands Cancer Center, University of Florida, Jacksonville, Florida; the M. D. Anderson Cancer Center, University of Texas, Houston; and the University of Pennsylvania, Philadelphia, has demonstrated that these complex, expensive facilities can be built on time and on budget and can rapidly reach targeted patient treatment capacity in order to satisfy ambitious clinical programs and business plans (Smith 2006).

References

Allgower, C. E., Schreuder, A. N., Farr, J. B. et al. 2007. Experiences with an application of industrial robotics for accurate patient positioning in proton radiotherapy. *Int. J. Med. Robotics Comput. Assist. Surg.* 3: 72–81.

Ambartsumian, V. A. 2008. A life in science. Astrofizika Journal Editorial Board. *Astrophysics* 51: 280–293.

Becquerel, H. 1900. Deviation du rayonnement du radium dans un champ electrique. *Comptes Rendus de l'Academie des Sciences,* 130: 809–815.

Castro, J. R. 1995. Results of heavy ion radiotherapy. *Radiat. Environ. Biophys.* 34: 45–48.

Chadwick, J. 1932. Possible existence of a neutron. *Nature* 129: 312.

Cheney, M. 2001. *Tesla: Man Out of Time,* New York: Simon & Schuster.

Cheney, M., Uth, R. and Glenn, J. 1999. *Tesla, Master of Lightning,* New York: Metro Books/Barnes & Noble Publishing.

Crane, L. August 13, 2010. Image of the Month: The left hand of Anna Roentgen. *Wellcome Trust Blog,* United Kingdom.

DeKosky, R. K. 1983. William Crookes and the quest for absolute vacuum in the 1870s. *Ann. Sci.* 40: 1–18.

Filler, A. G. 2009. The history, development, and impact of computed imaging in neurological diagnosis and neurosurgery: CT, MRI, DTI. *Nature Precedings* DOI:10.1038/npre.2009.3267.4.

Flanz, J., Durlacher, S. and Goitein, M. 1997. The Northeast Proton Therapy Center at Massachusetts General Hospital. *Proceedings of the Symposium of Northeastern Accelerator Personnel.* Woods Hole, MA. World Scientific 51–58.

Gaida, R. et al. 1997. Ukrainian physicist contributes to the discovery of X-rays. Mayo Foundation for Medical Education and Research. Archived from the original on May 5, 2008.

Haberer, T., Debus, J., Eickhoff, H. et al. 2004. The Heidelberg Ion Therapy Center. *Radiother. Oncol.* 73(Suppl. 2): S186–S190.

Heeg, P., Eickhoff, H. and Haberer, T. 2004. Conception of heavy ion beam therapy at Heidelberg University (HICAT). *Z. Med. Phys.* 14: 17–24.

Hishikawa, Y., Oda, Y., Mayahara, H. et al. 2004. Status of clinical work at Hyogo. *Radiother. Oncol.* 73(Suppl. 2): S38–S40.

Jäkel, O., Karger, C. P. and Debus, J. 2008. The future of heavy ion radiotherapy. *Med. Phys.* 35: 5653–5663.

Kevles, B. 1996. *Naked to the Bone, Medical Imaging in the Twentieth Century,* Camden, NJ: Rutgers University Press, pp. 19–22.

Kjellberg, R. N., Sweet, W. H., Preston, W. M. et al. 1962. The Bragg peak of a proton beam in intracranial therapy of tumors. *Trans. Am. Neurol. Assoc.* 87: 216.

Klein, H., Baumgarten A., Geisler, A. et al. 2005. New superconducting cyclotron driven scanning proton therapy systems. *Nucl. Instrum. Methods Phys. Res.* B 241: 721–726.

Koehler, A. M. 1987. Preliminary design study for a corkscrew gantry. *Proceedings of the Fifth PTCOG Meeting: International Workshop in Biomedical Accelerators,* Lawrence Berkeley Laboratory, Berkeley, CA, pp. 147–158.

Larsson, B., Leksell, L., Rexed, B. et al. 1958. The high-energy proton beam as a neurosurgical tool. *Nature* 182: 1222.

Lawrence, J. H., Tobias, C. A., Born, J. L. et al. 1958. Pituitary irradiation with high-energy proton beams: A preliminary report. *Cancer Res.* 18: 121–134.

Leicester, H. M. 1971. *The Historical Background of Chemistry,* New York: Dover Publications, pp. 221–222.

Li, Q., Dai, Z., Yan, Z. et al. 2007. Heavy-ion conformal irradiation in the shallow-seated tumor therapy terminal at HIRFL. *Med. Biol. Eng. Comput.* 45: 1037–43.

Lomax, A. J., Bohringer, T., Bolsi, A. et al. 2004. Treatment planning and verification of proton therapy using spot scanning: Initial experiences. *Med. Phys.* 31: 3150–7.

Lyman, J. T., Chen, G. W., Kanstein, L. et al. 1979. Heavy particle facilities at the Lawrence Berkeley Laboratory. *First International Seminar on the Use of Proton Beams in Radiation Therapy*, Moscow, Atomistadt, pp. 122–135.

Maughan, R. L., Van Den Heuvel, F. and Orton, C. G. 2008. Point/Counterpoint: Within the next 10–15 years protons will likely replace photons as the most common type of radiation for curative radiotherapy. *Med. Phys.* 35: 4285–4288.

Mazal, A. et al. 1997. Robots in high-precision patient positioning for confirmal radiotherapy. *Proceedings of the World Congress on Medical Physics and Biomedical Engineering, Medical and Biological Engineering and Computing*, Chicago.

Moyters, M. F., Pouliot, J. and Orton, C. G. 2007. Point/Counterpoint: Proton therapy is the best radiation treatment modality for prostate cancer. *Med. Phys.* 34: 375–378.

Myers, W. G. 1976. Becquerel's discovery of radioactivity in 1896. *J. Nucl. Med.* 17: 579–582.

Pedroni, E., Bacher, R., Blattmann, H. et al. 1995. The 200-MeV proton therapy project at the Paul Scherrer Institute: Conceptual design and practical realization. *Med. Phys.* 22: 37–53.

Peter, P. 1995. W.C. Roentgen and the discovery of x-rays. In *Textbook of Radiology*. GE Healthcare (ed.) (retrieved May 5, 2008 from Medcyclopedia.com).

Petrucci, R. H., Harwood, W. S. and Herring, F. G. 2002. *General Chemistry, Principles and Modern Applications*. Upper Saddle River, NJ: Prentice Hall, p. 41.

PTCOG (Particle Therapy Co-Operative Group). 2011a. Hadron Therapy Patient Statistics. http://ptcog.web.psi.ch/Archive/Patientenzahlen-updateMay2011.pdf (accessed and retrieved June 4, 2011).

PTCOG. (Particle Therapy Co-Operative Group) 2011b. Particle therapy facilities in operation. http://ptcog.web.psi.ch/ptcentres.html (accessed and retrieved November 21, 2011).

Schultz-Ertner, D. and Tsujii, H. 2007. Particle radiation therapy using proton and heavier ion beams. *J. Clin. Oncol.* 25: 953–964.

Schulz, R. J., Smith, A. R. and Orton, C. G. 2007. Point/Counterpoint. Proton therapy is too expensive for the minimal potential improvements in outcome claimed. *Med. Phys.* 34: 1135–1138.

Slater, J. M., Archambeau, J. O., Miller, D. W. et al. 1991. The proton treatment center at Loma Linda University Medical Center. *Int. J. Radiat. Oncol. Biol. Phys.* 22: 383–389.

Smith, A. 2006. Proton therapy. *Phys. Med. Biol.* 51: R491–R504.

Smith, A. R. 2008. Innovations and technical developments for particle therapy in the United States. *Proceedings of the 21st International Symposium: Modern Radiation Oncology: Innovative Technologies and Translational Research.*

Smith, A. R. 2009. Vision 20/20: Proton therapy. *Med. Phys.* 36: 556–568.

Stanton, A. 1896. Wilhelm Conrad Rontgen on a new kind of rays: Translation of a paper read before the Wurzburg Physical and Medical Society, 1895. *Nature* 53: 274–276.

Suit, H., Goldberg, S., Niemierko, A. et al. 2003. Proton beams to replace photon beams in radical dose treatments. *Acta Oncol.* 42: 800–808.

Thomson, J. J. 1897. Cathode rays. *Philosophical Magazine 44.* United Kingdom: Taylor and Francis, p. 293.

Thomson, J. J. 1903. *The Discharge of Electricity through Gases.* Charles Scribner's Sons, pp. 182–186.

Thomson, J. J. 1906. *Nobel Lecture: Carriers of Negative Electricity,* The Nobel Foundation.

Torikoshi, M., Minohara, S., Kanematsu et al. 2007. Irradiation system for HIMAC. *J. Radiat. Res. (Tokyo)* 48(Suppl A): A15–A25.

Trenn, T. J. 1976. Rutherford on the alpha–beta–gamma classification of radioactive rays. *Isis* 67: 61–75.

Tsujii, H. et al. 2007. Clinical results of carbon ion radiotherapy at NIRS. *J. Radiat. Res. (Tokyo)* 48(Suppl A): A1–A13.

Tsujii, H. et al. 1993. Clinical results of fractionated proton therapy. *Int. J. Radiat. Oncol. Biol. Phys.* 25: 49–60.

Wilson, R. 1997. *Astronomy through the Ages: The Story of the Human Attempt to Understand the Universe.* London: Taylor and Francis.

Wilson, R. R. 1946. Radiological use of fast protons. *Radiology,* 47: 487–91.

Wyman, T. (Spring 2005). Fernando Sanford and the Discovery of X-rays. *Imprint.* Associates of Stanford University Libraries: 5–15.

Zeeman, P. 1907. Scientific worthies. *Nature* 77: 1–3.

Section

2. Accelerators for Proton and Ion Therapy

Jacob Flanz

2.1 Introduction

From 1895, when Roentgen discovered X-rays, to 1913, when Coolidge developed the vacuum X-ray tube, it has been shown that energetic particles can be useful for medical applications. Table 2.1 summarizes some of these applications, including the particles used and the energy range required. It is clear that there is a wide range of applications and, therefore, also a wide range of requirements for the particles that must be considered.

2.2 Equipment Requirements

Particle therapy technology serves to create a beam that can achieve the clinical goals. This is the simple axiom that should be adhered to. The device that increases the charged particle energy (at relativistic speed) to the desired value of range penetration is called the accelerator. Different technology can result in different beam properties. Understanding this can greatly help in determining how the system can be used to deliver a useful treatment.

The equipment that is used to facilitate a patient treatment with the particle beam modality can be divided into a number of components, including

- Particle accelerator
- Beam transport system
- Beam delivery system
- Patient positioning system
- Patient alignment system

All of these systems are integrated to some extent, and their requirements are interrelated. Above all, the overall system design and design of the specific components must be safe and must satisfy the required clinical requirements. The following chart indicates the flow of requirements, in some semblance of order for the accelerator technology.

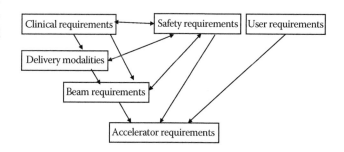

Note that the clinical requirements are buffered, in part, by the delivery modality, because the delivery modality will affect the type of beam that is needed to

Proton and Carbon Ion Therapy Edited by C.-M. Charlie Ma and Tony Lomax © 2013 Taylor & Francis Group, LLC. ISBN: 978-1-4398-1607-3.

Chapter 2

Table 2.1   Medical Applications of Accelerators

Particle	Energy	Application
Electrons	Tens of keV-MeV	X-rays
	6–25 MeV	Electron/photon therapy
	2–10 MeV	Sterilization, food preservation
	Hundreds of MeV to GeV	Synchrotron radiation–angiography
Protons	Tens of MeV	Neutron production/therapy
	10–100 MeV	Radioisotope production
	Tens–250 MeV	Proton radiotherapy
	500–700 MeV	Meson radiotherapy
Deuterons	7–20 MeV	Radioisotope production
Heavy Ions	70–400+ MeV/nucleon	Heavy ion radiotherapy

Note: MeV, megaelectronvolts; GeV, gigaelectronvolts.

achieve a desired clinical goal and, therefore, what the beam requirements will be. Furthermore, at each step of the analysis, the safety requirements must be evaluated to ensure that they are all captured before designing the component, such as the accelerator.

At the highest level, the goals of radiotherapy are to

- Deliver the required dose
- Deliver that dose with the prescribed dose distribution
- Deliver that dose in the right place

Clinical beam parameters, such as dose, dose rate, range, distal falloff, penumbra, and degree of dose conformity, will be associated with beam parameters such as beam current, beam energy, beam shape and size, and beam position. When discussing the beam production and delivery technology, it is always important to remember and associate the beam parameters with the clinical parameters; however, the association can be one to many or many to one. In addition (perhaps even more important), the tolerances associated with each of these parameters are critical. A change in the beam range, position, or shape could deposit a dose outside the target, especially in the case of beam scanning.

In addition to these raw beam properties, which are determined by a combination of the accelerator, beam transport, and beam-spreading mechanisms, another less considered quantity that will affect the treatment modality is the beam time structure.

2.3   Accelerator Physics

Virtually all the physics involved in accelerator and beam transport design is embodied within the Lorentz force law:

$$\mathbf{F} = q\mathbf{E} + q\mathbf{v}\mathbf{X}\mathbf{B}. \tag{2.1}$$

In this equation, the vector force $\mathbf{F}$ on a charged particle with charge q is determined by the electric field $\mathbf{E}$ and the magnetic field $\mathbf{B}$, in which the charge is introduced. In the case of an electric field, the force is in the direction of the electric field, in which case the charged particle is accelerated in its direction of motion, and the particle energy is increased. In the case of a magnetic field, the force is perpendicular to both the direction of the magnetic field and the velocity $\mathbf{v}$. In that case, because the force is perpendicular to the direction of initial motion, this force does not accelerate the particle in that direction of motion but adds a component of motion transverse to the initial direction, or it bends the particle trajectory but does not change the energy of the particle.

The technology of delivering a charged particle of the appropriate energy to a patient involves accelerating that particle and then bending and focusing a beam of the particles in the direction of the target in the patient. A beam of particles is a collection of particles within a small range of position, angle, and velocity. Most of the particles in the beam arrive at the target.

The shape of this beam distribution can have an impact on the beam delivery technique. Figure 2.1 represents the beam phase space [beam parameters associated with the

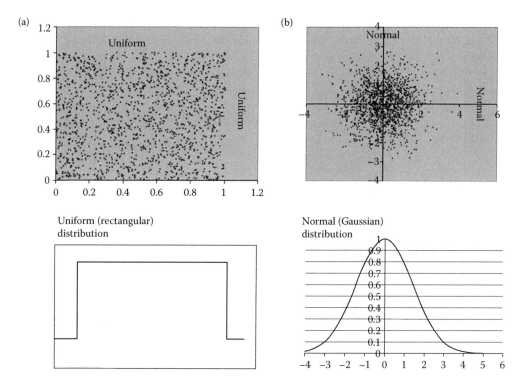

FIGURE 2.1 Rectangular (a) and Gaussian (b) beam phase space distributions. The 2-D spatial projection of the number of particles is shown below the phase space plots.

Hamiltonian canonical variables position and transverse momentum (or angle)] distribution for two different beams: one beam shows a Gaussian particle density, and the other beam shows a uniform distribution. Note that Gaussian beams can be more easily matched edge to edge, but the sharper the edge, the more critical the tolerance for the spacing between these beams. If two rectangular beams are shifted away from each other, there will be a deep crevice in the middle, but there is a wide range of positioning that can be tolerated between Gaussian beams before the summation of the two, in the middle, results in a nonuniform distribution.

Part of the reason for the size of the particle accelerator and beamline systems results from the Lorentz force law. If

it is necessary to change the direction of a particle or focus it, it must be bent. The radius of curvature of the bend follows from the Lorentz force law and can be simplified as

$$B \text{ (kg) } \rho \text{ (m) } = 33.356 \, P \text{ (GeV/c).} \qquad (2.2)$$

In this equation, a particle with a charge of 1 and momentum P (in GeV/c) is bent in a magnetic field B (in kilogauss) through a radius of curvature of ρ (in meters). Particles of therapeutic energies must have a range of about 30 cm in water. For protons, this is close to a proton kinetic energy of 200 MeV, and for carbon ions, this is close to a kinetic energy of 440 MeV per nucleon (which, in the case of carbon, has 12). Figures 2.2a and 2.2b provide

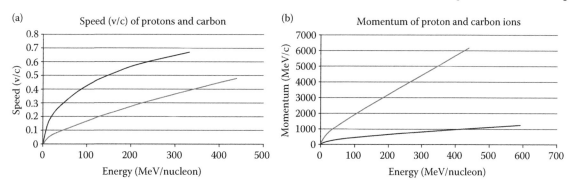

FIGURE 2.2 (a) Plot of the speed (in units of c) for protons (upper curve) and carbon ions (lower curve) as a function of the kinetic energy per nucleon. (b) Plot of the momentum of protons (lower curve) and carbon ions (upper curve) as a function of the kinetic energy per nucleon.

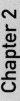
Chapter 2

a sense of the numbers. Figure 2.2a is a plot of the speed of the ion (in units of the speed of light *c*) as a function of the kinetic energy of the ion. We can see that, for therapeutic energies, the particle speed is relativistic. Figure 2.2b is a plot of the momentum of proton (blue) and carbon (pink) ions over a range of therapeutic ion energies. If the bending radius scales with the momentum, we can understand how the scale of ion therapy equipment (heavier than protons) grows. A normal electromagnet can reach a field of about 16 kG before saturation. According to Equation 2.2, a 230-MeV proton in a 16-kgauss field has a bending radius of 1.45 m, and a carbon ion of 440 MeV/μ has a bending radius of 12.8 m. Of course, superconducting technology provides the option to reach magnetic fields of several tesla (1 tesla = 10 kgauss) with a commensurate reduction in bending radius.

2.4 Accelerator Technology

Accelerators are devices that produce and shape an electric field to accelerate the charged particle. Based on the Lorentz law, we see that the electric field is key. However, an electric field can be formed in different ways.

- E ~ dB/dt—Maxwell's equations indicates that a changing magnetic field can produce an electric field. This was used for the betatron, but beam physics limitations restrict this technique to lower energies and light particles.
- E ~ applied voltage—An applied voltage can be either AC or DC; however, for clinical energies of hundreds of millions of volts, obtaining these fields involves serious engineering challenges.
- We can create an electric field or basically reform the electric fields that exist in atoms due to the existence of the charged particles in the atom.

We basically have two choices when accelerating a particle: we can send the particle through the accelerator once (single-pass schemes) or multiple times. A linear accelerator (linac) shapes the electric field to accelerate a beam in a linear path (just once), and thus, the length is proportional to the strength of the electric field and the energy gain desired. The power required is also related to the length. Conventional linear accelerators (certain radiofrequency types) do not produce sufficient electric field strength to build a compact heavy particle system, although nonconventional techniques are being investigated. Of course, the machine does not have to be strictly linear, but the overall length of the accelerating system is still determined by the acceleration gradient (in megaelectron-volts per meter).

One way of reducing the size of the machine and power required to accelerate charged particles is the efficient reuse of the electric field (multipass acceleration schemes). Therefore, circular machines such as cyclotrons, synchrotrons, or related devices are used in these environments. In general, due to the energy of the particles required for clinical use (for example, 250 MeV for protons and 440 MeV/nucleon for carbon ions), these accelerators can be large (or larger than a conventional photon linac).

The time during which the patient occupies the room depends on the choice of the activities defined by the clinic but usually includes in-room alignment, which could take several minutes. In this case, the accelerator usage could be a small fraction of the required room time. Thus, for maximal use of the machine, it could be more economical for it to feed up to several rooms.

There is also the opportunity to combine elements of different types of accelerators in order to achieve a balance among beam performance, cost, and size. Effort has been applied in the areas represented in Figure 2.3. As an example, note that elements of linacs and cyclotrons are combined in a cyclinac solution. In addition, within each type of system, there are variations on the theme. For example, the fixed-field alternating gradient (FFAG) accelerator can be implemented in a scaling or a nonscaling method, and there are pros and cons to both.

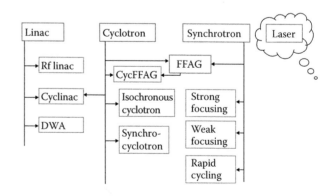

FIGURE 2.3 Types of accelerators and combinations thereof that are in use or under investigation for charged particle therapy uses.

2.5 Multiple-Pass Acceleration Schemes

2.5.1 Cyclotron-Based Systems

The charged particle beam path in a simple cyclotron is shown in Figure 2.4. An electric field is created at the center, and on either side, a magnetic dipole is positioned. The beam is injected into the center of the cyclotron and accelerated by the electric field at the center. When the beam leaves the electric field region, it enters the magnetic field region and is bent 180° and reenters the electric field region at the correct time to be accelerated in the opposite direction. The size and weight of a cyclotron is primarily determined by the strength of its magnetic field.

The C230 cyclotron built by Ion Beam Applications S.A. (IBA) is shown in Figure 2.5a. It is a room-temperature device with a magnetic field as high as 3 tesla in some places (Flanz et al. 1995). Extracted currents higher than 300 nA are possible. The overall weight of the iron core and the copper coils is 220 tons and stands in a footprint of a diameter of 4m. The cyclotron opens at the center and allows for maintenance, which is an important component of the required availability. The first such cyclotron was constructed for the Proton Facility at MGH and became operational in 1997.

The Still River cyclotron shown in Figure 2.5b is a superconducting accelerator that weighs about 20 tons due to the 9-tesla magnetic field strength (Matthews 2009), and it represents the other extreme (with today's technology). The current-carrying coils are in a superconducting cryostat that allows high electric currents to be reached. The beam properties are similar to conventional cyclotrons; however, the beam time structure is pulsed due to the modulating acceleration electric field frequency.

In between the IBA and Still River cyclotrons lies the Accel/Varian cyclotron, weighing in at 80 tons.

It is interesting to note that the technology for producing proton beams of sufficient energy for clinical therapy has advanced over the last decades. In fact, as shown in Figure 2.6, the weight of a cyclotron required for this has been reduced logarithmically.

The frequency of the electric field needed to synchronize the time it takes for the charged particle to follow its path with the phase of the electric field is given by the cyclotron equation

$$\omega = qB/m \qquad (2.3)$$

where ω is the angular frequency of the electric field, q is the charge of the particle, m is the mass of the particle, and B is the magnetic field of the cyclotron magnet. This

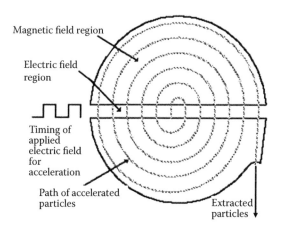

FIGURE 2.4 Pictorial description of the beam and electric field in a cyclotron.

FIGURE 2.5 (a) IBA C230 Cyclotron. (Courtesy of IBA s.a.) (b) Still River Superconducting Synchrocyclotron. (Courtesy of Ken Gall, Still River Systems.)

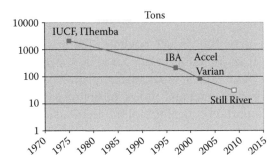

FIGURE 2.6   Progress in the reduction of the weight (and cost) of cyclotrons used for proton therapy.

FIGURE 2.8   IBA Ion Cyclotron solution for charge particle therapy.

magical relation indicates that, if the mass and charge are constant, then the frequency and magnetic field are constant and independent of the particle energy. When the energy is sufficient, the particle reaches the outer radius and is extracted from the cyclotron and directed to the treatment room. However, if the energy increase is large enough, then the effective mass increases due to relativistic effects, and the magnetic field and/or frequency cannot remain constant and still contain the accelerating particle. In that case, the cyclotron can be built with the appropriate magnetic field pattern to be "isochronous" (time independent), or the frequency can be modulated as the energy increases, thus forming a synchrocyclotron.

One method of increasing the magnetic field with radius is represented in Figure 2.7, with a magnetic gap that decreases with larger radius. This, however, curves the magnetic field (black arrows), resulting in components of the magnetic field in the radial direction (horizontal red arrows). Unfortunately, based on the Lorentz force law, we can see that the radial field defocuses ions that are not in the median plane, and without mitigation, the ions would not be contained during acceleration. This results in requiring

additional focusing, which led to the development of spiral sectors in a high-energy cyclotron.

In the isochronous cyclotron, a beam can be accelerated and extracted continuously. In fact, if there is sufficient stability, the beam current injected to the cyclotron can be modulated, which will result in a modulated output only delayed by the transit time, which is on the order of tens of microseconds. The synchrocyclotron implementation results in a short pulse of beam that can be accelerated. This pulse must track with the frequency modulation; otherwise, the beam will be mismatched with the accelerating voltage and will not be accelerated. Thus, the output beam is pulsed, with a repetition frequency given by the technology of the system used to vary the radio frequency (RF).

As the particle momentum increases, including the beam energy and the mass of the particle, the bending radius of the cyclotron must increase. Using superconductivity has led to a reduction in the overall weight and size for protons, as shown in Figure 2.5b; however, until recently, the use of a cyclotron for the acceleration of heavier particles for therapy has been thought to be too expensive. Recently, IBA has designed a cyclotron that would weigh about 600 tons but is only about 6.3 m in diameter. This is represented in Figure 2.8. Note the spiral-shaped sectors required for focusing the beam in this continuous, isochronous cyclotron.

2.5.2   Synchrotron–Based Systems

The charged particle beam path in a synchrotron is shown in Figure 2.9. The synchrotron is a ring (or some closed shape) of magnets. The beam is injected from outside the synchrotron and then circulates around the ring repeatedly through the accelerating structure. In order to keep the beam within the closed ring, the magnetic field of the magnets must increase in strength in conjunction with the beam energy increase. Thus, the beam is contained within the ring as its energy increases. When the beam reaches the desired energy, it is extracted. The time it takes to circulate one turn

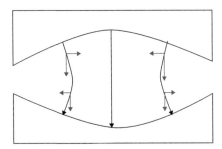

FIGURE 2.7   Possible modification of the poles of a cyclotron (exaggerated) showing how the gap is reduced toward the outer radius. This, however, results in a curved magnetic field (dark arrows from top to bottom). The curved field lines can be decomposed into vertical and radial (see the horizontal light arrows) directions.

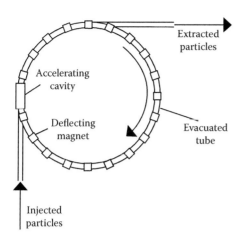

FIGURE 2.11 ProTom Synchrotron being assembled. (Courtesy of ProTom International.)

FIGURE 2.9 Schematic showing the simplified operation of a synchrotron accelerator, including the beam direction and electric field accelerating cavity.

in the ring depends, of course, on the particle velocity and, therefore, its energy; however, in the range of particle therapy and in the compact-sized synchrotron used, it is usually less than 1 μs. Therefore, if the beam was extracted in one turn, the pulse length would be less than 1 μs. In such an application, a rapid cycling synchrotron has been proposed, but as yet, it has not been used for particle therapy (Peggs et al. 2002). For clinical use, it is advantageous to have an intensity-modulated beam of time length in the hundreds of milliseconds or a beam lasting many seconds. Thus, the beam is extracted through techniques such as resonant extraction and variants thereof. Several synchrotrons have been built and are operating for particle therapy uses. The Hitachi synchrotron shown in Figure 2.10 is representative of the size of current synchrotrons used

in medical facilities for proton therapy (Hiramoto et al. 2007).

Concepts for smaller devices have been advanced, such as the one developed at the Lebedev Physical Institute, Russian Academy of Sciences, and licensed by ProTom International. Figure 2.11 shows the innovative design that represents a more cost-optimized approach to accelerator engineering while maintaining the ability to reach the desired specifications.

A significant advancement in the operation of the synchrotron was designed in the Hitachi machine and incorporated into other machines. Typically, the synchrotron operation is cyclic. However, it is sometimes useful to synchronize the beam production with the respiration cycle of a patient, particularly in such an accelerator that is not continuous. In this case, the Hitachi synchrotron allows for an arbitrary start of the acceleration cycle and is capable of an extended extraction sequence. Figure 2.12 shows the oscilloscope

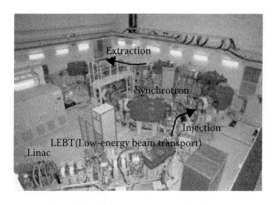

FIGURE 2.10 Hitachi Medical Synchrotron. The synchrotron is shown in the upper right, with the injector at the lower left. The extraction takes place at the upper left. (Courtesy of Hitachi Ltd.)

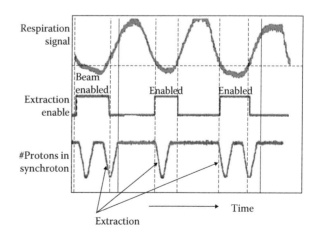

FIGURE 2.12 Time relationships in the Hitachi synchrotron among the respiration signal (upper trace), extraction enabling signal (middle trace), and synchrotron energy level (lower trace). (Courtesy of Hitachi Ltd.)

Chapter 2

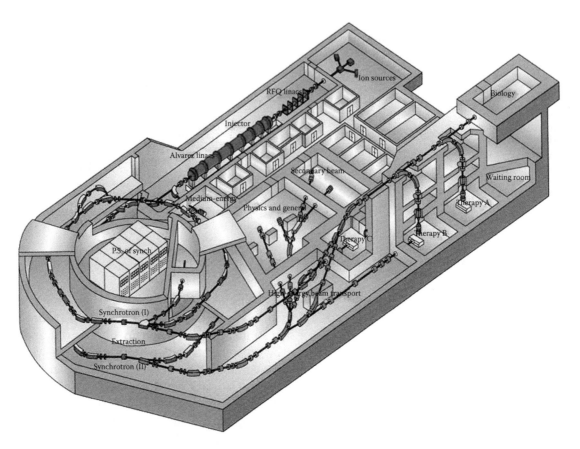

FIGURE 2.13   Overview of the Heavy Ion Medical Accelerator (HIMAC), in Chiba, Japan. It shows the two synchrotrons and multiple fixed-beamline treatment rooms. (Courtesy of Koji Noda and Hirohiko Tsujii.)

traces for this system. The upper trace shows the respiration signal coming from the patient, and the trace just below indicates that time allowed for the beam delivery to the patient. The lower trace indicates the extracted beam from the synchrotron coming, in time, during the allowed period.

The synchrotrons used for heavy ion treatments are larger than proton synchrotons, as shown in the concept drawing in Figure 2.13, e.g., the HIMAC facility, which began operation in 1994. In that facility, the maximum beam energy is 800 Mev/µ. This gives a range of 30 cm in water for Si ions. This accelerator system consists of two synchrotrons. The diameter of the synchrotron is about 40 m. One criticism of synchrotrons is the capability to control the extracted current. Techniques are being developed, such as the one used at HIMAC, and can obtain good intensity modulated extraction, as shown in the scope trace in Figure 2.14.

However, since then, designs have evolved to reduce the size of a carbon ion accelerator, leading to the design of the Heidelberg Ion Therapy Center, which began

patient treatments in 2010 (shown in Figure 2.15), and Gunma University Heavy Ion Medical Center, Gunma, Japan (shown in Figure 2.16), completed in mid-2010. The synchrotron in this facility has a diameter of 20 m and can accelerate carbon ions to 400 MeV/µ.

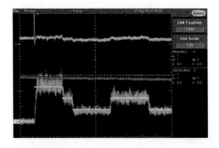

FIGURE 2.14   (See color insert.) Scope picture of the extracted beam (lower trace) from the HIMAC synchrotron: Modulated beam current during an extraction cycle and the difference possible from one cycle to the next. (Courtesy of Koji Noda.)

FIGURE 2.15 Schematic of the Heidelberg Ion Therapy (HIT) Facility, Heidelberg, Germany, the first heavy ion therapy facility with a gantry (shown on the right side of the drawing) in the world. (Courtesy of Thomas Haberer and HIT.)

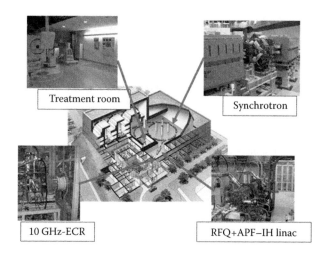

FIGURE 2.16 Overview of Gunma University Heavy Ion Medical Center, Gunma, Japan. (Courtesy of Mitsubishi Electric and Gunma University.)

FIGURE 2.17 Proof of principle (POP) FFAG proton accelerator at the KEK Laboratory in Japan.

2.6 Hybrid Concepts

2.6.1 FFAG Accelerator

The FFAG accelerator (Craddock 2004), first developed more than 50 years ago for the Midwestern Universities Research Association (MURA) project, is meant to merge the best aspects of cyclotrons and synchrotrons. Figure 2.17 shows an FFAG accelerator. The system comprised a ring of magnets such as a synchrotron; however, instead of the magnetic field ramping with the increasing particle energy as it is accelerated, the magnet system is designed to accommodate a large range of beam energies, and the fields are fixed as in a cyclotron. Current work is ongoing to determine whether this concept can be applied for particle therapy.

This system can be expandable and possibly used for heavy ion acceleration. Figure 2.18 shows a concept of a nested series of these accelerators enabling the acceleration of carbon ions.

2.6.2 Cyclinac

A cyclinac uses a cyclotron as an injector to an RF linear accelerator (Amaldi et al. 2004). This reduces the size and complexity of both the cyclotron and the linac. This system offers the advantage that it can use the electronic aspects of a linac to control the beam energy on a pulse-to-pulse basis, increasing the flexibility of a "cyclotron-only" system, which produces a

Chapter 2

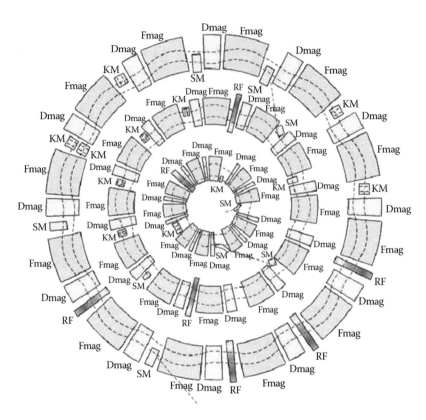

FIGURE 2.18 Concept for a high-energy heavy-ion FFAG accelerator.

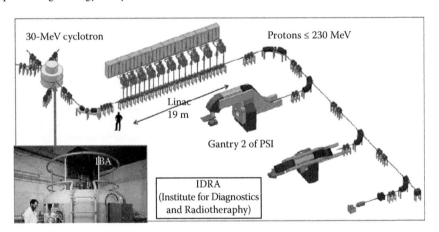

FIGURE 2.19 Representation of a cyclinac concept showing how the lower energy beam from the cyclotron is boosted by the linac and then delivered to the treatment rooms. (Courtesy of Ugo Amaldi.)

fixed beam energy. In addition, it would be possible to use the cyclotron for positron emission tomography (PET) isotope production, thus increasing the functionality of the facility. An example of a concept for this facility is shown in Figure 2.19.

2.7 High-Gradient Techniques: Single-Pass Acceleration

2.7.1 High-Gradient Linear Accelerator

One reason that a linear accelerator has not heretofore been used in a medical environment is the size, complexity of the RF system required, and beam characteristics such as duty factor. However, one concept currently under consideration is that of a high-gradient electrostatic accelerator. New dielectrics can hold off very high voltages without arcing. Gradients more than 100 MeV/m are sought. Such an accelerator

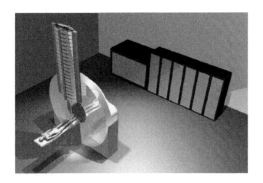

FIGURE 2.20   Artist's conception of an implementation of an dielectric wall accelerator by tomotherapy and the Compact Proton Accelerator Corporation.

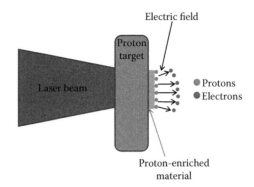

FIGURE 2.21   Diagram showing the mechanism for accelerating protons with a laser system. The electrons (little circles to the right of the figure after being ejected from the material), are released from the target by the laser power and set up an electric field which causes the protons (little circles near the proton enriched material section) to be accelerated.

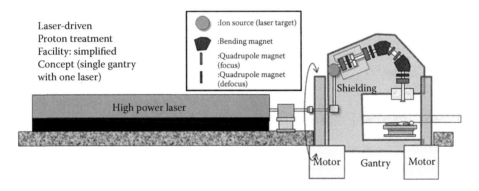

FIGURE 2.22   Concept of a laser accelerator based proton therapy facility with a gantry. (Courtesy of Paul Bolton.)

is called a dielectric wall accelerator (DWA) (Sampayan et al. 1995) and could produce a short-pulsed beam of clinically useful energy, with pulse-by-pulse control of beam energy, size, and intensity. It could be mounted on a robotic arm such as the CyberKnife system or a small gantry. One such concept is shown in Figure 2.20.

2.7.2   Laser Wakefield Particle Acceleration

An article in *Nature* discusses the "dream beam" (Faure et al. 2004). Experiments show that it is possible to create a huge electric field by using an intense laser pulse (intensity $\sim 10^{20}$ W/cm^2) to knock out electrons from a thin target, as depicted in Figure 2.21. This can result in an electric field of tera volts (TV)/m, and beam energy in the range of several tens of megaelectronvolts has been reached [more than 60 MeV was reached in 2009 with the Los Alamos National Laboratory (LANL) Trident laser]. Therefore, with

a target thickness of a few micrometers, therapeutic beam energies can be reached. The variation in the electric field is typically large, resulting in a large range of accelerated proton energies. Currently, issues of sharpening the beam energy spread and emittance are investigated. However, some tests show that, by concentrating the proton source at the center of the electric field axis, the energy spectrum can be improved. One of the more realistic concepts of such a facility is shown in Figure 2.22 for the laser-driven ion beam radiotherapy concept, which is a joint development in Japan. This system begins with the laser, whose light beam is deflected onto a rotating gantry. On this gantry is placed the proton target and a magnetic analysis system used to select only the energy spectrum that will be used for treatment. Some studies have been made on a compact particle selection and beam collimating system by Charlie Ma (this book's editor), who uses a 100-TW laser for experimental studies (Figure 2.23).

FIGURE 2.23 C.-M. Charlie Ma, Ph.D. (this book's editor), working on components for use with the laser accelerator technique.

2.8 Summary

engineering, and some are a result of necessary beam modifications external to the accelerator. Table 2.2 summarizes some of the parameters that can be used to determine appropriate beam delivery techniques for the various accelerators discussed in this chapter.

The identification of the constancy or dynamic behavior of the system can help determine the beam delivery method and provide an insight into the complexity of the operation of the system. The time structure of the beam is an important although misunderstood quantity, and some parameters associated with the pulse structure of the various accelerators are included in the table. Finally, a relative comparison of the raw emittance extracted from the accelerator (including any energy modification devices required) is provided. The emittance will, in some situations, determine the beam size at the target, although devices in the beam path such as ionization chambers, vacuum windows, and air will modify that and, in some cases, make the comparison in the table invalid.

There are many types of accelerators, and each produces beams with different properties. Clever scientists have found methods of effectively tailoring various beams for use in medical particle beam delivery. Some of the parameters are inherent in the accelerator physics and

2.9 Context within the Particle Therapy Facility

The technical equipment consisting of the accelerator, beamlines, and gantries must be integrated into a medical facility. This should be done with respect to the appropriate requirements of the facility. The equipment containing the beam has the potential for producing radiation during the time the beam is active,

and therefore, radiation shielding is necessary. In typical cases, depending on the location of the building, walls from 4 to 14 ft thick are needed, and ceilings as thick as 10 ft may be indicated. These are additional considerations that will be affected by the choice of accelerator.

Table 2.2 Comparison of Some Parameters among the Accelerator Types

Parameter	Cyclotron CW	Synchrocyclotron	Synchrotron	Cyclinac	FFAG	DWA	Laser
Magnetic field	Constant	Constant	Changing	n/a	Constant	Constant	n/a
Radio frequency	Constant	Changing	Changing	Variable	Changing	n/a (DC)	n/a
Energy change	Degrader	Degrader	Cycle time or within cycle?	~200 Hz	msec	Pulse-to-pulse	Pulse-to-pulse
Current	Hundreds of nA	Hundreds of nA	Several nA	nA	~Hundreds of nA	?	~10 Hz
Pulse frequency/ length	Continuous	10/1 ns	0.5/30 Hz Microseconds to seconds	200 Hz μs	Hundreds of hertz? 1 μs	Hundreds of hertz? Nanoseconds	ps
Scanning type	All (spot and continuous)	Spot	Spot and possibly continuous	Spot	Spot	Distal edge tracking	Distal edge tracking?
Emittance	Larger	Larger	Medium	Small	Medium	Small	Medium

Note: nA, nanoampere; ps, picosecond.

References

Amaldi, U. et al. 2004. The cyclinac: An accelerator for diagnostics and hadrontherapy. PTCOG meeting Paris.

Craddock, M.K. 2004. The rebirth of the FFAG. *CERN Courier* 44.

Faure, J. et al. 2004. A laser-plasma accelerator producing mono-energetic electron beams. *Nature* 431: 541–544.

Flanz, J. et al. 1995. Overview of the MGH-Northeast proton therapy center plans and progress. *Nucl. Instrum. Meth. in Phys Res B.* 99: 830–834.

Hiramoto, K. et al. 2007. The synchrotron and its related technology for ion beam therapy. *Nucl. Instrum. Meth in Phys Res. B* 261: 786–790.

Matthews, J. 2009. Accelerators shrink to meet growing demand for proton therapy. *Physics Phys. Today*, p. 22.

Peggs, S. et al. 2002. The rapid cycling medical synchrotron. RCMS, *Proceedings of EPAC*, Paris, France, pp. 2754–2756.

Sampayan, S. et al. 1995. High-gradient insulator technology for the dielectric wall accelerator. *Proceedings of the Particle Accelerator Conference*, Dallas, TX, UCRL-JC-119411.

3. Beam Delivery Systems for Particle Therapy

Marco Schippers

3.1 Introduction

The beam transport system between an accelerator and a patient consists of different stages and depends on the type of accelerator used. In most of the existing treatment facilities, one accelerator (cyclotron, synchrocyclotron, or synchrotron) is used, and the beam transport system guides the accelerated beam to one treatment device (or room) at a time. A synchrotron produces the desired particle energy directly so that the extracted beam can be transported straight to the treatment room of choice. However, in case a (synchro) cyclotron is used as an accelerator, one has to take into account that these machines operate with a fixed-particle energy. Within the patient, the maximum depth of a particle treatment is thus determined by this fixed energy. In order to match the particle energy to a shallower depth, this energy is reduced in the beam transport system. This is performed in a degrader and beam analysis section, together usually assigned as an energy selection system

(ESS). After the beam analysis, a transport system with a switch yard is used to bring the beam to the requested treatment area. Most floor plans of particle therapy facilities have a configuration similar to that shown in Figure 3.1.

Downstream an eventual energy selection, all beamline components must be designed to transport, intercept, or detect 0.1–3-nA proton beams between ~70 and 230–250 MeV (therefore within a magnetic rigidity range of 1.2–2.4 Tm) or 0.02–1 particle-nA ion beams between 125 and 400–450 MeV/nucl. (3.3–6.8 Tm) to any target station. The basic concept of the ion optics is an *achromatic* beam transport, which means that the beam transport is independent of slight deviations of the intended beam energy. Wherever possible, the beam transport system consists of symmetric optical building blocks. The symmetry ensures a reduction of optical aberrations, reduces the phase-space dependence, and minimizes the effects of small errors in the settings of beamline components.

Before the beam enters the beam delivery system, it can be an advantage to perform certain online

Proton and Carbon Ion Therapy Edited by C.-M. Charlie Ma and Tony Lomax © 2013 Taylor & Francis Group, LLC. ISBN: 978-1-4398-1607-3.

Chapter 3

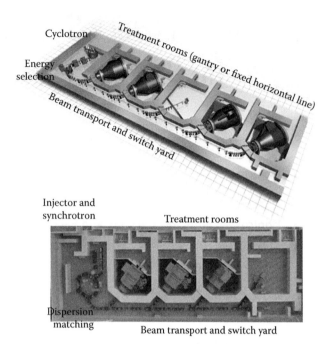

FIGURE 3.1   Typical layout of a particle therapy facility with a cyclotron and energy analysis (top) (Varian) and one with a synchrotron (Hiramoto et al. 2007). Treatment rooms are usually located next to each other and connected to the accelerator with a long beam transport line.

verifications. Such verifications could also be made at approximately the entrance of a treatment vault. At such *checkpoints*, measurements can be performed of the beam intensity, beam position, and beam size not only to ensure a proper matching to the beam delivery device but also for safety-related verifications. At the entrance of the treatment vault, the emittance of the beam must be matched to the gantry (rotating beam transport system, which allows irradiation of the patient from different directions) or fixed beamline. Finally, careful tuning of the beam transport in the gantry is needed to ensure angle-independent beam properties at the patient. The *nozzle* is the part behind the last bending magnet of a gantry or the last part of a fixed beamline. It typically consists of components for shaping the beam, energy modulation, range shifting, and dose monitoring, as well as X-ray equipment, light field equipment, and lasers for alignment.

In this chapter, the most important parts of the beam delivery system between the accelerator and the patient will be discussed in detail.

3.2   Energy Selection Systems

3.2.1   General Description

After acceleration, the particles need to be transported to the treatment room, where the patient is accurately positioned on a patient positioning system, and a beam delivery system accurately points the beam to the target region (tumor). The beam energy that is needed at the beam delivery system can be selected in the accelerator when a synchrotron is used. The energy spread in a synchrotron beam is very small; therefore, no further analysis of the beam energy is needed. In addition, in the case of a synchrotron for heavy ions (helium-carbon), the selectivity of the acceleration process is good enough to prevent contamination of the beam by other particles that have the same ratio of charge and mass number.

Cyclotrons and synchrocyclotrons do not possess the possibility of changing the energy in the accelerator. This implies that the beam energy must be reduced to match the penetration depth of the particles in the patient to the desired value. In most of the cyclotron-based particle therapy systems, the beam is focused on a *degrader*, a piece of light material in the beam path with an adjustable thickness. The energy of the particles leaving the degrader depends on the adjusted amount of material that they have traversed. When energy modulation (moving the Bragg peak back and forth over the tumor thickness) is performed in the nozzle, the ESS selects the energy needed to reach the most distal layer of the target volume. Figure 3.2 shows an overview of the ESS at the Paul Scherrer Institute (PSI), Villigen, Switzerland (Schippers et al. 2007). The degrader is followed by a collimation system to suppress the particles that have scattered out of the acceptance of the following beam transport system. A beam analysis system encompassing magnets and a slit selects the energy and energy spread (momentum and momentum spread) of the beam sent into the transport system. After a discussion on some general aspects, these respective components will be discussed in detail in this section.

In advanced systems (Schippers et al. 2007), it may be possible to perform energy modulation with the degrader as well. This method has the advantage that the transversal beam sharpness at the isocenter can be limited better, which is especially of advantage when pencil beam scanning is applied in the beam delivery

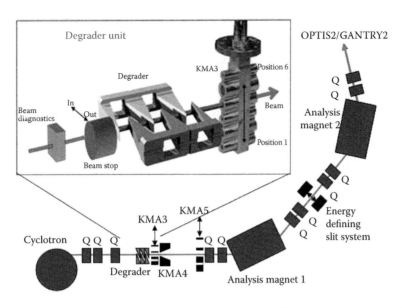

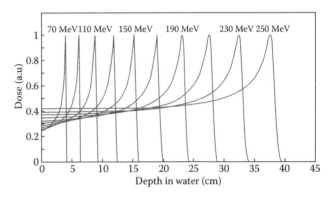

FIGURE 3.2 Schematic of the beamline at PSI. Solid rectangles are quadrupole (Q) and dipole magnets. Locations of collimators (KMA) and the energy defining slit system are indicated.

system. Another advantage is that this method creates a much lower neutron flux in the nozzle, as will be discussed in more detail in Section 3.5. Typically, the energy step size chosen to obtain a spread-out Bragg peak (SOBP) with good homogeneity corresponds for protons to a change in a (water-equivalent) range of approximately 5 mm. When *volumetric rescanning* is desired, in which the target volume is repeatedly covered over its full depth, the change of energy must be done as quickly as possible. Typically, one would like to make the 5-mm range step as fast as possible. The fastest system, which uses 50–80 ms to make such a range step, is currently achieved and used at PSI (Hug et al. 2008; Schippers et al. 2008). This, however, sets high constraints on the power supplies, the degrader electromechanics, the lamination of the beamline magnets and the control system, because all magnets downstream the degrader need to be synchronized with any step made by the degrader.

Another complexity, which also occurs with low-energy beams from a synchrotron, may arise at energies below approximately 110 MeV and with all ion beams of all energies. As shown in Figure 3.3, in these cases, the Bragg peak of the beam at the isocenter becomes sharper with decreasing beam energy, and at energies below ~110 MeV, it is very sharp. For treatments requiring a very sharp Bragg peak, such as treatments of eye melanoma, this can be advantageous. However, to obtain, for example, a homogeneous depth dose distribution in an SOBP, two to three times more Bragg peaks are needed to prevent ripples in the SOBP. This

FIGURE 3.3 Shape of the Bragg peaks for range modulation in an upstream degrader and beam analysis system. (Courtesy Terence Böhringer, PSI.)

would increase the treatment time, which is obviously not desired. Other methods that may be applied are to increase the accepted energy spread at the slit system [the usefulness depends on the energy (momentum) acceptance of the following beam transport system and gantry], to use ripple filters in the nozzle (see, for example, the work of Bourhaleb et al. 2008), or to insert a range shifter into the nozzle and entering the treatment room with energies above 110 MeV. The disadvantage of the last two methods is the increase of the lateral penumbra due to multiple scattering in the filter or range shifter.

3.2.2 Degrader

The lowest energy proton that is usually transported in the beam lines is 70 MeV. Lower energies would introduce complications due to the scattering, losses, instabilities in power supplies, and nonlinear effects in detectors and give Bragg peaks that are too sharp for direct use. The degrader should consist of light (low mass number) material to prevent too much increase of the beam emittance by multiple scattering. To degrade protons from 250/230 to 70 MeV, a degrader length of about 20/17 cm is needed. Multiple scattering causes an increase not only in beam divergence but also in beam size at the degrader exit. Often, graphite compressed to a high density (at PSI 1.88 g/cm³ is used) is used. In cases where the emittance increase must be minimized, beryllium may be an alternative (Anferov et al. 2007). However, one should take into account the higher neutron production per transmitted proton and the chemical toxicity when beryllium is used (van Goethem et al. 2009).

As shown in Figure 3.4, different degrader designs exist, which can be distinguished in discrete designs, made of a set of plates that are inserted into the beam, or shapes based on a wedge shape that allow for a continuous energy selection. In the case of a wedge shape, one always needs a counterwedge to obtain an equal amount of traversed material over the beam cross section. This, however, causes a minimum value of the energy loss of a few MeV in the degrader, because the wedges need a transversal overlap at least as large as

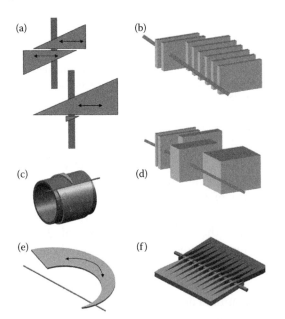

FIGURE 3.4 Several design concepts of a degrader in the beam transport system. (a) Two versions of a system based on two or one adjustable wedges. (b) Insertable slabs of graphite or Plexiglass. (c) Rolled-up wedge. (d) Insertable blocks with different thicknesses. (e) Rotatable Plexiglass curved wedge. (f) Adjustable multiwedge design. The beam is indicated at each degrader type.

the beam diameter. To also transmit energies between the accelerator energy and the maximum energy out of the degrader, an additional dedicated degrading system (e.g., a stack of thin plates or a thinner wedge) is needed.

Important specifications of the degrader are

- Speed (setting limits on the weight and size)
- Thickness (minimum energy to be reached)
- Transversal size (access)
- Accuracy of positioning (range accuracy)
- Material (the choice of which should be made by considering aspects of activation, contamination, mechanical stability, and whether the degrader is located in vacuum or in air)
- Choice between discrete or continuous energy steps
- Scattering (when multiple material layers are traversed, any distance between these layers will increase the scatter cone of the particles leaving the degrader)
- How near a collimator can be mounted at the beam exit side (affecting transmission efficiency)
- Reliability and service aspects (access, easy replacement, lifetime of moving components)

The degrader in the IBA systems (IBA) consists of a single carbon wedge rolled on a cylinder to limit the lateral dimensions (Figure 3.4c). A small fixed counterwedge is mounted close to the beam exit side of the degrader. The system is mounted in air.

At PSI and in the Varian systems (van Goethem et al. 2009; Varian Medical), the degrader is optimized for speed and consists of two multiwedge blocks opposite each other (Figure 3.4f) so that the travel distance between the maximum and the minimum energy setting is short. This allows fast energy selection without the need for very high velocity movement of the wedges. The degrader is mounted in the vacuum of the beamline (see Figure 3.5). Reasons for mounting the degrader in vacuum are radiation protection (to avoid contamination risks and reduce air activation; per area ≤10 nA.m is allowed at PSI) and beam diagnostics using high voltages mounted in close vicinity of the degrader.

3.2.3 Collimation Systems for Emittance Matching

Due to multiple scattering in the degrader material, the beam divergence and the beam size increase at the

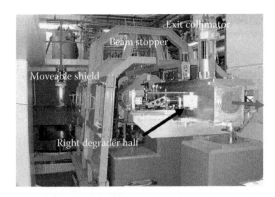

FIGURE 3.5 Vacuum box for the beam stopper, beam diagnostics, degrader, and collimation system at PSI. The cart with the lead shield has been rolled stream upward to allow access during the beamline installation. The beam is indicated by the gray arrow.

exit of the degrader when the exit energy decreases. In a high-density graphite degrader of 200-mm length, multiple scattering adds about 8 mm (1 sigma) to the initial beam size and about 40 mrad (1 sigma) to the divergence when degrading from 250 to 70 MeV. The degrader smears out part of the emittance of the incoming beam and can be regarded as the proton source for the following beam transport system. However, this smearing process does not completely decouple this beam from how the beam has left the cyclotron: the emittance of the incoming beam adds partly to the emittance (as discussed later), beam direction, location, and intensity can still affect the beam position and intensity at the entrance of the treatment areas.

The collimators, which limit the emittance of the beam leaving the degrader, are usually set up as a "beam-size defining" collimator immediately following the degrader and a "divergence" defining collimator, at a distance of approximately 1–1.5 m behind the degrader. In order to shorten the beamline length by 1m, it is sometimes chosen to mount the second collimator behind the first quadrupole set that follows the degrader. The disadvantages are, however, a high activation of the quadrupole(s) and that one has a rather complex relation between the apertures of the collimators and the accepted beam emittance, which may not allow all possible phase space configurations.

In order to have a well-defined beamline acceptance, the collimators that define the beam shape should be as short as possible but still thick enough to stop the protons at maximum energy. Therefore copper, brass, steel, or lead can be taken. However, as will be discussed in Section 3.2.5, depending on the collimator aperture, up to 80% of the incoming beam

intensity can be intercepted by the first collimator. Therefore, activation of the collimator material will be an issue, and it makes sense to choose light materials (e.g., graphite) for parts of the collimator system that are further away from the beam axis than the aperture-defining part. At PSI, therefore, as shown in the insert in Figure 3.2, the beam size that defines the collimator consists of a stack of copper rings (tubes) with different apertures and a thin wall. The inside of the tube acts as a collimator, whereas the particles that pass the tube at the outside are stopped in a graphite ring mounted behind the collimator stack.

The vacuum chamber housing the degrader has been made of aluminum to limit the medium short-term dose rate due to activation. The dose rate that is typically observed 30 min after switching the beam off is 4–5 mSv/h, at a 20-cm distance from the degrader. This decreases to about 1.5 mSv/h in 3 h. Further measures for reducing the exposure of service personnel have been the installation of local shielding, which is especially important at PSI due to the limited space to pass these systems. This shielding is visible in Figure 3.5. It is rolled stream upward from the degrader when the vault is not accessed by personnel to prevent activation of the lead. The shield has access ports to allow exchange of parts while shielding the personnel.

3.2.4 Energy Analysis System

The width of the energy spectrum of the beam leaving the degrader increases with the energy reduction in the degrader, as shown in Figure 3.6. This is due to

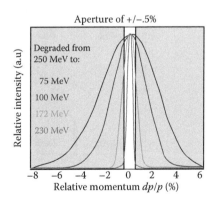

FIGURE 3.6 Momentum spectrum at the exit of the degrader, plotted for a different amount of degrading a 250 MeV proton beam (van Goethem et al. 2009). The schematically shown aperture selects ±0.5% momentum spread. When degrading only a little, most of the particles will be accepted. However, when degrading to low energy, the momentum spectrum is much broader than the accepted momentum fraction.

the statistical differences in the path length of the particles when traversing the degrader. The full width at half maximum (FWHM) of the spectrum at 70 MeV can be up to 13% in energy or 6.7% in momentum.

Beams with a momentum spread larger than about 1% cannot be transported in typically used beam transport systems, and furthermore, this spread decreases the distal sharpness of the Bragg peak. Therefore, a momentum selection (*beam analysis*) is performed. A beam analysis system consists of two large dipole magnets, each bending the beam over typically 45°–50°. Figure 3.7 shows the beam envelopes for 0%, 0.5%, and 1.0% momentum spread, as well as the *dispersion trajectory* (the distance *D* in millimeters to the beam axis of a particle with 1% more momentum than the reference beam, starting on the beam axis) for the beamlines at PSI to the eye treatment room OPTIS2 and to Gantry-2 (Schippers et al. 2007; Rohrer 2006a). Using quadrupoles before and behind the first dipole magnet, "Ana1," the beam optics is designed to have a *dispersive focus* halfway between the two bending magnets, at which the beam size–defining aperture immediately

behind the degrader is imaged. In Figure 3.7 this can be clearly recognized from the small horizontal beam size of the beam envelope with d*p*/*p* = 0% at this location.

For beams with some momentum spread, the beam spot is spread out in the horizontal plane, and a correlation exists between the energy of the protons and their distance to the optical axis, as shown in Figure 3.8. At the dispersive focus, a slit system positioned symmetrically left and right from the beam axis selects the beam momentum (which is linearly dependent on the strength of the magnetic fields), and the aperture of the slit system selects the momentum spread that is accepted for the beam transport to the treatment room. The second bending magnet, "Ana2," compensates for the dispersion again and makes the beam transport almost independent of the beam energy (*achromatic:* *D* = 0).

For eye treatments, the distal edge of the dose distribution is usually desired to be as sharp as possible, requiring a beam momentum spread down to ±0.5% or ±0.25%. For larger and deeper seated tumors, the distal penumbra is dominated by the range straggling

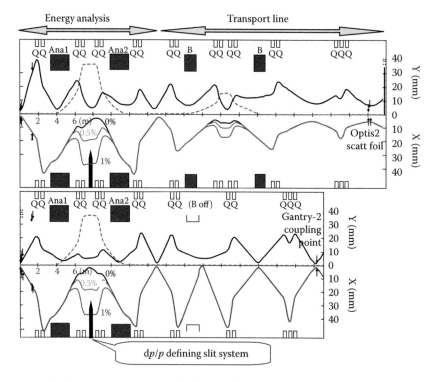

FIGURE 3.7 Beam envelopes (2-σ) to the eye treatment facility OPTIS2 (top) and to Gantry-2 (bottom) at PSI, both with d*p*/*p* = 0, ±0.5 and ±1.0%. The line above the horizontal axis indicates the vertical half-beam size, and the lines below the horizontal axis indicate the horizontal beam half-width for different momentum spreads. The dashed line indicates the dispersion trajectory; the negative dispersion is drawn above the axis. The magnets and momentum defining slit are also indicated along the 16-m-long trajectory. The calculation starts at the collimator at the degrader exit with an accepted emittance (2-σ) of ε*x* = 22π, ε*y* = 48π mm.mrad (high transmission for eye treatments) and ε*x* = ε*y* = 18π mm.mrad (symmetric for gantry) and ends at the checkpoints of OPTIS2 and Gantry-2. The aperture radii of the magnets are indicated in the plot. (The calculations have been performed with TRANSPORT.) (U. Rohrer 2006a, PSI Graphic transport framework based on a CERN-SLAC-FERMILAB version by K. L. Brown et al., http://pc532.psi.ch/trans.htm.)

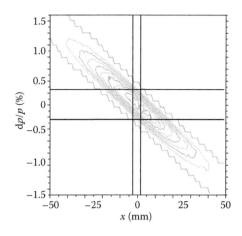

FIGURE 3.8 **(See color insert.)** Correlation between the relative beam momentum dp/p (in percentage) and the horizontal distance x (in millimeters) to the beam axis t the location of the dispersive focus in the energy selection system. The slope of the ellipse indicates the dispersion D of −33 mm/%. The vertical lines indicate the slit aperture at which the minimum momentum spread is obtained. Due to the convolution with the optical image of the aperture at the degrader exit, the correlation plot has a finite width. The minimal achievable momentum spread is indicated by the horizontal lines. The calculations have been performed with TURTLE. (U. Rohrer 2006b, PSI Graphic turtle framework based on a CERN-SLAC-FERMILAB version by K. L. Brown et al., http://pc532.psi.ch/turtle.htm.)

of the protons in tissue, and therefore, a larger momentum spread (±0.5%–±1%) can be accepted. Figure 3.9 displays the influence of momentum spread on Bragg peaks for eye treatments, as well as for treatments of deep-seated tumors.

When a small beam momentum spread is desired, one needs to resolve a momentum spread of 0.25% at the slits. In order to prevent spoiling of the beam by scattering in an excessively narrow slit opening and to obtain more spreading of the beam by dispersion than the beam size of the image (i.e., a high resolving power), a momentum dispersion D = 30–40 mm per percent

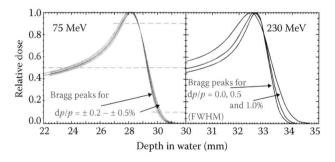

FIGURE 3.9 Left: Distal edge of the 75-MeV Bragg peak for momentum slit settings of ±0.2%–0.5%. Right: Dor 230-MeV protons, with dp/p = 0, 0.5 and 1% FWHM. (Data courtesy of Jens Heufelder, PSI.)

momentum spread is needed at the dispersive focus. This can be achieved with typically 45° of bending.

It is important to pay attention to this resolving power of the analysis system when a very small momentum spread is desired. The half-width x_{image} of the beam at the dispersive focus is given by

$$x^2{}_{\mathrm{image}} = (M.r_{\mathrm{col}})^2 + (R_{12}.x')^2 + (D.\sigma_p/P)^2 \qquad (3.1)$$

where M is the magnification of the image of the beam size–defining collimator, radius r_{col}, x' is the divergence of the beam behind the degrader, R_{12} is a coefficient representing a possible defocusing, D is the dispersion (in millimeters per percentage) at the dispersive focus, and σ_p/P is the relative width (in percentage) of the momentum distribution of the beam. The beam width is clearly not only dependent on the momentum distribution (last term) but also has spatial contributions (the first two terms). When the beamline has been set appropriately, the coefficient R_{12} is zero at the dispersive focus so that the second term vanishes.

The resolving power at the dispersive focus is defined as the amount of momentum change needed to achieve a displacement of the beam equal to $M.r_{\mathrm{col}}$, the imaged radius of the collimator. The accepted relative momentum spread is proportional to the aperture of the slit. However, as shown in Figure 3.8, the consequence of the finite resolving power is that a slit aperture smaller than the image of the collimator does not yield a further reduction in momentum spread. The minimal achievable momentum spread is given by the intersection of the ellipse with the vertical axis in Figure 3.8 as

$$\mathrm{d}p/P = (\sigma_p/P)_{\mathrm{max}} \sqrt{(1 - r^2)} \qquad (3.2)$$

in which $(\sigma_p/P)_{\mathrm{max}}$ equals the maximum momentum spread (=1% in Figure 3.8), and r is the correlation coefficient between momentum spread and position (−0.973 in Figure 3.8). Therefore, the dispersion (D) must be large (e.g., 25–35 mm/%), and $M.r_{\mathrm{col}}$ must be small. In order to accept as many particles as possible, one strives to maximize r_{col} and, therefore, to minimize M (for example, to a value between 1 and 2). A collimator with r_{col} = 5 mm and a dispersion of D = 30 mm/% thus yields a minimum momentum spread of approximately 0.25%, which results for a 70 MeV Bragg peak in a distal falloff of approximately 1.3 mm (90%–10%).

When carbon or helium ions are degraded, a high resolving power is even more important. In the degrader, as a part of the incoming particles will fragment due to nuclear interactions with the degrader nuclei. These fragments typically have the same velocity as the beam particles, and several of the fragments will also have a charge/mass ratio of 1/2. Such particles are not allowed to reach the patient, but when they have the same velocity as the ion beam, they cannot be separated from this beam after leaving the degrader. However, because they are created along the trajectory through the degrader, some slowing down will take place before they leave the degrader. Fortunately, the slowing down of the lighter fragments is less than the original beam ion. This difference in velocity allows a separation by the beam analysis of the majority of the fragments from the beam. Only the fragments that were created just before the degrader exit will have a magnetic rigidity close to the one of the ongoing beam, which explains the need for a high resolving power for sufficient suppression of these fragments.

At the end of Section 3.2.1, it was discussed that the Bragg peak becomes sharp when the beam is degraded to low energy. It can easily be seen that this is due to the fixed momentum *fraction* of the degraded beam, as accepted by the aperture of the slit at the dispersive focus (see Figures 3.6 and 3.8). The relative momentum spread of the beam leaving the degrader increases with decreasing energy, so the absolute momentum spread of the accepted beam (and, thus, the range straggling) is decreased.

To obtain the desired energy behind the energy selection system, three parameters play an important role: 1) the energy of the extracted beam from the cyclotron; 2) the degrader setting; and 3) the magnetic field of the beam analysis system. Note that the absolute values are important in therapy, contrary to many systems for nuclear physics experiments. For quality assurance and patient safety, not only the current through the magnet coil but also the magnetic field should be monitored, e.g. with a Hall probe. In case momentum distributions have a larger width than the acceptance of the slit system, the field strength of the first magnet is the determining parameter that sets the energy that is going through. At 130 MeV protons, a field error of 0.5% will lead to a range error of 2.2 mm, but a slightly wrong setting of the degrader will only result in a transmission decrease. At high energies, however, both the setting of the degrader and the energy from the accelerator are the important parameters, because a large fraction of the momentum distribution is accepted by the slit aperture at high energies. In order to guarantee the correct settings, a redundant measurement of the degrader position is required,

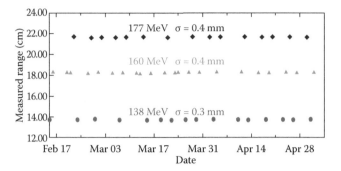

FIGURE 3.10 Results of daily range verification measurements at Gantry-1 at PSI for three different beam energies, evaluated over a period of 3 months. For each energy, the standard deviation of the measurements is given in the plot. (Data courtesy of Dölf Corray and Terence Böhringer, PSI.)

and a stabilization of the ambient temperature of the cyclotron and beam transport areas can be beneficial. Figure 3.10 shows how these uncertainties result in the spread of the measured range over a period of 3 months at Gantry-1 at PSI. The smallest range spread is found when the analysis system is the dominating system to set the energy (at low energy), which is consistent with what can be concluded based on Figure 3.6.

3.2.5 Transmission Determines the Dose Rate

The beam losses in the beamline section encompassing degrader, emittance matching, and energy analysis are very large. As a consequence of nuclear interactions, about 1% of the protons are lost per centimeter of traversed graphite. However, for protons, the increase of the emittance due to multiple scattering causes the largest losses. Of the remaining fraction of the beam intensity, up to another 90% can be lost in the energy selection. Because the combination of the different processes is rather complex to predict, Monte Carlo studies are needed for accurate predictions and optimizations of the collimator system downstream of the degrader. As an example, Figure 3.11 (top) shows the contribution of the individual components in the PSI degrader and collimator system to the loss percentage. Figure 3.11 (bottom) shows that the total transmission decreases strongly with the amount of energy lost in the degrader.

As a consequence, for example, at PSI, only 0.3% of the beam intensity from the accelerator reaches the entrance of the beam delivery system of the eye treatments when degrading from 250 to 76 MeV, with an acceptance of 44π mm.mrad in each transverse plane and a momentum acceptance of $\pm0.6\%$. The dose rates

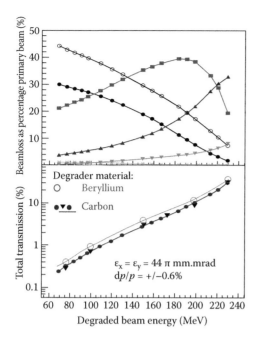

FIGURE 3.11 Top: Beam losses as a percentage of the primary beam as a function of the degraded beam energy for several beamline elements at PSI: Carbon multiwedge degrader (open circles); KMA3 set at 13 mm Ø (filled circles); KMA4 (squares); KMA5 set at 30 mm Ø (blue triangle pointing up); momentum defining slits at ±12 mm = ±0.3%dp/p (triangle pointing down), and losses in the magnets (stars). Bottom: Total transmission through a degrader and energy selection system with the same KMA3 and KMA5 settings, but with the momentum slits set at ±20 mm. (M. J. van Goethem et al. 2009. *Phys. Med. Biol.* 54: 5831–5846.)

typically needed for treatments of deep-seated tumors are only a few Grays per minute, and are usually not a problem to achieve. However, treatments of eye melanoma typically use 10–20 Gy/min, which require additional measures.

There are several possibilities of obtaining sufficient high dose rates in the treatment areas. First, one could simply increase the beam intensity from the accelerator. For a synchrotron, this is not always trivial, because the ring is usually filled to a maximum amount of particles, determined by repelling space charge forces when injecting into the ring. An injection system with a higher energy would be of advantage in this respect. One could also extract the beam faster. Although this helps, one will remain with the fixed dead time between the spills to fill the machine and accelerate to the required energy. In a cyclotron, one can simply increase the current from the ion source. When the machine has been designed well and losses during the acceleration and extraction process are limited, medical cyclotrons could extract up to a microamp or more, if necessary. However, the losses in

the accelerator and in the ESS may need extra shielding measures to prevent too large a neutron dose outside the vault. Therefore, it is often most effective to improve the transmission between extracted beam and patient. Apart from increasing the apertures of selected magnets in the beam transport system, which can be costly because more powerful power supplies are also needed then, the most straightforward method is to compromise on the distal fall of the Bragg peak. One can then increase the aperture of the momentum slit and/or transport a higher beam energy to the treatment room and perform a large range shift in the nozzle. However, one should realize that, apart from the decreased steepness of the Bragg peak, this also generates more neutrons in the treatment room. If these disadvantages are not desired, the nozzle can be optimized to make efficient use of the protons that enter the nozzle. A double-scattering system (Schippers et al. 2006; Chu, Ludewigt and Renner 1993; De Laney and Kooy 2007; Gottschalk, unpublished book; Koehler et al. 1977; Grusell et al. 1994; Pedroni et al. 1995; Schippers et al. 1992) or a scanning (wobbling) pencil beam can provide high efficiency. In some facilities, there is the possibility of using beryllium as degrader material (Anferov et al. 2007); due to its lower density, beryllium causes less multiple scattering. In Figure 3.11 (bottom), it is shown that, for the PSI geometry, this would increase the transmission with a factor of 1.2–1.3. However, because beryllium is brittle and toxic, special measures (for example, packing or mixing with aluminum) must be taken to use this material. In addition, one should not use beryllium in close proximity to the patient, because per proton reaching the patient, about 30% more neutrons are produced in beryllium compared to carbon (van Goethem et al. 2009).

An often underestimated contribution to the transmission is the phase space of the beam that enters the degrader. Van Goethem et al. (2009) have shown that, for moderate degradation to high energies, the transmission starting with a realistic phase space ($\varepsilon_x = 1.6 \pi$ mm.mrad and $\varepsilon_y = 3.4 \pi$ mm.mrad was used in the calculations) can be 30% lower compared to the transmission that starts with a zero-emittance pencil beam. Even for degradation from 250 to 76 MeV, the transmission can be 20% lower due to this emittance.

One can also take advantage of the effect that the multiple scattering contribution is added squared to the square of the divergence of the incoming beam. A very large angular spread in the incoming beam (i.e., very strongly focused) then causes a lower relative scatter contribution in the outgoing beam. This method

Chapter 3

needs very strong focusing systems before and behind the degrader. If this is done repeatedly and the degrading is performed in a few steps, one can gain a lot in transmission; however, one needs several meters of beamline length instead.

Last, note that a lower beam energy from the accelerator (e.g., 230 MeV instead of 250 MeV) also helps in limiting the transmission losses with a few percentage (relative). Of course, this small advantage has to be balanced against a 5-cm (water-equivalent) shorter range in tissue.

In the case of helium or carbon ions, the multiple scattering in a degrader is a factor 2 or 4 less, respectively. Therefore, the transmission will be about 4 to 16 times larger so that the transmission problems at low energies are much less severe.

3.3 Beam Transport Systems

The beam transport system between the energy selection system and the treatment rooms typically uses standard beam optics. The choice is not so critical, but important aspects to consider are maximum beam dimensions, locations of possible losses, locations of beam diagnostics, stability against beam shape change, modularity with periodicity and symmetry (to reduce higher order aberrations) in case of a long beamline, and a very good reproducibility of the settings. In addition to these requirements, note that a well-defined and transparent beam optics allows accurate interpolation between a limited amount of reference beamline settings without extensive quality control procedures for "each" energy. This also saves time during troubleshooting and is of benefit for the requirement of short switching times between treatment rooms or between beamline settings (energies). It must not be necessary to perform extensive beam tuning after a change of the setting, although one should develop a fast verification and correction procedure for a final accurate adjustment of the beam position. At Gantry-1 of PSI, which employs the spot scanning technique (Pedroni et al. 1995), the beam position at the patient is corrected (if necessary) after (part of) the first spot has been applied to the patient. This position is not always sufficiently reproducible due to the nonlaminated gantry magnets. Such a correction must be small and easy to perform. At OPTIS-2, where the beam position must be correct within 0.05 mm, a fast procedure is also applied, performed in between patient treatments. For Gantry-2, which has laminated magnets, the experience shows a sufficiently good reproducibility, so corrections are not expected to be necessary.

3.3.1 Layout and Beam Optics

Between the energy selection system and the entrance of the gantry rooms, the beam transport system is typically a telescopic system in which point-to-point imaging is applied with a unit magnification. The presence of intermediate foci can be conveniently used for beam diagnostics. When the beam hits a component that is unintended in the beam path, there is a big chance that this can be recognized from an increase of beam size. Examples of beam envelopes in the lines to OPTIS2 and Gantry-2 are shown in Figure 3.7. Of course, other schemes with a smoother beam transport than the one shown can be applied as well.

It is convenient to include a meeting point in the beamline design, at which the beam leaving the accelerator is aimed at by dedicated beam steering magnets and which acts as the start of the beam transport system. A steering magnet at the meeting point (working in both transversal directions) will match the beam direction to the beamlines. This alignment is important and prevents beam steering when other energies are set. When a degrader is used, the beam must be focused into the degrader to a small spot size to optimize the transmission.

Apart from using the accelerator or its injector to switch the beam on or off, a fast kicker magnet can be located in this beamline section before the degrader. At PSI, for example, this kicker is used to deflect the beam to a stopper to make the normally required beam interruptions, as shown in Figure 3.12. It is directly controlled by the therapy control system, and it acts as the first-level beam-interrupt system.

In order to prevent beam size increase due to energy spread in the beam, it is of advantage to design bending sections as achromats so that the dispersion is again zero behind the bend. This typically involves a split bending magnet with a quadrupole in between the two halves. This quadrupole acts like a mirror, reversing the direction of the dispersion created by the first magnet so that the second bending magnet brings the dispersion back to zero.

The beam that is extracted from the accelerator possesses a correlation between particle energy and distance to the optical axis of the beamline. Such a correlation must be removed for proper beam transport,

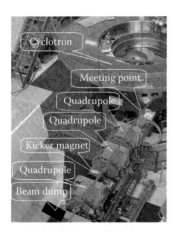

FIGURE 3.12 First beamline section during the installation of the cyclotron and beamlines at PSI. In the hole in the cyclotron yoke from which the beam exits, an *x/y* steering magnet will be installed, aiming the beam toward the meeting point. A second *x/y* steering magnet at the meeting point was not yet installed when this picture was taken. The kicker magnet at the bottom of the picture can deflect the beam on a beam dump (indicated by the black block) located in the vacuum box of the degrader. The third quadrupole is horizontally defocusing and helps in deflecting the beam.

especially when gantries are used. In a cyclotron beam (Schippers 1992), it will be mostly smeared out by the energy straggling in the degrader. In several synchrotron facilities, a matching section that consists of one or two bending magnets of about 30° is installed right behind the extraction (Meot 2008; Schönauer 2007; Benedikt 2005; Hiramoto et al. 2007). In this beamline section, also shown in Figure 3.1 (bottom), the dispersion of the bending magnets cancels the correlation.

Apart from dispersion matching, the phase space (emittance) of a synchrotron must be matched. The emittance and momentum spread can be typically up to a factor 10 smaller than the emittance of a cyclotron beam. Furthermore, in case of a horizontal extraction process, the horizontal emittance is not Gaussian shaped and, typically, a factor 10 smaller than the vertical emittance (Furukawa 2004; Furukawa 2006a; Ondreka et al. 2008).

A symmetry in phase space (or at least in beam size) is essential, especially at the entrance of a gantry system, because the shape and position of the beam at the patient and the transmission through the gantry should not depend on the gantry angle. Often, both the emittance and the beam position depend on the energy at which the beam is extracted from the synchrotron so that the beamline settings must include a beam energy dependent correction. Furukawa and Noda (2006) have proposed to employ a thin scatterer in the beam transfer line in order to compensate for

the asymmetry of the phase-space distribution. They showed that a thin scatterer, which introduces a scatter angle of 20 μrad and is set at an appropriate position in the beamline of the carbon-ion facility HIMAC (Koda et al. 2008), was sufficient to obtain a 2-D Gaussian distribution in the phase space and a symmetric beam condition for a rotating gantry. However, when the preservation of the small emittance is used in advantage to have a beam transport system with relatively small magnet apertures, other methods need to be used (Furukawa et al. 2006b; Tang et al. 2009). The method of using solenoids is very suitable to produce a coupling between the two transverse planes to compensate for the asymmetry. Tang et al. (2009) showed that a superconducting solenoid with a 1-m length and a field of 3.9 T can be used for 250-MeV protons. A design of a rotator section composed of quadrupoles is suggested by Benedikt et al. (1997, 1999). This method requires an additional quadrupole section, which is rotated mechanically over an angle proportional to the gantry angle.

The coupling of the beam transport system to a gantry can be made in different ways. Rotating vacuum connections are commercially available. In order to guarantee the emittance symmetry and zero dispersion, beam diagnostics is essential at this location. This need not be used during treatment routine but can be essential during the commissioning of the system. Very robust beam optics is obtained when the gantry images this coupling point to the patient. Then, slight differences in divergence between the horizontal and vertical planes are not visible at the isocenter. At PSI, a round collimator is located at the coupling point, with an aperture of the same size as the beam width. In this way, a possible misalignment of the beam results only in a decrease of the transmission but not in a change in spot position at the isocenter. In addition, a slight error or asymmetry in the beam size only leads to a small transmission loss at this collimator. However, a slight change in the shape of the beam at the isocenter is possible, which may have an effect on the dose distribution when the distance between the spots is too large. At PSI, the vacuum of the beamline has also been decoupled from that of the gantry. As shown in Figure 3.13, an air gap of 20 cm between the exit foil of the beamline and the entrance foil of the gantry is filled with the collimator, a fast (20-ms) mechanical beam stopper, and beam diagnostics that permanently measure the beam intensity, beam position, and spot size. In addition, a beam profile monitor can be inserted into the beam path.

Chapter 3

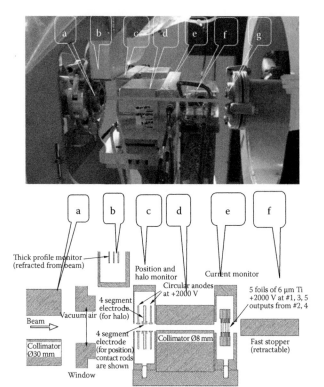

FIGURE 3.13 Coupling point of the gantry at PSI. The beam leaves the beamline vacuum at the left side of the picture (a). It then subsequently goes through a (b) moveable profile monitor, (c) beam position and beam size monitor, (d) collimator, (e) intensity monitor [parallel plate ionization chamber (IC)] and (f) fast mechanical beam stopper. It then enters the rotating beam tube of the (g) gantry entrance.

Because beam splitting would involve many complications due to the different and varying requirements in the treatment rooms, only one treatment room can be served at a time in the current systems. In order to prevent unnecessary waiting time, the time it takes to switch between rooms should be minimized and not take more than 0.5 min. Important parameters that determine this area switching time are magnet design (magnet yokes should be laminated; high voltages must be allowed between coil windings; there should not be too many turns to limit the inductivity), power supply design (allow high overshoot voltage), and control system issues (change all magnets all at the same time and perform fast verifications). When power supplies are shared between magnets, load switching must be done fast as well, which always requires extra time to ramp down to zero current when the switch is made. In addition, in all cases, a few seconds are needed at 100% current to be sure the magnet is always on the same *hysteresis curve* (the dependence between magnetic field strength and the current in the coils). Therefore,

this implies that typically less than 10–20 s should be specified to ramp from 0% to 100% of the current. Although this is already quite fast, a factor of 2–4 higher speed is needed when the degrader is used as a fast energy modulator (cyclotron case) and volumetric rescanning is desired. Because the magnet current is not varying periodically, one cannot rely on resonant circuits. Therefore, this speed requirement has some consequences on the cost.

3.3.2 Beam Diagnostics

Before and during the dose application at the patient, measurements of beam characteristics and device parameters are made. In general, it is a good practice to distinguish between measurements that are needed to tune the machine or warrant machine safety and diagnostics that are used to assure a safe and correct patient treatment. It may reduce complicated efforts to ensure a rigorous separation between these two systems. Such a separation is necessary, because systems involving the patient safety and/or dose monitoring are subject to special regulations and may need special designs and/or redundancy. For systems that are used only for machine verification, the requirements can often be less stringent.

The required speed to change beam energy or treatment room does not allow long tuning procedures. Therefore, one must strongly rely on the reproducibility of the beam transport system and the robustness of its ion optics design to small changes. Oftentimes, only a few beam parameters are monitored during treatment. The results of these measurements are beamline setting dependent and continuously compared with precalculated or measured reference values. The choice of the measured parameters and real-time information of device parameters and statuses must be made such that any deviation in the beamline setting will be detected.

At PSI, for example (Dölling et al. 2007), real-time intensity, position, and profile measurements are done continuously just before the degrader and intensity and position measurements are done at the entrance of each gantry (or treatment room) and in the gantry nozzle (through the dose monitors). Along the beam transport system, no further measurements of the beam are made during treatments.

In principle, standard methods for beam measurements can be used. However, apart from specific physics specifications, robustness is the most important. Dealing with small signals, avoiding microphony due to pumps, cooling water, and, sometimes, high background radiation levels and allowing many (sometimes fast) mechanical motions of actuators and proper mounting of end switches requires high-level engineering. Because the availability of the facility is of utmost importance, a certain level of redundancy is needed for the beam diagnostics. This does not imply doubling the equipment, but rather sufficient measurement points along the beamline as a whole to draw unique conclusions, even with one faulty device.

FIGURE 3.14 Thin profile monitor (partly assembled) for permanent installation in the beamline at PSI. Titanium foils of 6-μm thickness are used as electrodes to avoid excessive scattering of the beam and to prevent long-term radiation damage as expected for Kapton or Mylar foils. The detector is enclosed in a hood with two 50-μm titanium windows. (Dölling, R. et al. 2007. Proc. AccApp'07, Pocatello, Idaho, pp. 152–159.)

In order to have quick access and service, one can consider placing the electronics modules outside the vaults. The long cable lengths and possible electromagnetic pickup necessitate well-shielded and separate cabling in that case. In general, cables transporting (small) analog signals need to be mounted on separate cable trays. To prevent microphonic pickup, care should be taken of mechanical insulation from pumps and water tubes. A clear and rigorously pursued grounding philosophy must also be used to avoid ground loops and interference between devices.

Contrary to the usual need for nondestructive measurements in accelerator laboratories, we can allow (partially) beam destructive monitoring for verifications during beam tuning. Making sure that the device is out of the beam path when treating patients can even add to inherent safety. Of course, monitors that are always in the beam must be made as thin as possible to reduce emittance growth.

When fast tune changes need to be observed, the signal processing should be of large bandwidth (up to 10 kHz), and many channels need to be processed in parallel. In equipment specifications one should also take care of the large dynamic range (10 pA–1 μA) of the beam intensities and possible saturation of signals.

Typically, one needs to measure beam profiles, positions, intensities, energies, and losses. Beam profile monitors are extremely useful to detect setup and calibration errors. To obtain optimal information and beam envelop reconstruction possibilities, these should be located at beam waists and at locations

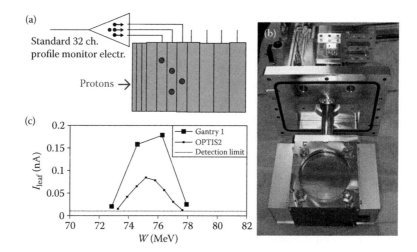

FIGURE 3.15 Beam energy is derived from range measurements in a stack of insulated copper plates [multileaf Faraday cup (Paganetti et al. 2003)], inserted in the beamline. (a) and (b) Operation principle and the device used in the PSI beamline. A version with thin plates and optimized around 75 MeV is used in the beamline dedicated for eye treatments, and a version with thicker plates to measure over the full range of 70–250 MeV is installed in the beamline to the gantry. (c) Beam energy distributions derived from the range distributions measured with both versions at a beam current of 0.46 nA. (Dölling, R. et al. 2007. Proc. AccApp'07, Pocatello, Idaho, pp. 152–159.)

Chapter 3

where the beam is relatively large, for example, before quadrupoles.

Often, ionization chambers flushed with ambient (dried) air or with nitrogen are used for profile (thus, also position) and intensity measurements (for example, see Figure 3.14). At the higher currents in front of the degrader, saturation effects can alter the signals. Hence, at PSI, secondary emission monitors are used there in addition. However, monitors based on secondary emission may suffer from performance changes due to deposits or damages of the electrode surface. Therefore, frequent checks of these monitors are necessary.

Beam stoppers with an electron suppression electrode and parallel plate ion chambers can be used for intensity measurements, whereas beam energy and momentum spread can be determined by multileaf Faraday cups (Figure 3.15; Paganetti et al. 2003). A profile monitor at the dispersive focus can also be used for relative energy

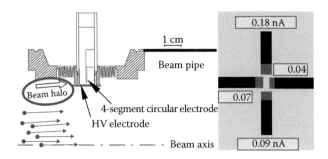

FIGURE 3.16 Beam halo monitor. The device is folded around a specially modified bellow in the beam pipe. The insertion slightly protrudes into the beam vacuum. Protons traversing the air volume in the chamber will generate a signal on the electrode. Signals below 1 pA can be detected. The electrode is segmented in four quadrants. The signal on the right shows the largest signal on the upper electrode.

measurements. Beam losses can be detected by ionization chambers next to the beamline or with beam halo monitors (Dölling et al. 2007; Figure 3.16).

3.4 Gantries

A gantry consists of a mechanically supported beam transport system that is able to rotate around the patient so that the tumor can be irradiated from many directions. When designing a gantry for particle therapy, one faces several practical problems, both in the mechanics (a 100–200 ton device, with pointing accuracy of 0.1 mm) and in the beam optics. An overview of different gantry configurations presently in use is given in Figure 3.17 (IBA; Koehler 1987; Optivus; Pedroni et al. 2004; Hitachi; Norimine et al.).

The possibilities of designing the beam optics are constrained by the limited number of components (space), the limited size of the magnets (weight), and, sometimes, relatively long distances between certain optical actions. First, the beam must be bent away from the original direction (which is the rotation axis of the gantry). Then, the beam is bent to a direction perpendicular to the original beam direction. This last bending can be achieved by one magnet (e.g., gantries of IBA and Varian) or by multiple magnets (e.g., PSI, Hitachi, Loma Linda, Heidelberg).

The first gantries that have been installed are of the corkscrew type, as shown in Figure 3.17a. This gantry design originates in the work of Andy Koehler (Koehler 1987). At the first hospital-based proton therapy facility at Loma Linda, CA, three of these gantries, which are 12 m in diameter, have been installed (Slater et al. 1992). In this gantry, basically a bend of 90° is made, followed by a 270° bend. As shown in Figure 3.17a, this bend is in a plane perpendicular to

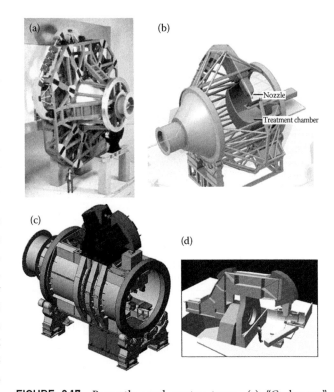

FIGURE 3.17 Presently used gantry types. (a) "Corkscrew" design at Loma Linda (Koehler 1987; Slater et al. 1992). (b) Gantry type for the example used in Boston, Jacksonville, and Oklahoma, with a 45° + 135° bending magnet (IBA). A similar scheme is used in the gantries installed in Munich (Varian Medical; http://medgadget.com/archives/2009/02/europe_approves_varians_proton_therapy_system_a_cancer_zipping_cyclotron.html). (c) Gantry used for example in Tsukuba and Houston (http://www.pi.hitachi.co.jp/rd-eng/product/industrial-sys/accelerator-sys/proton-therapy-sys/index.html; Norimine, T. et al. 2002), in which the 135° magnet has been split in 45° + 90°. (d) Gantry-2 at PSI, dedicated for parallel pencil beam scanning (Pedroni et al. 2004).

the original beam direction, resulting in a very short system. Both bends are achromatic, the first one by splitting the bend into two times 45° and the second bend by splitting in two times 135°. With quadrupoles in splits, the dispersion is guided to zero each time. Although this gantry has a large radius, the beam optics and the low emittance from the synchrotron allow relatively narrow magnet gaps and small magnets widths.

It is important to design an achromatic beam transport within the gantry. This is needed to prevent beam size increase at the isocenter (important for pencil beam scanning) and to reduce the dependence of the beam at the isocenter from the gantry angle. In the IBA (Figure 3.17b) and Varian gantries, only two bending magnets are used: 1) a 45° magnet to bend out of the plane and 2) a single 135° magnet to bend toward the patient. As shown in Figure 3.18, which shows a possible beam optics in the IBA gantry in Boston, the five quadrupoles between the bending magnets have a combination of functions, focusing the beam in both planes and matching the dispersion at the entrance of the last bending magnet. Due to the large bending angle of this magnet and the need to have zero dispersion at the isocenter, a quite-large dispersion and a large angular dispersion are needed at this point. As shown in Figure 3.18, this implies that the dispersion in the last quadrupole before the 135° magnet must be about 35 mm/%, resulting in a large beam size in the bending plane at this location. The strength of this last quadrupole is determined only by the goal to have zero dispersion behind the last bending magnet.

The focusing of the beam at the isocenter, therefore, is mainly achieved with the previous four quadrupoles, with the constraint imposed by the limitation of the dispersion function. The four quadrupoles at the entrance of the gantry are matching the phase space to the gantry and also contribute to the focusing of the beam at the isocenter. Because the quadrupoles that perform the focusing of the beam at the isocenter are located at a large distance from the isocenter, it is not possible to obtain a small beam size at the isocenter. For scattered beams, this is not a problem, but for scanning beams, either quadrupoles must be added behind the last bending magnet or severe limitations on the used beam emittance must be accepted (leading to lower beam transmission between the cyclotron and the isocenter).

The last dipole in the gantries of Hitachi, Tsukuba, Japan (Figure 3.17c) has been split into a 45° and a 90° part. This may have mechanical advantages and also allows extra elements for additional focusing or steering. However, the dispersion function at the entrance of this split magnet is up to 50 mm/% (Norimine et al. 2002). Because these gantries are usually applied in a synchrotron-driven facility with small emittance beams with very low momentum spread, this is not a problem.

Behind the second bending magnet in the PSI gantries (Figure 3.17d) and in the gantry for carbon ions in Heidelberg (Figure 3.19; Haberer et al. 2004; Kleffner 2008; Weinrich 2006), the beamline runs horizontally over a longer distance. In this section, several quadrupoles are installed to match the dispersion, to confine

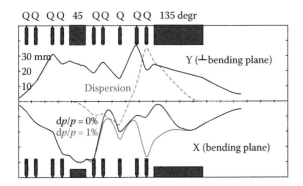

FIGURE 3.18 Possible beam envelopes in the IBA gantry, as derived by U. Rohrer, PSI (Rohrer, U., private communication). The dispersion trajectory is plotted with a dashed line. In the bending planes, beam envelopes are plotted for 0% and 1% momentum spread. The envelopes have been derived from data published by J. Flanz et al. [J. B. Flanz, *Nucl. Instrum. Meth. B* 99 1995 830–834 and J. B. Flanz, Particle Accelerator Conference (1995).]

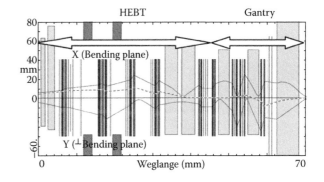

FIGURE 3.19 Beam envelopes in the Heidelberg gantry for carbon ions in Heidelberg. In this plot, the horizontal (in bending plane) envelopes are plotted above the axis. The dispersion is scaled as millimeters per percentage. HEBT = High-energy beam transport line between the synchrotron and the gantry. (From C. Kleffner, Proc. EPAC 2008, 1842–1844.)

Chapter 3

the beam size within the gantry, and to focus the beam at the isocenter to a size of a few millimeters (pencil beam). The scanning magnets are located just before the 90° bending magnet. The dispersion in these systems also reaches maximum values of 40–50 mm/%. The ion optics in most gantries is designed such that the coupling point at the entrance of the gantry is imaged to the scatter foils in the nozzle or, in case of pencil beam scanning, to the isocenter.

At the time of writing, pencil beam (spot) scanning is applied only in a few facilities: PSI, Villigen; German Cancer Research Center (DKFZ), Heidelberg, Germany; Rinecker Proton Therapy Center (RPTC), Munich, Germany; and M. D. Anderson Cancer Center, Houston, Rexas. In addition, the first patient at Massachusetts General Hospital (MGH), Boston, has been treated with spot scanning. When pencil beam scanning is employed, the scanning magnets are usually located downstream the last bending magnet, which thus needs a large gantry radius to have enough distance to the isocenter. The center of the scanning magnets is the virtual source from which the diverging pencil beam pattern is emerging. When the scanning magnets are not combined into a single magnet, the treatment planning must take care of the different locations of the virtual sources from which the pencil beams emerge.

In the PSI gantries (Pedroni et al. 1995, 2004) and in the gantry for carbon ions at DKFZ (Haberer et al. 1993; Figures 3.19 and 3.20), the scanning magnet(s)

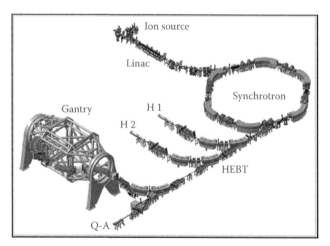

FIGURE 3.20 Layout of the carbon-ion facility at DKFZ. The beam is accelerated in a synchrotron. There are two treatment rooms with fixed horizontal beam lines, equipped with a nozzle for pencil beam scanning. One treatment room is using, so far, the only gantry in the world for carbon ion (Haberer et al. 2004; Kleffner 2008; Weinrich 2006; Haberer et al. 1993).

are located before the last (90°) bending magnet. At PSI, the ion optics of this magnet is made such that the pencil beams are displaced parallel to each other at the isocenter; the virtual source is located at infinity. Parallel pencil beam displacement has advantages with respect to treatment planning and the design of patched fields. In addition, the skin dose can be slightly reduced compared to divergent scanning.

New gantry designs that are presently being studied aim at a reduction of the weight and size, which is especially important when ions are used. Moreover, when energy modulation is performed upstream, the associated fast changes in magnet setting put costly requirements on the last bending magnet and on their power supplies. The FFAG-type gantry (Trbojevic et al. 2007) uses ion optics allowing relatively small (but many) magnets. The very large energy acceptance would allow that the magnetic fields need not be changed when the energy is changed before the gantry. The design made at Kernfysisch Versneller Instituut (KVI), Groningen, The Netherlands (Vrenken et al. 1999), uses one scanning magnet before and one behind the 90° bending magnet. This reduces the needed gap compared to the PSI design, but it does not have parallel scanning as a consequence.

To reduce costs, several designs have simplified (shortened) the beam transport system so that the gantry is no longer isocentric. However, as a consequence, these need a moving treatment table or "treatment room" (Reimoser et al. 2000; Kats 2008). The simplest version of this is Gantry-1 and has been used at PSI since 1996, at which the patient table is connected to the gantry mechanics. A disadvantage of such a system is the effort needed to reach the patient quickly at any gantry angle. This applies especially when beams from below are used, because then the patient table is quite high above the floor.

The application of superconductivity is investigated in several designs, with the aim of reducing the size of the gantry by using strong magnetic fields. This could especially be beneficial in the case of ions. Rotating cryogenic systems have been demonstrated at the Harper Cyclotron (Blosser 1989), for example. Novel cooling techniques or high-Tc conductors that do not use liquid helium will be advantageous for this application. Fast changes in magnet current are a problem for most superconductors; therefore, upstream energy modulation may not be an option in this case. In addition, the strong magnetic fields may give rise to strong stray fields, causing angular dependent effects.

3.5 Nozzle

The last part of the beam transport system shapes the beam to the desired shape and hosts the final dose monitoring and is called the treatment nozzle. It typically consists of components for shaping the beam [collimator(s), scatter foils, and/or scanning magnets], energy modulation, range shifting, dose monitoring, beam property verification, X-ray equipment, light field equipment, and lasers for alignment. Many reviews are available on beam shaping, as well as in other chapters in this book, and developments to improve techniques or reduce costs are in progress. Furthermore, in facilities that employ both pencil beam scanning and scattering, flexibility is required to switch between scanning and scattering or to perform pretreatment or aftertreatment verifications on the patient position.

The required flexibility, space constraints, diversity of the devices, and requirements on beam shape (sharpness) often lead to conflicting requirements and impose complex mechanical solutions. Collimators, compensators, and range shifters need to be located as close as possible to the patient to maintain a sharp lateral dose falloff but cannot always be close to the skin of the patient (e.g., in the neck region). However, these (often bulky) devices also need to be changed per field so that remote handling would save effort and time.

Especially for systems using a scattered beam, it should be realized that a considerable fraction of

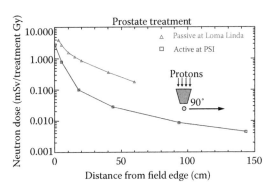

FIGURE 3.21 Neutron dose per gray of proton treatment as a function of the distance to the proton beam axis in the isocenter. A passive scattering system is compared with a scanning system. (From A. Wroe et al. 2007. *Med. Phys.* 34:3449–3456.)

the protons that enter the nozzle are stopped in collimators. These will create a neutron dose, which should be taken into account for certain treatments (Agosteo et al. 1998; Schneider et al. 2002; Schneider et al. 2006; Fontenot et al. 2008; Zheng et al. 2008). As shown in Figure 3.21, pencil beam scanning offers clear advantages here, because this technique has only a little material in the beam path (Schneider et al.; Wroe et al. 2007). For a scattered proton beam, it was shown that a reduction of the beam size at the patient collimator, performed by upstream precollimation, also yields a lower neutron dose at the patient (Taddei et al. 2008).

3.6 Control Systems, Mastership, and Facility Mode

The development, commissioning, and, sometimes, certification of the control and safety systems of a particle therapy facility are extremely challenging tasks, and the needed effort has been underestimated in almost all facilities that used a certain hardware configuration for the first time. This also has as a consequence that the effort to change existing and certified control systems are often considered too large, expensive, or complicated. A separation of functions, however, reduces the risks and complexity that might occur in the case of a system in which the design is based on one combined operation and safety system in which "everything is connected to everything else." Of course, well-designed systems with a global function approach to the facility can be conceived without this separation, but the separated function approach leaves more freedom for further technical developments.

Therefore, the modularity of the system is extremely important. A clear and consequent separation is needed between systems that control the machine (e.g., setting the current of a power supply, inserting a beam monitor, measuring the beam intensity) and systems that control safety. The only interactions between these safety systems and the control system consist of receiving and sending status information.

The concept of the control system architecture is related to the goals and design of the safety systems. Questions such as who is in control in a facility with multiple treatment rooms (mastership), who can do what (machine access control) and when (facility mode), and how a separation of (safety) systems is guaranteed need to be considered in any design. There are many different system concepts in use; therefore, it is not possible to give a complete description here.

Chapter 3

In order to illustrate the basic principles, the concepts used at PSI will used as an example in this section.

3.6.1 Control Concept

At PSI, a rigorous separation has been implemented between the responsibilities of cyclotron and beam transport lines and those related to the treatment equipment. This has decoupled the tasks and responsibilities of the machine as a beam delivery system and a user who asks for a beam with specified characteristics and decides whether the beam is accepted for treatment.

This separation of responsibilities is present both in the hardware and in the control system architecture (see Figure 3.22). A machine control system (MCS) controls the accelerator and beamlines, and it only controls the machine performance itself. Each treatment area has its own therapy control system (TCS). Each TCS communicates with the MCS through a beam allocator (BALL), a software package that grants the TCS of the requesting area exclusive access (the *master status*) to the corresponding beamline up to the accelerator. In addition, it grants the master TCS a selected set of actions. This includes control of the degrader, beamline magnets and kicker, and the right to give beam on/off commands. The BALL acts on the demand of the TCS so that no mixing of responsibilities between machine and user occurs. The master TCS will ask the MCS through the BALL to set the beamline and accelerator according to a predefined setting list. Independent of the MCS, the master TCS will start, verify, use, and stop the beam to perform the patient irradiation (Pedroni et

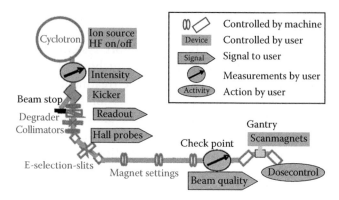

FIGURE 3.23 Dedicated measurements or actions (indicated as text on gray background) are performed by the TCS and safety systems, independent of the system that controls the accelerator and beamlines (indicated as open symbols in the beam line).

al. 1995). Dedicated measurements, verified by the TCS and safety systems, are made through their "own" beam diagnostics, as illustrated in Figure 3.23. The results of these measurements indicate whether the status with respect to the beam is alright (i.e., whether the beam complies with specifications on energy, position, direction, emittance, and intensity).

3.6.2 Facility Modes

In order to organize when certain operator actions are allowed, different (at least two) facility modes must be defined. A *therapy mode* is used for patient treatment, and a *machine, physics,* or *experimental mode* is used for the daily setup of the machine and allows beam tests to be made with the accelerator and the energy degrader. It can be convenient to define an intermediate mode (at PSI: *diagnostic mode*) for tuning a beamline that is allocated to an area with a master status. The facility mode determines the safe state of the configuration of all signals and status information for the safety systems. In the machine mode, beam blockers prevent the beam from being directed to a user area. Only the operator of the treatment area that has obtained mastership can set the facility mode to the therapy or the diagnostic mode. Switching from an area that is in the diagnostic mode to an area in the therapy mode requires a procedure that first forces the beamline and beam current setting into a safe state.

3.6.3 Hardware and Software

In general, one should strive to use well-proven components and systems. Aspects to consider when

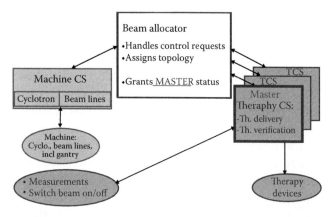

FIGURE 3.22 Concept of the different control systems at PSI. Each treatment area is controlled by a TCS. Only one of these systems has mastership over the facility and can set beamline components through the BALL. Necessary measurements and beam on/off are done directly by the master TCS.

selecting hardware are robustness, fail-safe design, which transient states are possible, what happens if the device is switched off or cables are not connected, robustness and signaling of overflow or signal saturation, time response (speed as well as reproducibility), possible *safety integrity level* (SIL) qualification (the robustness of such a device or measure), and certification by the manufacturer. Programmable logic controllers (PLCs) can be used for user interface applications and general control functions. In general, however, PLCs are not allowed to be used in safety systems. Therefore, some companies have developed dedicated and certified safety PLCs. To reach the required level of safety, special concepts (for example, redundancy) have been integrated into the PLC design. One part of these concepts is a rigorous test program that is to be performed after any small change in a program of the PLC. When speed or a reproducible time response is an issue (e.g., in switch-off systems), advanced logic components and/or digital signal processors (DSPs) are preferred.

The separation of the safety systems and the control systems extends to the cabling of the hardware and, if possible, to the hardware itself (for example, the use of dedicated ionization monitors per safety system). Each system should, for example, also have its own signal cables and limit switches.

In order to achieve enough speed for energy changes or switching treatment area, as much intelligence should be installed locally (i.e., at the device level). This is often performed by using dedicated VME boards (a standard for electronic processing units), equipped with field-programmable gate arrays (FPGAs; programmable component in an electronic circuit) or local processors (CPU). This way, beam diagnostic modules, for example, can perform onboard (fast) profile analysis, check intensities against limits, or generate interlock signals based on the characteristics of a beam profile. Power supplies only need very basic commands (on/off, requested current, or field), because they "know" their $I(t)$ curves and generate an OK signal when a new current is within limits. VME crates with PowerPC processors can also be used for interlock/warning functionality.

All the devices, except for time-critical safety systems, are connected, for example, through a switched high-speed Ethernet network. Operator workstations and servers for dedicated functions (database, different servers, archiving, and alarm handling) are also connected to this network. For dedicated components (for example, vacuum and cooling systems), a PLC is often used, which is then also connected to the Ethernet so that it can also be controlled by a user interface on an operator workstation. Time-critical safety systems must be hardwired.

In general, software applications running on servers and used through the operator workstations, provide user interfaces not only to control the machine but also to perform the communication with the BALL. Interpreting safety system signals and interlocks is an important task needed to maintain a high availability of the system. The higher level of the servers and workstations is also suitable for applications that perform automated tasks such as ramping magnets and measuring a set of parameters while changing another parameter or continuous online evaluation of measured signals. Many applications are needed, which encompass standard tools such as synoptic displays, control loops, device control, archiving, and data retrieval and analysis.

3.7 Conclusion

Currently, the beam delivery systems for particle therapy have reached a mature and well-demonstrated state. Due to local boundary conditions, the layout of a new treatment facility may be different from the usual one, which is a good moment to reconsider and optimize the currently used schemes. When the planned treatment program is expected to be limited to certain indications (for example, refraining from eye treatments), one may simplify or leave out parts of the generally used beam delivery components.

In the last few years, many studies have been performed to improve the accelerators and gantry systems, but real improvements can only be achieved when the facility as a whole is considered, including even the most trivial and straightforward parts. One of the most important specifications, availability, can be seriously compromised when easy access, relaxed design parameters, and quick exchange are not an intrinsic part of the design process. For all new development and simplifications, the first key questions should always be whether one can provide at least the same quality of planned treatments with the newer version. This may seem trivial, but we should be aware that economic pressure may unwillingly have negative consequences in this respect.

Chapter 3

3.8   Acknowledgment

Much information presented here has been obtained by fruitful discussions with many colleagues in particle therapy during visits and conferences. In this respect, the author would especially like to mention the experience shared with our colleagues at PSI: Damir Anicic, Frank Assenmacher, Terence Boehringer, Dölf Coray, Rudolf Dölling, Marc-Jan van Goethem (also for the discussions and reading this paper), Martin Grossmann, Tony Lomax, Anton Mezger, Eros Pedroni, Hans Reist, and Jorn Verwei.

References

Anferov, V. A. et al. 2007. *Proc. 18th Int. Conf. Cyclotrons Appl.*, 231–233.

Agosteo, S. et al. 1998. *Radiother. Oncol.* 48: 293–305.

Benedikt, M. et al. 1997. Proc. Particle Accelerator Conference, New York, p. 1379.

Benedikt, M. et al. 1999. *Nucl. Instrum. Meth.* A 430: 523.

Benedikt, M. 2005. *Nucl. Instrum. Meth.* A 539: 25–36.

Blosser, H. 1989. Proc. PAC 1989 (IEEE), pp. 742–746.

Bourhaleb, F. et al. 2008. *J. Phys. Conf. Ser.* 102: 012002.

Chu, W. T., Ludewigt, B. A. and Renner, T. R. 1993. *Rev. Sci. Instr.* 64: 2055–2122.

De Laney, T. F. and Kooy, H. M. (eds.). 2007. *Proton and Charged Particle Radiotherapy*. Philadelphia: Lippincott Williams & Wilkins.

Dölling, R. et al. 2007. Proc. AccApp'07, Pocatello, Idaho, pp. 152–159.

Flanz, J. et al. 1995. *Nucl. Instrum. Meth.* B 99: 830–834.

Flanz, J. B. 1995. Particle Accelerator Conference.

Fontenot, J. et al. 2008. *Phys. Med. Biol.* 53(6): 1677–1688.

Furukawa, T. et al. 2004. In Proceedings of the PAC, Knoxville, Tennessee, p. 910.

Furukawa, T. and Noda, K. 2006. *Nucl. Instrum. Meth.* A 565: 430–438.

Furukawa, T. et al. 2006. *Nucl. Instrum. Meth.* A 560: 191.

Furukawa, T. et al. 2006. *Nucl. Instrum. Meth.* A 562: 1050–1053.

Gottschalk, B. Unpublished book available in pdf format at http://huhepl.harvard.edu/~gottschalk/.

Grusell, E. et al. 1994. *Phys. Med. Biol.* 39: 2201.

Haberer, T. et al. 2004. *Radiother. Oncol.* 73: S186–S190.

Haberer, T. et al. 1993. *Nucl. Instrum. Meth.* A 330: 296–305.

Hiramoto, K. et al. 2007. *Nucl. Instrum. Meth.* B 261: 786–790.

Hug, E. et al. 2008. *PSI Scientific Report*, 56–59.

IBA. http:// .iba.be/healthcare/radiotherapy/particle-therapy/particle-accelerators.php.

Kats, M. M. 2008. Proc. RuPAC 2008, pp. 291–294.

Koda, K. et al. 2008. Proceedings of EPAC08, Genova, Italy, TUPP125.

Kleffner, C. 2008. Proc. EPAC 2008, pp. 1842–1844.

Koehler, A. M. et al. 1977. *Med. Phys.* 4: 297–301.

Koehler, A. M. 1987. Proc. 5th PTCOG Meeting: Int. Workshop on Biomedical Accelerators, Lawrence Berkeley Lab. Berkeley, CA, pp. 147–158.

Méot, F. 2008. *Nucl. Instr. Meth.* A 595: 535–542.

Norimine, T. et al. 2002. Proceedings of EPAC 2002, Paris, France, pp. 2751–2753.

Ondreka, D. et al. 2008. Proc. EPAC08, Genova, Italy, TUOCG01.

Optivus Proton Therapy Inc., http://www.optivus.com/conforma-3000.html.

Paganetti, H. et al. 2003. *Med. Phys.* 30: 1926–1931.

Pedroni, E. et al. 1995. *Med. Phys.* 22: 37–53.

Pedroni, E. et al. 2004. *Z. Med. Phys.* 14: 25–34.

Reimoser, S. A. et al. 2000. Proc. EPAC 2000, pp. 2542–2544.

Rohrer, U. 2006a. PSI graphic transport framework based on a CERN-SLAC-FERMILAB version by K. L. Brown et al. http://pc532.psi.ch/trans.htm.

Rohrer, U. 2006b. PSI graphic turtle framework based on a CERN-SLAC-FERMILAB version by K. L. Brown et al. http://pc532.psi.ch/turtle.htm.

Schippers, J. M. 1992. *Proc. 13th Int. Conf. Cyclotrons Appl.*, 615–617.

Schippers, J. M. et al. 2006. *PSI Scientific Report*, 62–63.

Schippers, J. M. et al. 2007. *Nucl. Instr. Meth.* B 261: 773–776.

Schippers, J. M. et al. 2008. *Proc. 18th Int. Conf. Cyclotrons Appl.*, 2007, 15–17.

Schneider, U. et al. 2002. *Int. J. Radiat. Oncol. Biol. Phys.* 53: 244–251.

Schneider, U. et al. 2006. *Strahlenther. Onkol.* 182(11): 647–652.

Schönauer, H. 2007. *Talk at Joint Symp. on Medical Accelerators* Wiener Neustadt (unpubished).

Slater, J. M. et al. 1992. *Int. J. Radiat. Oncol. Biol. Phys.* 22: 383–389.

Tang, J. Y. et al. 2009. *Phys. Rev. Spec. Top.-Ac.* 12: 050101-9:050101-9.

Taddei, P. J. et al. 2008. *Phys. Med. Biol.* 53(8): 2131–2147.

Trbojevic, D. et al. 2007. Proc. PAC07, pp. 3199–3201.

van Goethem, M. J. et al. 2009. *Phys. Med. Biol.* 54: 5831–5846.

Varian Medical. http://www.varian.com/media/oncology/proton/images/mpc_gesamt.jpg.

Vrenken, H. et al. 1999. *Nucl. Instrum. Meth.* A 426: 618–624.

Weinrich, U. 2006. Proc. EPAC 2006, TUYFI01.

Wroe, A. et al. 2007. *Med. Phys.* 34:3449–3456.

Zheng, Y. et al. 2008. *Phys. Med. Biol.* 53(1): 187–201.

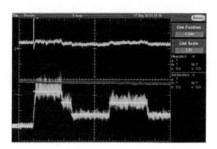

FIGURE 2.14 Scope picture of the extracted beam (lower trace) from the HIMAC synchrotron: Modulated beam current during an extraction cycle and the difference possible from one cycle to the next. (Courtesy of Koji Noda.)

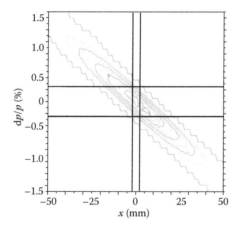

FIGURE 3.8 Correlation between the relative beam momentum dp/p (in percentage) and the horizontal distance x (in millimeters) to the beam axis t the location of the dispersive focus in the energy selection system. The slope of the ellipse indicates the dispersion D of -33 mm/%. The vertical lines indicate the slit aperture at which the minimum momentum spread is obtained. Due to the convolution with the optical image of the aperture at the degrader exit, the correlation plot has a finite width. The minimal achievable momentum spread is indicated by the horizontal lines. The calculations have been performed with TURTLE. (U. Rohrer 2006b, PSI Graphic turtle framework based on a CERN-SLAC-FERMILAB version by K. L. Brown et al., http://pc532.psi.ch/turtle.htm.)

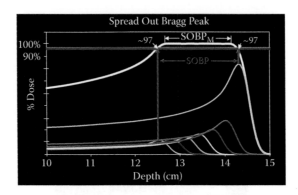

FIGURE 4.1 Depth dose curve for an SOBP (white) and depth dose curves of the Bragg peaks (in color) used to produce the SOPB. The vertical dotted red lines show the position on the SOBP of the maximum of the most distal and most proximal Bragg peaks. In this example, these positions correspond to approximately the 97% distal and proximal points of the SOBP. The distance between these points is defined as the SOBP. The maximum of the SOBP is that depth over which the dose is uniform, or nearly so, and is defined as SOBP$_M$.

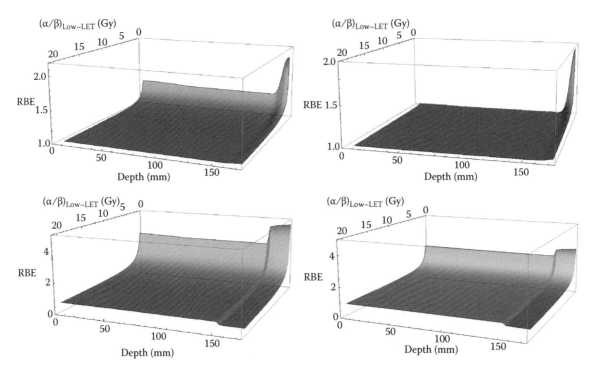

FIGURE 5.26 Proton (top) and carbon (bottom) RBE versus depth (combined dose and LET$_d$ effect) and $(\alpha/\beta)_x$ for a SOBP normalized to 2 Gy (left) and 15 Gy (right).

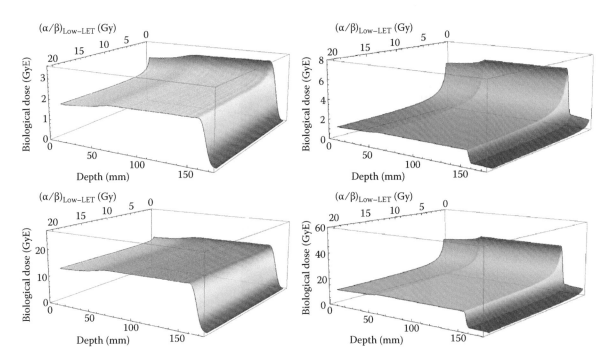

FIGURE 5.27 Top left: Biological proton SOBP at 2 Gy for different $(\alpha/\beta)_{X\text{-ray}}$. Top right: Biological carbon ion SOBP at 2 Gy for different $(\alpha/\beta)_{X\text{-ray}}$. Bottom left: Biological proton SOBP at 15 Gy for different $(\alpha/\beta)_{X\text{-ray}}$. Bottom right: Biological carbon ion SOBP at 15 Gy for different $(\alpha/\beta)_{X\text{-ray}}$.

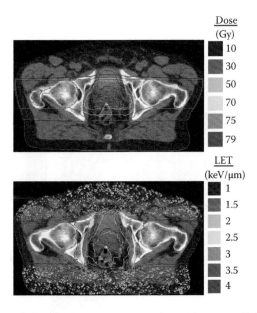

FIGURE 5.28 Dose (top) and LET_d (bottom) distributions in a prostate plan with two parallel opposed beam.

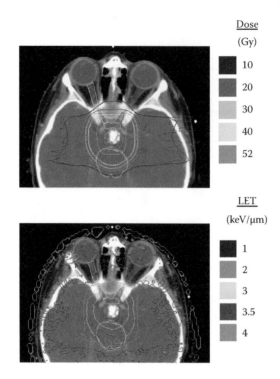

FIGURE 5.31 Dose (XiO, solid line; Monte Carlo, dashed line) and LET$_d$ distributions in the treatment of a craniopharyngioma.

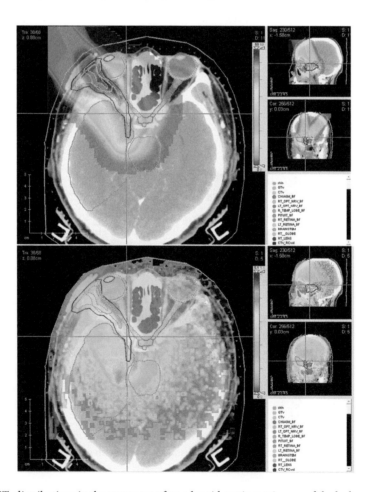

FIGURE 5.33 Dose and LET$_d$ distributions in the treatment of an adenoid cystic carcinoma of the lachrymal gland.

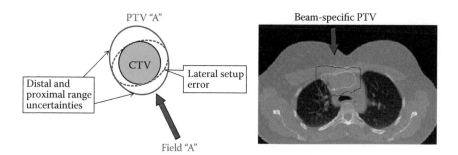

FIGURE 7.4 Beam-specific PTV concept. To accommodate both setup errors and range uncertainties, a beam-specific PTV can be used for setup verification. Because the charged particle beam will not be affected by a setup error parallel to the beam direction, setup error will only contribute to a lateral margin, as indicated by the dotted red line. However, range uncertainties due to organ motion in the beam path as well as CT imaging uncertainties will make up the distal and proximal margins. The combined PTV is shown in the solid red line for Field A. The beam-specific PTV also depends on local tissue heterogeneities. An example of beam-specific PTV for a lung cancer patient is shown to the right. The beam angle is shown by the red arrow. The CTV is shown in green and the beam-specific PTV is shown in red.

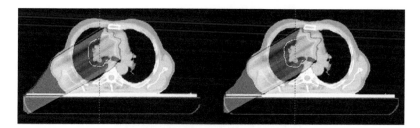

FIGURE 7.5 Example showing the dosimetric impact of the couch edge if the patient is not exactly positioned on the same location of the couch. One of the treatment beams go through the edge of the couch before reaching the tumor target. Left: Nominal (planned) position with beam central axis going through couch edge. Right: Same beam setup as before but with the patient shifted 1 cm laterally on the couch towards the beam. As a result, the portion of the couch in the beam is 1 cm short, resulting beam overshoot in the distal end. The isocenter is set to the center of the CTV in both situations.

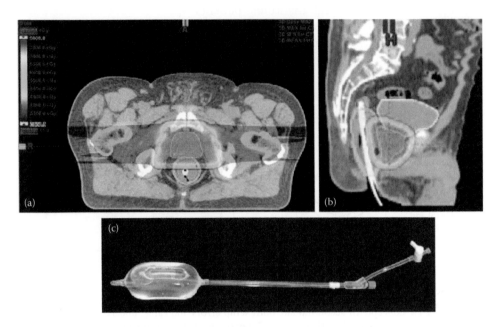

FIGURE 7.7 Typical treatment plan for prostate treatment using bilateral passively scattered proton therapy. The posterior rectum was displaced away from the high dose region by the rectal balloon, as shown in the axial (a) and sagittal (b) planes. The rectal balloon was specially designed for prostate immobilization with asymmetric flat surface and groove (c).

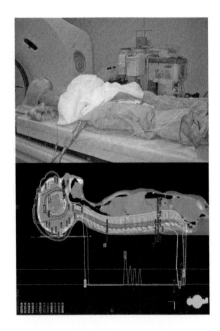

FIGURE 7.9 Setup technique for craniospinal irradiation for pediatric patients. Top: Patient is immobilized with a vacuum bag for the legs and a head holder and thermoplastic face mask for the head and neck. Bottom: Dose distribution is highly conformal to the treatment site while delivering essentially no dose to the tissues anterior to the vertebral bodies.

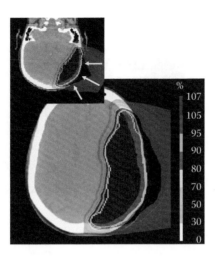

FIGURE 8.4 Example dose distribution from the PSI spot scanning system for a meningioma. This is a three-field plan, with the field arrangement as indicated in the insert. All fields are incident from the same quadrant such as to utilize the BP fall-off in order to spare the uninvolved (contralateral) part of the brain as much as possible.

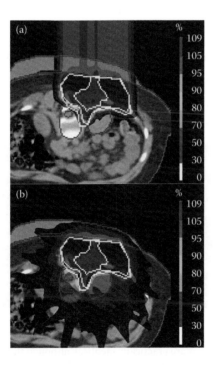

FIGURE 8.5 (a) Single-field proton plan used to irradiate this large superficial desmoids tumor in a 12-year-old boy. Note the use of a techPTV (interior yellow contour) to spare the spinal cord and partially the kidney. For comparison, a nine-field photon IMRT is shown at the bottom (b). Proton plan delivers six times lower dose to the normal tissues than the proton plan.

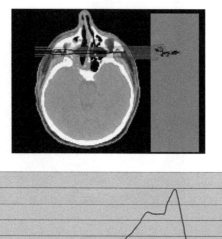

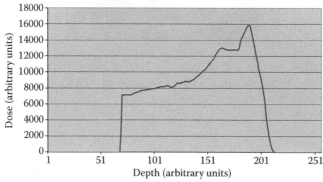

FIGURE 8.6 Effect of density heterogeneities on the shape of proton pencil beam (top) and integral BP (bottom). In many cases within the patient, the actual delivered depth dose curve will be quite different from the normal textbook BP.

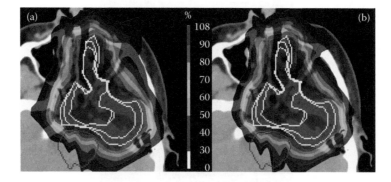

FIGURE 8.11 (a) Three-dimensional IMPT and (b) DET plans to a large chondrosarcoma. Both plans consist of coplanar five fields, equally distributed around the patient. Both plans are practically identical, although the plan on the right delivers BPs only at the DET, whereas the 3-D-IMPT plan delivers BPs in three dimensions throughout the volume.

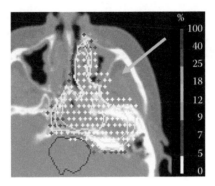

FIGURE 8.12 Example slice through the same case as shown in Figure 8.11, but showing the positions and relative weights of the delivered BPs for one field of the plan (direction shown). Each cross marks the position of a delivered BP and the relative BP peak weight (fluence) is color-coded using the scale on the right.

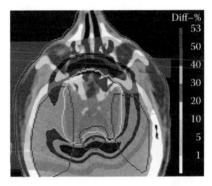

FIGURE 8.13 Dose variance distribution calculated based on the technique described in Morávek et al. (2009) for a SFUD proton plan. Here the colors indicate the possible spread of doses at each point under a given set of error conditions, in this case setup errors along each of the cardinal axes. This figure shows that within the target volume, the dose is very robust (small variance), but at the lateral edges of the field, dose can vary considerably due to the sharp dose gradients in these regions.

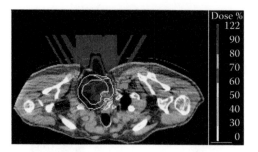

FIGURE 8.14 Typical narrow-angle IMPT plan for a spinal axis chordoma. This is a three-field plan consisting of a posterior field and two posterior-lateral obliques, angles 30° away from posterior. Although this choice of angles nicely spares the abdominal organs, the choice is primarily driven by the requirement to bring the incident fields predominantly along the principle axes of patient motion, which for a patient in a prone position such as this is along the anterior-posterior direction.

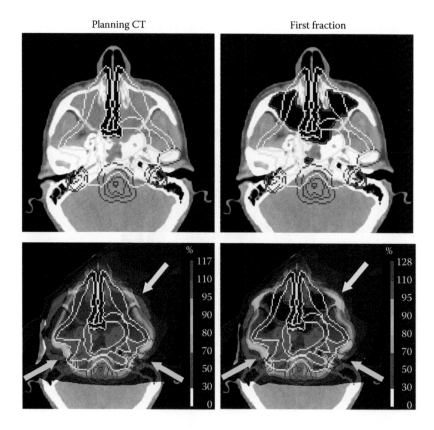

FIGURE 8.16 An example of extreme changes in nasal cavity fillings between the planning CT (right column) and the first treatment fraction (left). Nevertheless, despite these massive changes, because of the field arrangement chosen (indicated by the arrows), the differences in calculated dose are still rather small (see bottom two images).

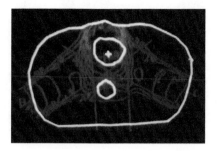

FIGURE 9.1 Therapy planning image from 1973, Loma Linda University Medical Center. The dose distribution and lines of attenuation in tissue were developed from ultrasound scans.

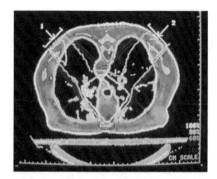

FIGURE 9.2 Therapy planning image from 1978, Loma Linda University Medical Center. This planning image was developed from CT scans. In addition to displaying the anatomy, CT scan-based planning images allowed assessment of density variations as X-ray beams passed through tissue.

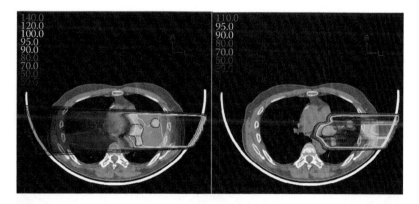

FIGURE 9.3 Single-beam dosimetry of a lateral photon beam (left) and lateral proton beam (right). These images demonstrate how lateral proton beams stop at the surface of the mediastinum, while photons do not.

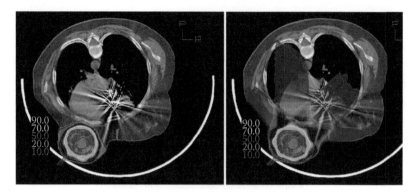

FIGURE 9.5 Left: Typical proton dose distribution utilizing four equally weighted beams (lateral, anterior, right, and left anterior oblique). Right: Same beam arrangement using 6-MV X-rays. Although this photon treatment plan would not likely to be used for treatment, the comparison demonstrates the normal tissue-sparing effects of proton beams; note that the dose distribution from the proton beams avoids the heart, lungs, and esophagus.

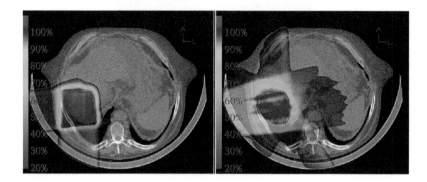

FIGURE 9.6 Treatment planning comparison of a two-field proton plan (left) and a five-field IMRT plan (right) for a large hepatocellular carcinoma (outlined in red).

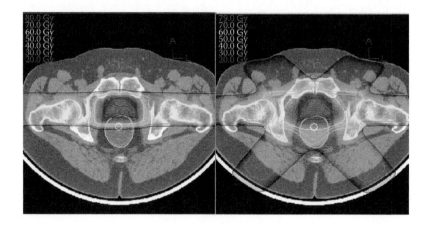

FIGURE 9.7 Treatment planning comparison of a two-field proton plan (left) and a six-field IMRT plan (right) for carcinoma of the prostate (outlined in red).

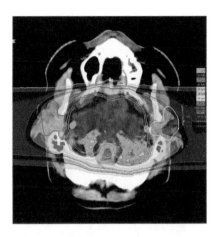

FIGURE 10.1 Sixty-three-year-old female with extensive chordoma of the skull-base; axial dose distribution, prescribed dose: 66 GyE carbon ions in 3 GyE per fraction.

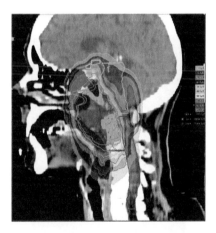

FIGURE 10.2 Sixty-three-year-old female with extensive chordoma of the skull-base; sagittal dose distribution, prescribed dose: 66 GyE carbon ions in 3 GyE per fraction.

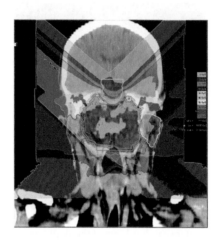

FIGURE 10.3 Sixty-three-year-old female with extensive chordoma of the skull-base; coronal dose distribution, prescribed dose: 66 GyE carbon ions in 3 GyE per fraction.

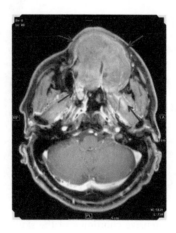

FIGURE 10.4 Sixty-nine-year-old patient with large adenoid cystic carcinoma, pretherapeutic contrast-enhanced MRI.

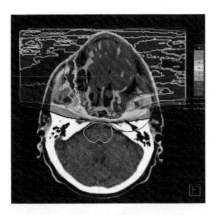

FIGURE 10.5 Sixty-nine-year-old patient with adenoid cystic carcinoma; axial carbon ion boost dose distribution (24 GyE C12 + 50 Gy IMRT).

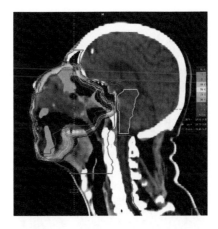

FIGURE 10.6 Sixty-nine-year-old patient with adenoid cystic carcinoma; sagittal carbon ion boost dose distribution (24 GyE C12 + 50 Gy IMRT).

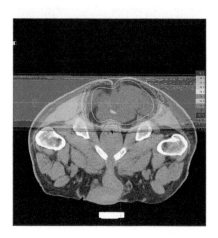

FIGURE 10.8 Sixty-nine-year-old patient with sacral chordoma; axial carbon ion boost dose distribution (24 GyE C12 + 50 Gy IMRT).

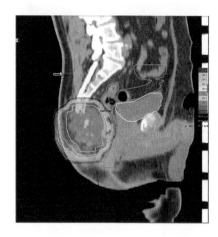

FIGURE 10.9 Sixty-nine-year-old patient with sacral chordoma; sagittal carbon ion boost dose distribution (24 GyE C12 + 50 Gy IMRT).

4. Requirements for Particle Therapy Facilities

Alfred R. Smith and Oliver Jäkel

4.1   Introduction

Modern particle therapy is increasingly being delivered in hospital-based facilities or free-standing facilities closely associated with a hospital or cancer center. The transition from research- to hospital-based particle therapy has prompted the development of standards, systems interfaces, and clinical processes and procedures that enable the integration of particle therapy into multidisciplinary cancer care environments. As a result,

modern particle therapy facilities provide cancer therapy in a highly efficient and integrated manner, similar to that provided in conventional photon and electron radiation oncology facilities.

A particle therapy facility should contain state-of-the-art technologies that have been tested and proven to be safe, robust, and reliable; however, an upgrade path should be provided for the implementation of new technical developments that advance the field of particle therapy without undue cost or interference with the operations of the overall facility. Analogous to conventional radiation oncology, a variety of facility

Proton and Carbon Ion Therapy Edited by C.-M. Charlie Ma and Tony Lomax © 2013 Taylor & Francis Group, LLC. ISBN: 978-1-4398-1607-3

Chapter 4

configurations, equipment options, and treatment delivery methods are available for particle therapy. Particle therapy facilities contain one treatment room or multiple treatment rooms, and a large facility may include ancillary capabilities for research and development.

When the decision has been made to pursue the development of a particle therapy facility, one of the early project tasks is to develop the clinical and technical specifications, collectively called hereafter the "requirements," for the major particle therapy technologies. These requirements will ultimately determine the quality of treatments and the efficiency of clinical operations in the new facility. The goal of this chapter is to provide guidance in establishing the overall requirements related to the production, transport, and delivery of treatments using protons and carbon ions. The requirements described in this chapter should be thought of only as guidelines; a particular facility will have requirements dictated by local conditions, clinical needs, and technology preferences.

There are both clinical and cost implications for each requirement. In addition, there may be compromises and trade-offs in the requirements for a particular facility in order to achieve the best possible clinical system compatible with project constraints. In general, more than one technical implementation may satisfy the requirements, so various solutions should be examined in order to optimize the balance between overall requirements and project constraints.

Requirements for particle therapy should be provided in a clear concise language and, when appropriate, accompanied by specific numbers and units (and variances if applicable), e.g., "the beam penetration in water should be ≥ 32 g/cm^2, and the penetration resolution should be ± 0.1 g/cm^2." Compliance with each numerical requirement should be demonstrated by measurements obtained during acceptance tests. However, note that some requirements cannot be expressed in numerical terms (e.g., "each treatment room should have a robotic patient positioner").

In this chapter, the discussion of the requirements for particle therapy facilities is divided into two sections: 1) requirements for a proton therapy facility and 2) requirements for a carbon ion therapy facility. There are, of course, some overlapping requirements for the two particle types. A carbon ion therapy facility, in general, can also accelerate, transport, and deliver protons; however, a carbon ion facility can be developed only for carbon ion therapy. If the intent is to develop a facility that delivers both proton and carbon ion therapy, the requirements provided in this chapter can be combined to obtain the requirements needed for a dual-particle facility.

4.2 Requirements for a Proton Therapy Facility

Clinical and technical requirements will depend on the treatment delivery technique used, which include the general categories passive scattering and pencil beam scanning. Pencil beam scanning techniques can further be divided into the following subcategories: 1) Intensity modulated proton therapy (IMPT); 2) single-field uniform dose (SFUD) therapy; and 3) uniform field scanning. Treatment delivery techniques are discussed in Chapter 3 and in the literature (Smith 2009).

Beam penetration in the patient or phantom provides an excellent example of the dependence of technical implementation on delivery technique. For the same energy extracted from the accelerator, passive scattering techniques will result in less beam penetration than pencil beam scanning techniques. The reason is that, in passive scattering techniques, the beam interacts with scattering materials in the nozzle that spread the beam laterally, whereas in pencil beam scanning techniques, the treatment field is spread laterally by magnetic deflections of the pencil beam. In addition, in passive scattering, the beam is spread in depth by the use of a ridge filter or rotating modulation wheel, whereas in pencil beam scanning, this is achieved by sequentially changing the beam energy. Therefore, in passive scattering, the interactions with the scattering materials in the nozzle will absorb energy from the beam (the amount of energy absorbed will depend on the extent of the lateral scattering required to achieve a given field size and the extent to which the high dose region is spread in depth); therefore, the beam exiting the nozzle will have a significantly lower energy than the beam entering the nozzle, reducing the beam penetration accordingly. Consequently, for a given specification for maximum penetration, passive scattering techniques will require a higher energy than pencil beam scanning techniques to be extracted from the accelerator.

When selecting a vendor for proton therapy equipment, it is usually good practice to engage in a competitive procurement process beginning with the issuance of a request for proposal (RFP) to equipment vendors. Accordingly, in developing requirements for a competitive procurement process, requirements that are specific to a particular vendor or equipment type should not be developed. In general, multiple vendors and technology implementations can satisfy the requirements for a high-quality proton therapy system, and we can negotiate better pricing and enhanced equipment packages by making vendors compete with one another. However, if a specific technical implementation is required, it should be clearly stated in the requirements.

Only a few sources that provide comprehensive requirements for particle therapy facilities are currently available. Specifications for proton therapy facilities have been provided in more detail than those for carbon ion facilities (Chu et al. 1993b; Gall et al. 1993). In some respects, these specifications need to be updated to account for advances in technology; however, they are still relevant.

4.2.1   General Requirements for Proton Therapy Facilities

In this section, we provide the requirements that generally apply to all proton therapy facilities and proton treatment techniques.

Accelerators should be capable of producing proton beam energies in the range of 70–230 MeV; however, it is preferable to have an upper energy of 250 MeV, especially for passive scattering techniques in which significant energy losses occur in the scattering system. Recently, synchrotrons with an upper energy of 330 MeV have been developed, with the intent of using higher energies for proton tomography techniques. The energy spread ($\Delta E/E$) of the beam entering the treatment delivery nozzle should be on the order of ±0.1% so that range straggling in the patient can be minimized (Section 2.1.2).

The dynamic range for beam intensity should be ≥10:1. The stability of the extracted proton beam, measured at the accelerator extraction point, should not vary by more than ±1 mm or by ±1 mrad during beam extraction.

The accelerator beam intensity should be high enough to deliver a dose rate of 2 Gy/min/L to a cubic target (e.g., 10 cm × 10 cm × 10 cm) at any depth in the patient for both passive scattering and pencil beam scanning techniques. The efficiency of a passive scattering system is approximately 40%–50%; therefore, for the same dose rate in a given target, the extracted

beam intensity for passive scattering techniques would need to be correspondingly higher than for pencil beam scanning techniques.

4.2.1.1   Beam Transport Systems

The beam transport system includes a common beam transport line consisting of dipole (bending) and quadrupole (focusing) magnets and beam analysis instrumentation: the proton beam travels inside a vacuum tube running through the cores of the magnets. Chapter 3 provides a detailed description of beam transport systems. According to the treatment room beam schedule, the beam transport line carries the beam to a branch point (bending magnet) for the scheduled treatment room, and the bending magnet switches the beam into that room. Inside the treatment room, the beam is transported to either an isocentric gantry or a fixed beamline, which then transports the beam to the treatment delivery nozzle. Magnets in a beam transport line should be laminated in order to increase the efficiency of magnet operation and to decrease the time required to change magnet set points. The latter is especially important in pencil beam scanning techniques in which the beam energies and, therefore, the magnet currents, are changed frequently.

As aforementioned, the beam transport system is continued inside the treatment room by either an *isocentric gantry* or *fixed beamlines*. There are several types of isocentric gantries (e.g., those that have full 360° rotations and those that have limited rotations). For gantries that have less than 360° rotation, for example, 180°, combinations of gantry rotations and robotic couch movements can be utilized to achieve multiple coplaner and noncoplaner beam entry angles and, accordingly, optimized treatment plans. Most proton gantries have an effective source axis distance (SAD) of ≥2m. Fixed beam treatment rooms can have either one horizontal beamline or a combination of a horizontal beamline and another fixed beamline at an angle (e.g., at 45 or 90°) to the horizontal beamline, with all beamlines having the same isocenter. The number of treatment rooms having gantries or fixed beamlines depends on local cost constraints and clinical needs.

Isocentric gantries should have a maximum rotational speed of at least 1.0 RPM and, preferably, one or two selectable lower speeds. Emergency stops from 1.0 RPM should occur in ≤5°. The gantry angle accuracy should be ±0.5° of the selected value. The mechanical isocenter accuracy should be contained within a sphere of uncertainty of ≤1 mm diameter during a full rotation. Some systems use a gantry error correction table

Chapter 4

to correct the treatment couch position for residual gantry isocenter errors at each gantry angle in order to increase the accuracy of the clinical isocenter.

4.2.1.2 Patient Support Systems

Patient support systems have traditionally consisted of a robotic couch having 6 degrees of freedom [three isocentric rotations (a ±90° rotation about the vertical axis plus pitch and yaw rotations of ≥±3°) and three translations in a horizontal plane containing the isocenter]. The treatment couch may have exchangeable couch tops to accommodate a variety of patient sizes and treatment sites. Recently, commercial robots have been modified for use in particle therapy applications and are being used with increasing frequency. One advantage of robots is their ability to transfer a patient from a moveable transport cart, position the patient at the isocenter for treatment, and then transfer the patient back to the transport cart after treatment. This allows treatment setup rooms to be located outside the treatment room, which means that patients can be imaged and set up in the treatment position before being transported to the treatment room, thus allowing increased efficiency and greater patient throughput. Robots can also facilitate the use of in-room computed tomography (CT), where the robot moves the patient between the treatment position and the CT scanner. Couch tops should have minimum proton beam attenuation; they are usually are made of a carbon fiber sheath with a low-density foam core. Couch tops should safely and accurately hold patients up to 6 ft in height and weighing up to 400 lbs.

The combination of gantry rotations and robotic couch movements should allow 4-π solid angle beam delivery to a patient placed in a prone or supine position on the patient support couch. Patient support systems often have a correction table that corrects for positioning errors; the positioning accuracy should allow a solid phantom placed on the couch top to be moved from an arbitrary position in the couch movement space to the defined treatment isocenter within a sphere of uncertainty of a 0.5-mm diameter. The target isocenter should remain within a sphere of uncertainty of ≤1 mm diameter for all isocentric rotations of the combined gantry and patient support system. The absolute accuracy should be ≤1 mm at all positions.

4.2.1.3 Patient Treatment Positioning System

Each treatment room should have a patient treatment positioning system to ensure that the patient is correctly positioned for treatment. Currently, orthogonal lasers with crossbeams are used to place the patient in the approximate treatment position. Imaging systems (X-ray tubes and flat panel imagers) located in the gantry treatment enclosure are then used to obtain digital radiographs (DRs) that are compared to digitally reconstructed radiographs (DRRs) generated by the treatment planning system. DRRs are derived from the treatment planning CT scans. The image analysis system compares the DRs and DRRs and calculates the couch correction (translations and rotations) necessary to place the patient in the correct treatment position according to the initial treatment planning CT scans. Couch corrections can be transferred automatically from the imaging system to the couch controller via the therapy control system (TCS); however, the actual couch correction should not be made until the radiation therapist requests it using the couch pendant control when he/she is near the patient and can observe all couch movements.

Increasingly, facilities are likely to have either in-room CT or robotic C-arm cone beam CT (CBCT) imaging systems, which can be used to image and position patients for treatment. In the future, these imaging systems may be used for adaptive proton therapy or treatment dosimetric verifications. Data from the imaging system can be transferred to a four-dimensional console that is integrated with the treatment planning system and the oncology information system (OIS).

4.2.1.4 TCS

The proton therapy system should have an integrated and comprehensive TCS. The TCS must interface with the overall system's various internal components and monitor and control their functions. The TCS must also interface externally with the OIS to enable the exchange of information between the OIS and TCS. This function is essentially identical to that for linac-based therapy systems: treatment delivery parameters generated during treatment planning are transmitted to the OIS, which then transfers these parameters to the TCS; the OIS performs "record and verify" functions; and the TCS transmits final treatment data to the OIS to be archived in the electronic medical chart. The TCS also contains connections to various door interlocks, crash buttons, and radiation monitors in the facility.

The TCS interfaces with the treatment safety system (TSS) and monitors and controls the functions of the accelerator, beam transport (beamlines and gantries), treatment delivery (e.g., beam modifying and dose monitoring), and patient positioning components of the overall system. The TCS usually has two or, sometimes, three modes of operation, which generally fall into the categories of "service," "physics," and "treatment." Sometimes, the functions of the operating modes are combined. In general, the control system performs the following top-level operations.

- Systems start-up and shutdown
  - Control and monitor accelerator functions, including acceleration, energy selection, beam extraction, beam intensity, and beam shutoff.
  - Set up the magnets in the beam accelerator and transport systems and monitor magnet currents. The magnetic field strength is monitored in selected locations.
  - Monitor the vacuum in the accelerator and beam transport lines.
  - Monitor the beam size and position at selected points in the beam transport line.
  - Schedule beam sequencing into the treatment rooms according to treatment room readiness signals and beam requests from treatment control rooms.
  - Automatically set up patient treatment parameters in the treatment rooms (excluding gantry, patient positioner, and beam modifying devices).
- Monitor dose delivery and treatment beam parameters (e.g., beam position and size entering nozzle, beam symmetry, and uniformity, SOBP width, beam energy exiting the nozzle, and delivered dose).
- Provide error messages and, when requested, system performance and status logs.
- Control treatment room hardware components such as the gantry, nozzle, image systems, treatment couch, and control pendant.

In modern proton therapy systems, it is essential that the overall TCS is designed such that the control of individual treatments is assigned to individual treatment rooms and associated treatment control rooms. In proton therapy, radiation therapists should exercise control of their treatment rooms in much the same way as they do for conventional linac-based treatment rooms. That is, during normal operations, the radiation therapist should not have to communicate with someone in the main control room during treatment sequences or rely on beam tuning, beam switching, or other actions by someone in the main control room.

Energy selection, beam transport, and beam scheduling for a particular treatment should be handled automatically by the TCS in accordance with the treatment parameters transferred from the OIS to the TCS for a particular patient.

Particularly for treatment rooms in which pencil beam scanning techniques are used, it should be possible to control the gantry, couch, and imaging operations from the treatment control room, making it possible to treat several fields in sequence without reentering the treatment room.

4.2.1.5 TSS

The safety of patients, operating personnel, and the equipment itself is of paramount importance. The overall safety system design should be driven by a rigorous risk analysis that includes all mechanical, treatment, and stray radiation hazards; it should be designed to mitigate all identified hazards. The TSS should be designed to have redundant safety systems for critical components and functions; potential failure modes with high-level criticality should have triple redundancy. Redundant safety systems must be completely independent of one another with respect to, for example, input and output signals and power supplies. There must be no single point of failure in any critical safety system. Automatic collision avoidance systems should be present in each treatment room to minimize the risk of collision between the patient or patient support system and the equipment (e.g., gantry, imaging devices, and the treatment nozzle and snout). If collision occurs, the collision detection and interlock system should shut off the power to all equipment in the treatment room; this system should have a fast recovery procedure to enable the operator to remove the patient and/or equipment from collision conditions.

Safety functions are performed by interlocking the operation of a limited number of important equipment components (e.g., accelerator power supplies, including those for the ion source and accelerating RF, beam stops, gantry and patient positioner and snout movement motors, gantry brakes, door interlocks, and area crash buttons) to a set of safety conditions that are specified by closed electrical contacts. The closed contacts, representing the fulfillment of the required safety conditions, are connected in a logic series; each element in the series is represented by a logic controller such as a programmable logic controller; the desired output is activated when all contracts in the series are closed, which means that all safety conditions are met. The status of interlocks are prerequisites for an action

Chapter 4

(e.g., irradiation) and do not involve a control process; an interlock for a device cannot be part of the control for that device. All materials that are placed in the beam path (e.g., beam stops, beam dumps, beam monitors, and beam-shaping devices) should be interlocked.

4.2.1.6   Reliability

Reliability at a very high level is required for the proton therapy facility to achieve its goals of patient throughput and financial viability. The overall system reliability should provide a treatment availability of ≥96% for scheduled patient treatments for up to 16 h per day, 5 days per week. An overall availability of ≥96% requires that common systems such as the accelerator have a reliability of ≥98%.

4.2.1.7   System Parameters

The proton therapy system should satisfy the following requirements.

1. The beam switching time, defined by the time between the completion of treatment in one room and beam readiness for treatment in another room, should be ≤1.0 min, preferably ≤30 s.

2. The time required for changing energies in the beam production system should be ≤2 s, preferably ≤1 s. The time depends, among other things, on the accelerator type; generally, more time is required to change energies in a synchrotron than in a cyclotron or synchrocyclotron using an ESS.

3. Patient treatments should commence within 30–45 min of a cold start-up of the accelerator. There should be no need for manual beam tuning after the morning warm-up and quality assurance procedures.

4. The time and number of operations needed for all the facility's components should be minimized so that operators need only to enter (or download) the minimum amount of information necessary to specify the desired state of the system.

5. The system should permit rapid manual and automatic (from prestored tables) setup of all parameters needed to treat a patient.

6. The system should be "fault tolerant" (i.e., it should not abort a treatment for noncritical error messages), and it should provide user-accessible tolerance tables for all major treatment devices.

4.2.1.8   Equipment

The equipment should be highly efficient and user-friendly in its operations and functions. Care should

be taken to minimize the number of actions that personnel must take to treat patients and to maximize the utilization factor (demand/capacity). Proton therapy facilities are scarce and expensive; therefore, resources and patient throughput should be optimized while maintaining safety and treatment accuracy.

4.2.1.9   Compliance with Regulations and Standards

The proton therapy system should meet all applicable local, state, national, and international regulations and standards. The equipment should have United States Food and Drug Administration (U.S. FDA) 510(k) certification or its equivalent in other countries. The interfaces that connect the major system components (treatment planning imaging system, treatment planning system, OIS, and TCS) should comply with the DICOM-RT-ION standard, which makes it possible to integrate components from multiple vendors and to replace equipment or implement new technologies.

4.2.1.10   Additional Technologies

The overall proton therapy system must include additional technologies to those described above: imaging systems [CT-simulation (CT-SIM) and access to positron emission tomography (PET)-SIM and magnetic resonance imaging (MRI)]; a treatment planning system with capabilities for photon, electron, and proton treatment planning; and an OIS. Proton treatment planning capability should include all treatment delivery techniques expected to be carried out in the facility, e.g., passive scattering, pencil beam scanning (i.e., IMPT, SFUD, and uniform scanning), eye treatments, and radiosurgery. All technology interfaces should comply with the DICOM-RT-ION standard. Treatment planning systems are discussed in detail in Chapter 8.

4.2.2   Requirements for Specific Proton Treatment Techniques

Proton therapy requirements depend on the treatment delivery technique used; in this section, we provide the specific requirements for the passive scattering and pencil beam scanning techniques.

4.2.2.1   Passive Scattering Systems

The first type of beam delivery technique is passive scattering. Passive scattering techniques use scattering devices in the treatment delivery nozzle (usually double scatterers for large fields and single scatterers for small, e.g., radiosurgery, fields) to spread the beam transversely. In the longitudinal dimension of the dose distribution a range modulation wheel or ridge filter (Chu et al. 1993a) is used to create a region of dose uniformity (i.e., SOBP) in the dose distribution. Between the nozzle exit and the patient surface, a treatment field–specific collimator is used to shape the field laterally to conform to the maximum beams-eye-view extent of the target volume, and a range compensator is used to correct for patient surface irregularities, density heterogeneities in the beam path, and changes in the shape of the distal target volume surface. The size of the SOBP is chosen to cover the greatest extent in depth of the target volume. Chapter 3 provides a complete description of passive scattering systems.

Until the mid-1990s, passive scattering was the only treatment mode used for proton therapy. In 1995, Pedroni et al. described a proton therapy system developed at the Paul Scherrer Institute, Villigen, Switzerland, which was the first to use pencil beam scanning techniques to treat cancer patients (Pedroni et al. 1995). In May 2008, the M. D. Anderson Cancer Center, University of Texas, Houston, began treating patients with scanned beams; this was the first commercial system to provide scanned beams for optimized treatments (Smith et al. 1999). During the next decade, pencil beam scanning techniques were expected to gain widespread acceptance and to rapidly dominate proton therapy; two vendors now offer scanning-only proton therapy systems. Therefore, we expect the widespread use of passive scattering techniques to substantially decrease over the next several years, except in facilities that currently use passive scattering systems.

Most of the requirements provided as follows can be met or exceeded by most vendors of proton therapy equipment; however, all vendors will not be able to meet all the requirements. Many of the definitions used in this section were taken from the International Commission on Radiation Units and Measurements (ICRU) Report No. 78 (ICRU 2007). This ICRU report also provides useful figures that explain the application of some of these definitions.

1. Treatment beam requirements for large-field treatment beams
   a. *Depth of penetration* is the depth (in grams per square centimeter) along the beam central axis, measured in water, to the distal 90% point of the maximum dose.

The maximum penetration is ≥ 30 g/cm^2 for field sizes ≤ 10 cm $\times$ 10 cm and ≥ 23 g/cm^2 for field sizes ≤ 25 cm $\times$ 25 cm.

The maximum penetration depends on the maximum energy available from the accelerator and field size. Thick scatterers required to spread the beam to large field sizes absorb beam energy and therefore decrease the maximum penetration.

The minimum penetration is ≤ 4 g/cm^2; penetrations shallower than 4 g/cm^2 can be achieved by using range shifters in the nozzle. The penetration resolution accuracy and reproducibility should be ± 0.1 g/cm^2.

   b. *Distal dose falloff* is the distance (in grams per square centimeter) in which the dose, measured in water along the beam central axis, decreases from 80% to 20% of the maximum dose value. Distal dose falloff is determined by two factors: 1) the energy spread of the beam ($\Delta E/E$) entering the phantom and 2) the proton range straggling in the phantom. It is desirable to keep $\Delta E/E \leq \pm 0.1\%$ at the entrance to the treatment delivery nozzle; however, this value will increase because of proton interactions in the scattering system and nozzle elements. A reasonable goal is to constrain the energy spread of the beam entering the phantom such that it contributes no more than 0.2 g/cm^2 to the distal dose falloff. The proton range straggling contributes approximately 1.1% of the proton range to the distal dose falloff. Using these factors, we can determine that a typical total distal dose falloff for a proton beam with a penetration of 24 g/cm^2 would be approximately 4.6 mm.

   c. *Range modulation or spread-out-Bragg-peak (SOBP) length* is the distance in water between defined distal and proximal dose percentage points of the extended, uniform maximum dose level. ICRU Report 78[7] defines the SOBP length as the distance between the distal and proximal 90% points of the modulated depth dose curve. However, there are other definitions in use, for example, the distance between the 90% distal point and the 97% proximal point, and the distance between the distal and proximal 97% points. The problem inherent with SOBP definitions that use a dose point ~90% either distally or proximally, or both is that there are often variations in the slopes of the proximal and distal dose that fall off of the

Chapter 4

maximum dose that will lead to ambiguities in the stated SOBP length. For example, the proximal and distal slopes of the maximum dose will, in general, vary with depth, SOBP size, and the design of range modulation device used to produce the SOBP. A nominal SOBP length of 5 cm for one situation would not necessarily be 5 cm for another; therefore, the definition given above may lead to different results in different situations, even within a particular system. Here, we use a definition for SOBP that can be used consistently across various systems and within a given system. The SOBP width is defined as the distance between the maximums of the most distal and proximal Bragg peaks used to form the SOBP. This concept is illustrated in Figure 4.1. The width of the maximum uniform dose region, which is the essential dimension for treatment planning, is defined as $SOBP_M$, as shown in Figure 4.1. The length of the maximum uniform dose ($SOBP_M$) is analogous to the quantity m_{100} defined by Gottschalk (2003) as the "modulation at full dose."[8]

The $SOBP_M$ length should be ≥18 g/cm² and should be variable in steps no greater than 1.0 g/cm². The practical limit of the $SOBP_M$ length for a given penetration is necessarily less than the full penetration of the beam, because the penetration is defined at the 90% distal dose level, and the slope of the dose profile

proximal to the $SOBP_M$ will limit how close to the surface the maximum uniform dose can be placed. In most proton therapy systems, a family of $SOBP_M$ sizes can be produced. The dose uniformity in the $SOBP_M$ should be ±2%.

d. *Field size* is the distance (in centimeters) between the 50% points of the maximum dose value, measured along the line perpendicular to the beam central axis, on the isocenter plane in air. Some vendors offer treatment snouts (an extendable device attached to the nozzle that holds the treatment-specific collimator and range compensator) that are designed to hold circular apertures and range compensators, whereas other vendors offer treatment snouts that accommodate square devices. Accordingly, field sizes are given in terms of both diameters of circular fields or squares. Here, we will use dimensions for square fields; the corresponding circular field can be obtained by calculating the diameter of a circle that will encompass the square field size.

The largest beam size currently available for passive scattering systems is 25 cm × 25 cm, which is the recommended maximum field size. Beam penetration is inversely proportional to the field size, because larger scatterers, required to produce large field sizes, absorb more beam energy. In addition, large field sizes require a large SAD, which affects the size of the gantry structure. Field-shaping collimators are usually made of machined brass; because collimators for 25 cm × 25 cm fields are too heavy for manual insertion, they must therefore be segmented into lighter, more manageable thicknesses. If a multileaf collimator (MLC) is desired, this may limit the maximum field size, because an MLC for a 25 cm × 25 cm field may be very massive and take up too much space between the nozzle and the patient. In addition, the size of the MLC may require a larger air gap between the MLC and patient surface, which may degrade the lateral penumbra of the treatment beam.

Treatment snouts should accommodate large, medium, and small field sizes to enable the use of collimators and range compensators that are sized to be a close match to the target volume being treated. The use of one large size for all treatments increases both the weight and the cost of materials for apertures and range

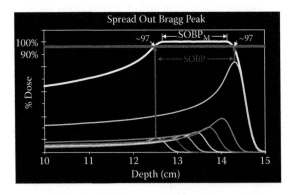

FIGURE 4.1   (See color insert.) Depth dose curve for an SOBP (white) and depth dose curves of the Bragg peaks (in color) used to produce the SOPB. The vertical dotted red lines show the position on the SOBP of the maximum of the most distal and most proximal Bragg peaks. In this example, these positions correspond to approximately the 97% distal and proximal points of the SOBP. The distance between these points is defined as the SOBP. The maximum of the SOBP is that depth over which the dose is uniform, or nearly so, and is defined as $SOBP_M$.

compensators, as well as the amount of neutron production close to the patient.

e. *Target volume dose uniformity* is the allowed variance in dose over the defined target volume as measured in a water phantom. The target volume is defined in various ways; however, we will define the target volume as the volume encompassed by the $SOBP_M$ in depth and laterally by two lateral penumbra widths from the 50% isodose levels of the lateral beam profile. The dose uniformity in the target volume should be $\leq\pm4.0\%$.

f. *Delivered dose reproducibility and repeatability* are the abilities to reproduce the delivered dose on the same day or to repeat the same dose within the period of a week. The reproducibility of the proton system dosimetry should be within $\pm1.0\%$ [2 standard deviations (SD)] over the span of one day and $\pm2.0\%$ (2 SD) over a period of 1 week when measuring treatment doses of 2 Gy.

g. The *dose rate* for a given beam intensity extracted from the accelerator depends on the target volume size. For average target volumes, the dose rate should be at least 2 Gy/min; for very large volumes, the dose rate should be ≥1.5 Gy/min; and for small (e.g., radiosurgery fields), the dose rate should be ≥10 Gy/min.

2. Treatment beam requirements for small beams used primarily for ocular treatments (e.g., treatment for uveal melanomas). Uveal melanoma is a relatively uncommon disease, so it is not required that every proton therapy facility has the capability to treat this disease. This capability should be reserved for large academic hospitals or facilities located near large eye clinics where uveal melanoma cases are frequently seen. Here, we provide a limited set of requirements for ocular treatments.

a. Beam energy at the nozzle entrance: ~80 MeV
b. Maximum field size: 3.5-cm diameter
c. Beam penetration: $0.7–3.8$ g/cm^2
d. Range adjustment resolution: 0.01 g/cm^2
e. Maximum SOBP width: 3.8 g/cm^2
f. SOBP size resolution: 0.2 g/cm^2
g. Dose rate: ≥12 Gy/min (fraction sizes are approximately 12 Gy)
h. Dose uniformity: $\pm3\%$ over a defined target volume
i. Dosimetry reproducibility and repeatability: $\pm0.5\%$ (2 SD) for 1 day and $\pm1.5\%$ (2 SD) for 1 week

4.2.2.2 Pencil Beam Scanning Treatment Systems

The second type of beam delivery technique is pencil beam scanning (also often called spot scanning). This technique uses two pairs of scanning magnets located at the nozzle entrance to position spots in the transverse dimension of the target volume and energy changes to cover the longitudinal dimension of the target volume. Energy changes in scanning techniques can be performed with various methods: 1) energy changes in the accelerator when a synchrotron is used; 2) energy changes with an ESS when a cyclotron is used; or 3) either of these two methods plus energy absorbers in the treatment nozzle. Beam scanning treatment techniques are described in detail in Chapter 3.

Pencil beam scanning techniques can be used to deliver at least three types of treatments.

1. *Uniform field scanning*—In this technique, pencil beam scanning is used instead of passive scattering to spread the beam laterally to generate large uniform treatment fields. In uniform field scanning, collimators and range compensators are still required to shape the treatment beam laterally and distally to the target volume. Uniform field scanning techniques produce dose distributions analogous to those produced by passive scattering techniques.

2. *SFUD*—In this technique, the results of optimization techniques in treatment planning are used to achieve a desired dose distribution (usually uniform) in the target volume. Objective functions are used to optimize spot patterns for single treatment fields. SFUD treatment plans consist of one or more individually optimized fields.

3. *IMPT*—This technique is analogous to intensity-modulated photon treatments that are usually called IMRTs but which we will call IMXT. IMPT is defined as the simultaneous optimization of all Bragg peaks from all fields with or without additional dose constraints to organs at risk (OARs) such that, when all fields are delivered to the patient, their combination results in a desired dose distribution in the target volume and OARs. All of these pencil beam scanning techniques are described in the literature and in Chapter 3 (Smith 2009).

As previously mentioned, the use of pencil beam scanning techniques for proton treatments did not begin until the mid-1990s. However, it is expected that scanning techniques will be used more often than passive scattering techniques to deliver proton therapy in

Chapter 4

the future primarily, because scanning techniques provide better dose localization and utilize proton beams more efficiently.

1. Treatment beam requirements for IMPT and SFUD techniques

    Unless otherwise noted, the definitions for the following treatment beam parameters are the same as those given above for passive scattering techniques.

    a. *Depth of penetration*—The maximum penetration is ≥ 32 g/cm^2. The minimum penetrations is ≤ 4 g/cm^2. Penetrations shallower than 4 g/cm^2 can be achieved by using range shifters in the nozzle. The penetration resolution accuracy and reproducibility should be ± 0.1 g/cm^2.

    b. *Distal dose falloff*—See requirements for passive scattering. The energy spread of the beam incident on the phantom (patient) is smaller than that for passive scattering techniques, because the beam intercepts less scattering material in the nozzle. A reasonable goal for a scanned beam is to constrain the energy spread of the beam entering the phantom to contribute no more than 0.1 g/cm^2 to the distal dose falloff. The proton range straggling contributes approximately 1.1% of the proton range to the distal dose falloff. Using these factors, ee can determine that a typical total distal dose falloff for a proton beam with a penetration of 24 g/cm^2 would be approximately 3.6 mm.

    c. *Lateral penumbra*—In the absence of additional collimating devices, the lateral penumbra is determined by the spot size in water. The relation between spot size in water and penumbra for pencil beam scanning techniques using uniform spot weights and a spot spacing of 1-σ is given by $P = \sigma * 1.68$, where P represents the 80%–20% penumbra, and σ represents the SD of the Gaussian that describes the single spot in water. If nonuniform spot weights are applied, the penumbra can be sharpened, and the relation becomes $P = \sigma * 1.3$.

    The lateral penumbra should not exceed the spreading in water caused by multiple scattering by more than 2 mm. This requirement will probably not be met for low energies; therefore, in such cases, it may be necessary to provide collimation for pencil beam scanning delivery.

    d. *Field size*—Maximum field sizes for scanned beams are usually larger than those for passive scattering, because the scattering material and its associated beam losses are not involved in spreading the beam. Factors that determine the maximum beam size in scanning techniques are the length of the SAD and the size of the gaps in the scanning and/or gantry magnets. The maximum field size should be 40 cm $\times$ 40 cm; however, for most clinical situations, 40 cm $\times$ 30 cm is satisfactory.

    e. *Target volume dose uniformity*—A convenient way of defining the depth dimension of the target volume is to define a length, in water, bounded by the positions of the peak doses of the most distal and most proximal Bragg peaks (i.e., the Bragg peaks used in the deepest energy layer and the most shallow energy layer) used in the treatment plan. For the lateral boundaries of the target volume, we use the same definition as used in Section 4.2.2.1, point f. To check the ability of the scanning system to deliver a uniform dose to a planned target volume, we recommend using a treatment plan in a water phantom for the largest volume that can be delivered by the scanning system, with the aim of delivering a uniform dose distribution throughout the volume: the dose uniformity for the target volume should be $\leq \pm 4.0\%$.

    f. *Pencil beam spot size (1-σ) and shape*—As measured in air at the isocenter the beam (spot) size should be ≤ 4 mm at energies with a penetration of ≥ 32 g/cm^2 and ≤ 8 mm at energies with a penetration of ≤ 4 g/cm^2. Ideally, the spot size should be controlled by a magnetic lens system, which would permit the spot size to be optimized over the entire energy range.

    The beam spot may have a different Gaussian profile in the *x* and *y* transverse directions, because spots may be elliptical in shape rather than circular. The degree to which the spot shape is not circular depends on the beam emittance. In addition, in some cases, the spot fluence at the isocenter may not have a pure Gaussian shape (i.e., it may require two or more Gaussians with different sigmas to model the beam profile).

    This effect has been found to be larger at greater depths in the phantom. The growth of the beam sigma at depth is determined by elastic multiple scattering and nuclear interactions. Thus, the total beam size and shape is determined by a combination of the Gaussian of the in-air fluence, the multiple scattering

Gaussian, and a Gaussian with a large sigma, which describes the large angle scattering due to nuclear interactions (Pedroni et al. 1995).

g. *Spot placement accuracy*—This is essential in order for the scanning system to deliver dose distributions that accurately correspond to the calculated treatment plan. The spot position is defined, in air, as the center of the (approximately) Gaussian fluence distribution in the plane, which is perpendicular to the central beam axis and contains the isocenter. The spot placement accuracy for each spot should be ≤± 0.5 mm (1 SD). To facilitate accurate placement of spots, the spot position monitoring system should have spot position monitors at the entrance and exit of the nozzle. Using dual monitors with a wide separation, the trajectory of the pencil beam to the isocenter plane can be accurately calculated, and if the projected position of the spot at the isocenter plane is out of tolerance, the treatment can be paused.

h. *Accuracy of delivered dose distribution*—The delivered proton dose distribution, which may be highly nonuniform for IMPT, should match the dose distribution calculated by the treatment planning system with a high degree of accuracy. The pencil beam scanning system should be capable of accurately reproducing the nonuniform proton dose distribution calculated by a treatment planning system, which has been configured and validated to accurately reflect the treatment delivery system's capabilities. The γ-evaluation method, which was introduced by Low et al. (1998), is a method that can be used to quantitatively compare dose distributions by combining dose-difference and distance-to-agreement criteria. Since then, Low et al.'s original method has been refined and modified by several authors, and it is now capable of analyzing both two-dimensional (2-D) and three-dimensional (3-D) dose distributions (Low and Dempsey 2003; Depuydt, Van Esch and Huyskens 2002; Bakai, Alber, and Nüsslin 2003; Stock, Kroupa and Georg 2005; Jiang et al. 2006; Spezi and Lewis 2006). The γ-evaluation method is highly recommended for comparing calculated and delivered IMPT and SFUD dose distributions. Although a 3-D analysis can be achieved by performing a 2-D γ comparison energy slice by energy slice and by forming a stack of axial 2-D γ maps, the analysis should be performed using a full 3-D γ evaluation in which all three dimensions of the dose distribution are included in the evaluation (Gillis et al. 2005). Until recently, 3-D γ evaluations have required long calculation times. However, these times have been significantly reduced by the development of advanced 3-D calculation methods (Wending et al. 2007).

A complete discussion of the γ-evaluation method is beyond the scope of this chapter, so we strongly encourage the reader to review the cited articles for a better understanding of the computational methods. Stated in straightforward terms, the dose variance between the measured dose at a point in the delivered dose distribution and the calculated dose can be stated either as the maximum permissible dose error at a point or as the distance of closest approach to a point that receives the planned dose.

- The delivered dose is within ±2.5% (1 SD) of the mean planned dose in the high-dose volume.
- The delivered dose is within ±1 mm (1 SD) of a point within the planned dose distribution that was specified to receive the planned dose.

i. *Spot dose resolution*—The dose monitoring system should be capable of delivering dose to a resolution of ±1 cGy.

j. *Effective dose rate*—The time needed to treat a volume of 10 cm × 10 cm × 10 cm located at any depth to a dose of 2 Gy should be ≤1.5 min; 1.0 min is recommended. The effective dose rate will vary somewhat on the method used for scanning (i.e., discrete spot scanning or dynamic beam scanning definitions are provided in Chapter 3); in general, dynamic beam scanning will have the highest effective dose rate.

k. *Bragg peak spreading*—For low-energy beams the Bragg peak width is very sharply peaked with a steep proximal and distal dose falloff. When energy stacking is used, the energy steps must be very small in order to achieve a uniform dose distribution in depth. In addition, for very narrow Bragg peaks, small errors in spot placement may result in significant dose errors. In order to decrease the number of beam tunes (energies) and to decrease the probability of dose errors, the width of the Bragg peak can be increased by using a ridge filter to produce a mini-SOBP.

Chapter 4

4.2.3 Typical Configurations for Proton Therapy Facilities

There are currently approximately 30 proton therapy facilities in operation worldwide. These facilities have a range of sizes, spatial configurations, and arrangements for the beam delivery beamlines. Facility sizes range from small treatment facilities having only one treatment room to large academic facilities having up to four treatment rooms. In single-room facilities, the accelerator can be located on the gantry inside the treatment room or in an adjacent room. In either case, when a rotating isocentric gantry is used, the height of the treatment room will be about three floors, and the footprint area will, in general, be larger than for a conventional linac-based radiation oncology treatment room.

In multiroom facilities, the typical layout is linear, with the accelerator located on one end and the beam transport line running linearly behind the row of treatment rooms (Figure 4.2). In situations where space is limited, the linear scheme may vary, and the accelerator may even be located on a different floor than the treatment rooms. In recent times, there has been considerable interest in smaller, less expensive proton treatment facilities, which have compact accelerators and compact isocentric gantries (e.g., gantries that rotate 180° from the vertical). With the use of robotic patient support systems, such gantries can deliver treatment beams from any angle. An example of such a compact treatment facility is shown in Figure 4.3.

Proton therapy facilities have a variety of operational arrangements: 1) integration into a physics research facility; 2) a free-standing facility (i.e., not physically attached to a hospital or cancer center); and 3) a hospital-based facility (i.e., operationally and physically integrated into a hospital or cancer center). All of these operational types have been relatively successful and are clinically and financially viable; however, the current trend is for hospital-based facilities, which are fully integrated into the hospital and radiation oncology operations. In some cases, new hospital or cancer center facilities have been built with full and complete integration of proton and conventional

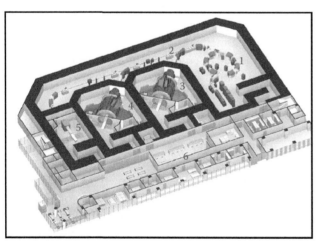

1. Accelerator	4. Gantry room
2. Beam transport line	5. Fixed beam room
3. Gantry room	6. Patient support area

FIGURE 4.2 Typical proton therapy facility layout showing accelerator on far right, beam transport line, two isocentric gantry treatment rooms, and a fixed-beam room. The figure has a synchrotron accelerator; cyclotrons have most often been used in proton therapy facilities, but either type of accelerator can provide the energies, beam intensities, and reliability required for proton therapy applications. The design of proton facilities can be quite flexible and can include features such as more or fewer treatment rooms, different combinations of gantries and fixed beamlines, multiple fixed beamlines in one room, a research/development room, and a custom beamline for treating eye tumors. New facilities may also have a CT or CBCT in the treatment room.

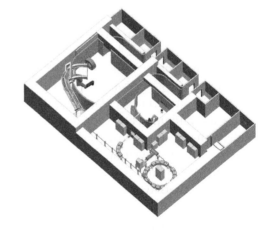

FIGURE 4.3 A compact proton therapy facility with a synchrotron accelerator, one fixed-beam room, and one isocentric gantry room. The gantry has a rotation range of 0°–180°. The robotic patient positioner, in combination with the gantry, enables treatment beams to be applied from 360° around the patient. A compact facility could also have only one treatment room or the facility shown in the diagram could have two isocentric gantries. The fixed-beam room could also have multiple fixed beams. The compact synchrotron requires much less space than synchrotrons that have been used to date. Due to very low proton losses in the accelerator and pulse-to-pulse energy changes without the use of an external beam selection system, significantly fewer neutrons are produced; hence, the shielding walls can be thinner than those for cyclotron-based facilities. This system is scanning only, which means fewer neutrons are produced in the treatment rooms than when passive scattering techniques are used. (Courtesy of ProTom International, Inc.)

radiation oncology operations. Because proton therapy becomes more widespread and mainstream, it can be expected that fully integrated hospital-based proton therapy facilities will become more common. Facilities can also be differentiated according to whether the facility is for profit or nonprofit and whether the facility is privately or publicly owned.

A few proton therapy facilities have special-purpose irradiation rooms (e.g., those used for ocular and radiosurgery treatments and those used only for research and development purposes). Within a multipurpose facility, there may also be treatment rooms dedicated to particular treatments, such as those used exclusively, or nearly so, for pediatric cancer treatments or for prostate cancer treatments.

4.3   Requirements for Carbon Ion Facilities

Similar to proton beam therapy facilities, the beam delivery technique (scattering versus scanning) will be an important factor in determining the clinical and technical requirements. Among the pencil beam scanning techniques described in Chapter 3, only the so-called intensity controlled raster scanning technique has been implemented for clinical use. Details about the treatment delivery techniques for carbon ions are discussed in the literature (Alonso 1989; Haberer et al. 1993; Kanai et al. 1999). An overview of the current status of ion beam therapy is found in the work of Jäkel, Karger, and Debus (2008).

General differences between proton and ion beam facilities arise from a number of factors.

- The energy loss of ions such as carbon is significantly higher than for protons, mainly due to their larger charge. Consequently, the energy of the particles needed to achieve the same penetration depth is higher. Based on the Bethe equation, it can be deduced that the energy loss of carbons is ~20 times larger than for protons at the same range.
- In order to achieve these large energies (typically, 5.4-GeV kinetic energy is needed for carbon ions, which corresponds to 450 MeV per nucleon, to achieve a depth of 35 cm in water), much larger accelerators are needed, and up to now, only synchrotrons have been implied in clinical facilities. Because the energy loss distribution (or the spectral distribution of secondary electrons) is very similar for protons and carbon ions with the same range, it is mainly the number of secondary electrons that increases when going from protons to carbons rather than their energy. This leads to much more dense ionization tracks of carbon beams and in addition to a significant change in the ionization density or linear energy transfer (LET) toward the Bragg peak.
- Because carbon ions are composed of six protons and six neutrons, nuclear interactions will be of higher importance. In particular, due to projectile fragmentation, lighter particles with a velocity

similar to the incoming primary ion are produced. Because these have lower energy loss, they have considerably longer ranges, leading to a dose tail beyond the Bragg peak.
- The change in LET together with the production of a rather complex spectrum of secondary charged particles leads to a complex dependency of the response of biological systems, as well as detectors to the beam energy, penetration depth, dose, and fluence.

Although the number of vendors of carbon facilities is smaller than for proton facilities, there still is a variety available. Thus, a similar procurement process should be followed as for protons. Because some of the vendors have not yet provided facilities, it may even be more important to define the performance characteristics and specifications that are needed.

Because all of the currently existing carbon ion facilities have been developed to a large extent by research laboratories, a new concept can be developed within a consortium of partners from research, industry, and health care providers. This was successfully done in Japanese facilities, at Heidelberg Ion Therapy Center (HIT), Heidelberg, Germany, at the National Center for Oncological Treatment (CNAO), Milan, Italy, and at MedAustron, Wiener Neustadt, Austria. Some specifications for these facilities can also be found in the literature (Amaldi 2004; Haberer et al. 2004) and on the Web sites of the corresponding research centers, e.g., Gunma University Heavy Ion Medical Center (GHMC) in Japan and the Helmholtzzentrum für Schwerionenforschung GmbH (GSI), which is the German national heavy ion research laboratory.

4.3.1   General Requirements for Ion Beam Therapy Facilities

In this section, we provide the requirements that generally apply to all ion beam therapy facilities and ion

Chapter 4

beam treatment techniques. It should be noted that many facilities offer proton and ion beams, so the requirements from both sections in this chapter apply.

4.3.1.1 Accelerators

All ion facilities currently in operation are based on synchrotrons, although cyclotrons, in principle, are a possible alternative for ions. New accelerator concepts such as the DWA or laser acceleration for ions will require considerably more time to be developed compared to protons. This is due to the much higher energy required. Within the next one to two decades, it is very unlikely that these techniques will be available for clinical application. An exception may be the FFAG concept, which may be especially useful to reduce the magnet size in synchrotrons and gantries. The same is true for superconducting magnet systems, which may allow smaller accelerators and gantries (Trbojevic et al. 2007). The advantage of using synchrotrons compared to cyclotrons for ion beams is that, first, there does not exist sufficient experience with high-energy cyclotrons and, second, the possibility of directly changing the extraction energy. This can be achieved by predefined settings for all beamline elements, even from pulse to pulse, such as in the machines at HIT and CNAO. Here, many energy levels are available, and no further range shifter is needed.

When only a few energy levels are available (e.g., at Heavy Ion Medical Accelerator in Chiba (HIMAC), Chiba, Japan, at Hyogo Ion Beam Medical Center (HIBMC), Hyogo, Japan, or Gunma), an additional range shifter for fine tuning of the beam range is needed, but its overall thickness is much smaller. In all existing ion facilities, a depth modulation over the full particle range is avoided.

For currently available proton cyclotrons, generally, no energy modulation is possible. Consequently, a range modulation that covers the full range of the particles is needed. For ions, this will result in a large loss of particles in the range shifter. This has to be compensated by a much higher beam current and also leads to a considerable activation of the range shifter and the production of fast neutrons and light fragments in the treatment beam. Activation and neutron production, in turn, request better shielding and make these parts of the machine less easily accessible for maintenance and more expensive. Nevertheless, there are plans of building a compact superconducting cyclotron for carbon ion beam therapy by one vendor (Jongen et al. 2006). A general discussion of

accelerators can be found in Chapter 2 and elsewhere (Boehne 1992; Alonso 2000).

Generally, for carbon ions, the available range of energies is between 80 and 430 MeV/u (or, somewhat less preferable, 400 MeV/u in some cases). As in proton machines, higher energies are preferable if passive elements are used for the beam delivery. Range straggling is much smaller for carbon beams compared to protons. On the other hand, synchrotrons have a very small energy spread ($\Delta E/E$) of typically less than ±0.1%, which is generally appropriate. In the case of an active beam delivery with very little material in the beam, it may even be advisable to introduce a so-called miniridge filter to increase the energy spread (Weber and Kraft 1999). By doing so, the number of energies needed to irradiate a target volume is reduced, and the irradiation times will be reduced.

The dynamic range for the beam intensity should be ≥10:1. The stability of the extracted ion beam, measured at the accelerator extraction point, should not vary by more than ±1 mm or by ±1 mrad during beam extraction. Beam shutoff in a synchrotron is usually done after extraction from the ring and should require ≤500 μs, depending on the number of particles per spill. If a beam scanning system is used to deliver the beam to the patient, much more relaxed conditions for the beam position stability may be sufficient, because the isocenter position of the beam can be corrected by a feedback loop to the scanning magnets.

The energy levels should be stable to within 0.5%, which corresponds to a range uncertainty of 0.5 mm. Due to the intrinsic energy selection of the synchrotrons, this is usually not considered a problem.

The beam width extracted from the accelerator should also be controlled to within 20%–30% of the nominal FWHM. This is important, especially if beam scanning is used, and no collimation of the beam is introduced.

Generally, the same beam intensity requirements as for proton beams can be used for ion beams. When scanning with active energy variation is used, the desired dose rates of 2 Gy/min/L to a cubic target will not always be reached. This is mainly due to the relatively slow change of the accelerator beam energy, which is typically done from one spill to another and thus takes 1–2 s. Some recent developments at HIMAC, Chiba, Japan, allow for an energy change during extraction, reducing the overall irradiation time significantly. Absorber-based energy variation systems also react faster by either inserting range shifters (within less than 100 ms, HIMAC) or by deflecting the beam on a wedge (PSI, 80 ms).

4.3.1.2 Beam Transport Systems

The beam transport system for ions is, in principle, designed the same way as for protons, with the main difference being that magnetic fields have to be significantly stronger due to the high magnetic rigidity of high-energy heavy ions. Bending radii are thus considerably larger for ion machines. See Chapter 3 for details. If active energy variation of the accelerator is provided, all the beamline elements have to be adjusted separately for each extracted energy. In order to do that within a minimum of time, special laminated magnets have to be designed in order to minimize arising eddy currents.

4.3.2 Patient Support Systems

The demands on patient support systems are the same as for protons, and no further requirements have to be made.

4.3.3 Patient Treatment Positioning System

The demands on treatment positioning systems are the same as for protons, and no further requirements have to be made. However, it should be mentioned that, when working with a gantry for heavy ions, movements of the gantry bearings and treatment room walls may not be excluded due to the large moving masses. This may require additional laser systems mounted outside the treatment area, where they will be less subject to these movements.

4.3.4 TCS

The demands on the TCS are the same as for protons, with only one further requirement: the selection of the ion species (if that functionality is provided by the facility) has to be controlled and displayed by the TCS.

4.3.4.1 Reliability

The requirements on reliability are the same as for protons, and no further requirements have to be made.

4.3.4.2 System Parameters

The same requirements as for protons should be satisfied.

4.3.4.3 Equipment

The same requirements as for protons should be satisfied.

4.3.4.4 Beam Gating Capability

For active beam delivery systems, the interplay between organ movement and the dynamic beam delivery poses a general problem. Currently, three lines of development are followed in order to mitigate the effects of target motion.

The most advanced method is beam gating, where the irradiation is gated by a suitable patient monitoring device. Another system under investigation is the fast rescanning, where a homogeneous target irradiation is restored by multiple scans, which leads to an averaging of the dose inhomogeneities introduced by interplay effects.

Another technique called target tracking aims at a real-time tracking of the target motion with the scanning beam (Bert et al. 2007; van de Water et al. 2009). Although this technique is still under development, new facilities should be prepared to use these techniques. Tracking is the most demanding technique, because it requires real-time adoption not only of the lateral beam position but also of the beam energy.

4.3.4.5 Compliance with Regulations and Standards

When combining medical products from different vendors, the compatibility of devices has to be assured by the manufacturer through conformance statements.

4.3.4.6 Additional Technologies

A special feature of heavier ions is the activation of tissue during irradiation, which offers the unique possibility of using *in vivo* monitoring of dose delivery. Dedicated in-room PET scanners may be considered, or a PET-CT scanner can be located close to the treatment room (half-lives of the main isotopes in the region of 10 min). To move a patient to an out-of-room scanner, the PET-CT should be compatible with a shuttle system.

4.3.5 Requirements for Specific Ion Treatment Techniques

Ion therapy requirements depend on the treatment delivery technique used; in this section, we provide the specific requirements for the passive scattering and pencil beam scanning techniques.

Chapter 4

4.3.5.1 Passive Scattering Systems

As for proton therapy, only passive systems were used for ion beam delivery until the 1990s. Although a pencil beam scanning system was developed in Berkeley in 1992, it was only used for a single patient (Alonso 1989). The first hospital-based ion beam center (HIMAC) in Japan, which started operation in 1994, still relies on passive scattering systems only. The same is true for a facility in Hyogo (HIBMC) that opened in 2002. Most of the requirements for protons can be also followed for ions, and consequently, we summarize only those issues where some differences may occur.

- *Lateral penumbra*—The multiple scattering for ions in the patient is less than 1% of the penetration depth; therefore, at a depth of 30 cm, the typical penumbra size would be approximately 3–5 mm, depending on the field size.
- *Field size*—The largest beam size currently available for passive ion scattering systems is 22 cm in diameter at the HIMAC facility; generally, a larger field size (at least in one dimension) of 25–30 cm is recommended. For ion beams, the air gap does not contribute significantly to the penumbra of the field. However, a retractable snout is still advisable in order to limit the penumbra resulting from the collimator and compensator.
- *Delivered dose reproducibility and repeatability*— The same specifications as for protons should be followed, but the delivered doses should be close to the clinical requirements (typically 0.5 Gy and 1Gy).
- *Dose rate*—For a given beam intensity extracted from the accelerator, the dose rate depends on the target volume size. For average target volumes, the dose rate should be at least 1 Gy/min; for very large volumes, the dose rate should be ≥0.5 Gy/min; and for small volumes (e.g., radiosurgery fields), the dose rate should be ≥5 Gy/min.

4.3.5.2 Pencil Beam Scanning Treatment Systems

In 1997, a pencil beam scanning system was introduced for carbon ions at GSI, Darmstadt, Germany (Haberer et al. 1993) and was used as the sole beam delivery system.

The first commercial ion scanning system was introduced at HIT in November 2009 and closely followed the design of the GSI development. At the new ion beam facilities at Gunma University and HIMAC, pencil beam scanning is also under development.

As with passive systems, we only summarize here the items where differences between protons and ions are arising.

1. *Field size*—The maximum field sizes for scanned beams of ions are currently smaller than for passive beams, because the high magnetic field requires fast deflection of the high energetic ion beams. The systems currently available allow for field sizes of 20 cm × 20 cm (HIT); larger field sizes are currently under development (e.g., 25 cm × 25 cm at HIMAC scanning beam). The treatment of larger fields may be facilitated by using translations of the treatment table.
2. *Pencil beam spot size (1-σ) and shape*—Especially for high energetic ion beam, the beam spot size may deviate significantly from a Gaussian profile and may exhibit an asymmetric form in the horizontal plane due to the extraction via a septum. It is then necessary to have sufficient overlap between adjacent beam spots in order to reach a homogeneous dose distribution.

   Even a perfect Gaussian-shaped beam will deviate from this idealization at larger depths due to nuclear fragments that are produced in depth and are ejected at larger angles. Thus, the beam may be parameterized by 2 or 3 Gaussians (Furukawa 2010).
3. *Spot placement accuracy*—The spot placement accuracy for each spot should be ≤±0.5 mm (1 SD). If a large overlap of many pencil beams is achieved like in the raster scanning technique used, e.g. at HIT, an accuracy of 10% of the FWHM of the beam is more appropriate.
4. *Effective dose rate*—The time needed to treat a volume of 10 cm × 10 cm × 10 cm located at any depth to a dose of 1 Gy should be ≤3 min; 1.0 min is recommended. The effective dose rate will vary considerably on the method used for scanning (i.e., discrete spot scanning or dynamic beam scanning definitions are provided in Chapter 3; in general, dynamic beam scanning will have the highest effective dose rate).

4.3.5.3 Layer Stacking Treatment Systems

As a form of a hybrid system, at HIMAC, the so-called layer stacking system has been developed for ion beam therapy. It includes a range modulator with a small depth modulation of ~1 cm, a range shifter, and an MLC in order to perform conformal irradiations. The irradiation is performed layer by layer, and the depth and the field size of each layer are adjusted by the range shifter and MLC accordingly. By doing so, the dose can be conformed better to the target volume than by the conventional fixed depth modulation and field size of passive systems (Kanematsu et al. 2002). This system is employed at HIMAC and also at Gunma University facility. Because it is a very specialized system, we will not give dedicated requirements for this system here.

4.3.6   Typical Configurations for Ion Therapy Facilities

There are currently three carbon ion therapy facilities in operation worldwide within a clinical setting and one within a research facility. About five more facilities are under construction. Most ion facilities in Japan are purely carbon facilities, whereas all European facilities provide protons and carbon ions. All of these facilities currently have three treatment rooms with one or two fixed beamlines and only one with a single isocentric gantry exists. Even without a gantry, the height of buildings will be much higher than for a linac-based facility due to the inclined beamlines.

A typical arrangement of a treatment facility with two horizontal beamlines and an isocentric gantry is shown in Figure 4.4.

As for proton facilities, the trend for ion facilities is also to integrate them into existing hospital structures in order to integrate ion beam radiotherapy into the oncology services. Currently, there is no rationale for treating eye tumors with carbon ions; therefore, no dedicated facility exists.

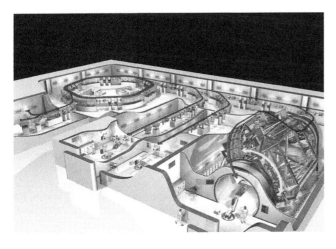

FIGURE 4.4 Heavy-ion facility with a synchrotron on the upper left, two fixed-beam rooms, and one isocentric gantry room. This facility, located at Heidelberg, Germany, was built primarily for the use of carbon ion and proton beams, but other heavy particles can be accelerated. (Courtesy of the University of Heidelberg.)

4.3.7   Radiation Shielding Considerations for Proton and Carbon Ion Facilities

The Particle Therapy Co-Operative Group (PTCOG) has published an excellent report that covers all aspects of radiation shielding for proton and carbon ions facilities that can be found on the Internet site for PTCOG 2010.

The requirements for radiation shielding for a particle therapy facility depend on a number of factors, including the type of particle (i.e., protons or carbon ions), the type of accelerator, the configuration of the facility, and the type of treatment techniques employed. The accelerators used so far in proton facilities have been cyclotrons, synchrocyclotrons, and synchrotrons; however, within the next decade, laser accelerators and dielectric wall accelerators may be used in proton facilities. To date, carbon ion facilities have only used synchrotrons; however, a high-energy superconducting cyclotron has recently been designed for carbon ion therapy.

A typical large particle therapy facility might consist of a beam production system (a cyclotron and ESS or a synchrotron and associated injector), a cyclotron or a beam transport line, and several treatment rooms with isocentric gantries or fixed beamlines. Some facilities also have a treatment room dedicated to eye treatments and/or an experimental room for research and development purposes. Recently, single-room therapy systems with a synchrocyclotron integrated into the gantry in the treatment room have been marketed, and the first such system may be in operation within a year. Several vendors offer two- or single-room systems with the accelerator outside the treatment room.

For radiation therapy applications, dose rates of 1.0–2.0 Gy/min are typically used, and treatment fields on the order of 12 cm × 12 cm are commonly used. A passive scattering system can generate treatment fields up to 25 cm × 25 cm; pencil beam scanning techniques can produce treatment fields up to 40 cm × 40 cm and larger. Beamlines devoted to eye treatments use dose rates of 12–18 Gy/min, but the field sizes used are on the order of 3-cm diameter. There are a few facilities that use radiosurgery techniques that have dose rates and field sizes between those for large field treatments and eye treatments.

During the normal operation of particle therapy facilities, secondary radiation is produced at locations where beam losses occur. Several equipment design features have a strong impact on the shielding requirements; these include techniques that

Chapter 4

cause large beam losses in the beam production system (such as inefficient extraction systems, degraders associated with fixed-energy accelerators, and emittance filters), beam stops or beam dumps, passive scattering systems, range degraders and range modulation devices in the treatment nozzle, and beam-shaping devices such as collimators and range compensators near the patient. In addition, particles stopping in the patient or dosimetry phantom also produce secondary radiation. Therefore, secondary radiation can be produced at many places in the facility, leading to the need to shield every area where such radiation is produced.

The interaction of protons and carbon ions with matter results in "prompt" and "residual" secondary radiation. Prompt radiation persists only during the time that the beam is present. Residual radiation from activation remains after the beam is shut off. For charged particle therapy facilities, neutrons dominate the prompt, secondary radiation dose.

The goal of shielding is to attenuate secondary radiation to levels that are within regulatory or design limits for individual exposure, and to protect equipment from radiation damage. This requires knowledge of the following parameters (PTCOG 2010):

1. Accelerator type, particle type, and maximum energy
2. Beam losses and targets
3. Beam-on time
4. Beam shaping and delivery
5. Regulatory and design limits
6. Workload, including the number of patients to be treated, treatment energies, field sizes, beam angles, and dose per treatment
7. Use factors
8. Occupancy factors

The attenuation length of neutrons in the shielding material determines the thickness of shielding that is required to reduce the dose to acceptable levels. Shielding for neutrons must be such that sufficient material is interposed between the source and the point of interest, and neutrons of all energies must be sufficiently attenuated. Dense material of high atomic mass such as steel meets the first criterion, and hydrogen meets the second criterion because of effective attenuation by elastic scattering. However, steel is transparent to neutrons of energy ranging from 0.2 to 0.3 MeV. Therefore, a layer of hydrogenous material must always follow the steel. Alternatively, large thicknesses of concrete or concrete with high-z aggregates can be used. This is the usual shielding solution for particle facilities (PTCOG 2010). In some cases, the use of local shielding materials close to the source of secondary radiation is useful for decreasing the thickness of exterior shielding walls and for protecting nearby electronic modules from neutron damage.

Monte Carlo calculation is the most efficient and comprehensive method for calculating shielding thicknesses and configurations; several Monte Carlo codes are routinely used, and these are described in the PTCOG document.

References

Alonso, J. R. 1989. Magnetically scanned ion beams for radiation therapy. *Nucl. Inst. Meth. B* 40/41:1340–1344.

Alonso, J. R. 2000. Review of ion beam therapy: Present and future. Report LBNL-45137. Lawrence Berkeley National Laboratory.

Amaldi, U. 2004. CNAO—The Italian Centre for Light-Ion Therapy. *Radiother. Oncol.* 73 (Suppl. 2):S191–S201.

Bakai, A., Alber, M., and Nüsslin, F. 2003. A revision of the γ-evaluation concept for the comparison of dose distributions. *Phys. Med. Biol.* 48:3543–3553.

Bert, C., Saito, N., Schmidt, A., Chaudhri, N., Schardt, D., and Rietzel, E. 2007. Target motion tracking with a scanned particle beam. *Med. Phys.* 34(12):4768–4771.

Boehne, D. 1992. Light ion accelerators for cancer therapy. *Radiat. Environ. Biophys.* 31:205–218.

Bryant, P. J., Badano, L., Benedikt, M., Crescenti, M., Holy, P., Maier, A. T., Pullia, M., Reimoser, S., Rossi, S., Borri, G. et al. 2000. Proton-Ion Medical Machine Study (PIMMS), CERN report CERN-PS-2000-007-DR, http://cdsweb.cern.ch/record/449577/files/, accessed June 21, 2012.

Chu, W. T., Ludewigt, B. A. and T. R. Renner. 1993a. Instrumentation for treatment of cancer using proton and light-ion beams. *Rev. Sci. Instrum.* 64:2055–2122.

Chu, W. T., Staples, J. W. and Ludewigt, B. A. et al. 1993b. Performance specifications for Proton Medical Facility. LBL Report No. 33749.

Depuydt, T., Van Esch, A. and Huyskens, D. P. 2002. A quantitative evaluation of IMRT dose distributions: Refinement and clinical assessment of the gamma evaluation. *Radiother. Oncol.* 62:309–319.

Furukawa, T. 2010. Performance of fast raster scanning system for HIMAC new treatment facility. Talk at PTCOG 49, Gunma, Japan.

Gall, K. P., Verhey, L. and Alonzo, J. et al. 1993. State of the art? New proton medical facilities for the Massachusetts General Hospital and the University of California Davis Medical Center. *Nucl. Instrum. Meth. B* 79:881–884.

Gillis, S., De Wagter, C. and Bohsung, J. et al. 2005. An intercenter quality assurance network for IMRT verification: Results of the ESTRO QUASIMODO project. *Radiother. Oncol.* 76: 340–353.

Gottschalk, G. 2003. On the characterization of spread-out Bragg peaks and the definition of "depth" and "modulation." http://huhepl.harvard.edu/-gottschalk/BGdocs.zip/SOBP.pdf.

GSI. http://www.gsi.de/en/start/beschleuniger/therapieprojekte.htm. Accessed June 21, 2012.

Gunma University Heavy Ion Medical Center (GHMC). http://heavy-ion.showa.gunma-u.ac.jp/en/facilities02.html. Accessed June 21, 2012.

Haberer, T., Becher, W., Schardt, D. and Kraft, G. 1993. Magnetic scanning system for heavy ion therapy. *Nucl. Instrum. Meth. A* 330:296–305.

Haberer, T., Debus, J., Eickhoff, H., Jäkel, O., Schulz-Ertner, D. and Weber, U. 2004. The Heidelberg Ion Therapy Center. *Radiother. Oncol.* 73 (Suppl. 2):S186–S190.

ICRU Report 78. 2007. Prescribing, recording, and reporting proton-beam therapy. *J. ICRU* 7(2).

Jäkel, O., Karger, C. P. and Debus, J. 2008. The future of heavy ion radiotherapy. *Med. Phys.* 35:5653–5663.

Jiang, S. B., Sharp, G. C. and Neicu, T. et al. 2006. On dose distribution comparison. *Phys. Med. Biol.* 51:759–776.

Jongen, Y., Beeckman, W., Kleeven, W. J. G. M. et al. 2006. Design studies of the compact superconducting cyclotron for hadron therapy. *Proc. EPAC 2006*, Edinburgh, Scotland, pp. 1678–1680.

Kanai, T., Endo, M., Minohara, S. et al. 1999. Biophysical characteristics of HIMAC clinical irradiation system for heavy-ion radiation therapy. *Int. J. Radiat. Oncol. Biol. Phys.* 44:201–210.

Kanematsu, N., Endo, M., Futami, Y., Kanai, T., Asakura, H., Oka, H. and Yusa, K. 2002. Treatment planning for the layer-stacking irradiation system for three-dimensional conformal heavy-ion radiotherapy. *Med. Phys.* 29:2823–2829.

Low, D. A., Harms, W. B., Mutic, S., and Purdy, J. A. 1998. A technique for the quantitative evaluation of dose distributions. *Med. Phys.* 25:656–661.

Low, D. A. and Dempsey, J. F. 2003. Evaluation of the gamma dose distribution compassion method. *Med. Phys.* 30:2455–2464.

Pedroni, E., Bacher, R., and Blattmann, H., et al. 1995. The 200 MeV proton therapy project at the Paul Scherrer Institute: Conceptual design and practical realization. *Med. Phys.* 22: 37–53.

PTCOG. 2010. Shielding design and radiation safety of 11 charged particle therapy facilities; PTCOG Publications Sub-Committee Task Group on Shielding Design and Radiation Safety of Charged Particle Therapy Facilities (ed.); http://ptcog.web.psi.ch/Archive/Shielding_radiation_protection.pdf. Accessed June 21, 2012.

Smith, A. 2009. Vision 20/20: Proton therapy. *Med. Phys.* 36:556–568.

Smith, A., Gillin, M., and Bues, M. et al. 2009. The M. D. Anderson proton therapy system. *Med. Phys.* 36:4068–4083.

Spezi, E. and Lewis, D. G. 2006. Gamma histograms for radiotherapy plan evaluation. *Radiother. Oncol.* 79:224–230.

Stock, M., Kroupa, B. and Georg, D. 2005. Interpretation and evaluation of the γ index and the γ index angle for the verification of IMRT hybrid plans. *Phys. Med. Biol.* 50:399–411.

Trbojevic, D., Parker, D., Keil, E. and Sessler, A. M. 2007. Carbon/proton therapy: A novel gantry design. *Phys. Rev. Spec. Top.-Ac.* 053503-1–053503-6.

van de Water, S., Kreuger, R., Zenklusen, S., Hug, E. and Lomax, A. J. 2009. Tumor tracking with scanned proton beams: Assessing the accuracy and practicalities. *Phys. Med. Biol.* 54: 6549–6563.

Weber, U. and Kraft, G. 1999. Design and construction of a ripple filter for a smoothed depth dose distribution in conformal particle therapy. *Phys. Med. Biol.* 44(11):2765–2775.

Wending, M., Zijp, L. J. and McDermott, L. N. et al. 2007. A fast algorithm for gamma evaluation in 3-D. *Med. Phys.* 34: 1647–1654.

Chapter 4

Section II

5. Radiobiology of Proton and Carbon Ion Therapy

Alejandro Carabe

5.1   Introduction

State-of-the-art radiation therapy enables the delivery of the prescribed dose to the target volume with unprecedented precision due to a deep understanding of the science and engineering required to achieve this ultimate goal in radiotherapy. However, the overall precision of a radiotherapy treatment requires mastering not only the physics and engineering aspects behind it but also the biology that determines the end point and result of such treatment. Huge efforts have been made by the proton and carbon ion radiotherapy communities to characterize the radiobiology of these two particles and determine how different their relative effectiveness in achieving an end point is compared to X-rays.

In the second half of this chapter, we will revise the concept of relative biological effectiveness (RBE) and the physical and biological factors on which it depends. The intention of this chapter is not so much to revise the status of mathematical modeling describing the RBE of these particles but to describe the available data and, perhaps, recognize caveats where more experimental work may be needed in order to further increase the precision in the delivery of these types of radiations in the context of cancer therapy from a radiobiological point of view.

Proton and Carbon Ion Therapy Edited by C.-M. Charlie Ma and Tony Lomax © 2013 Taylor & Francis Group, LLC. ISBN: 978-1-4398-1607-3.

Chapter 5

5.2 BP, Spread-Out Bragg Peak (SOBP), and Linear Energy Transfer (LET)

Protons and carbon ions are charged particles that deposit energy along their tracks with a maximum dose deposition at the end of their range, commonly known as the Bragg peak (BP). The increase in dose with depth, together with the abrupt drop in dose immediately after the BP, differs markedly from the dose-deposition profiles produced by conventional radiation sources (X-rays and electrons) currently used in radiotherapy. Because the treatment volumes commonly encountered in radiotherapy are generally deep seated (with the exception of superficial disease normally treated with orthovoltage X-ray or low-energy electron beams), the difference in the location of the maximum dose makes proton and carbon ion beams interesting radiation modalities for cancer treatment.

In addition to the favorable depth-dose properties mentioned above, protons and carbon ions are known to have an enhanced biological effect toward the distal portion of the depth-dose curve. This means that the advantage of using protons and, more particularly, carbon ions over photons and electrons is based not only on their dosimetric properties but also on their higher cell killing (per unit dose) within the target volume compared to that at smaller depths.

In this section, we will discuss the fundamental physical properties of protons and carbon ions and the rationale for their use in radiotherapy.

5.2.1 Formation and General Characteristics of the Proton and Carbon Ion Pristine Bragg Peak (PBP)

When a fast charged particle enters a medium, it interacts with the electrons and nuclei of the medium, losing part of its energy in each interaction. If we ignore nuclear forces and consider only the interactions arising from Coulomb forces with the orbiting electrons, the energy lost (or transferred) will result in *ionization* (production of ion–electron pairs) and *excitation* of the atoms present in the medium, and the rate of transfer of energy per unit distance (or *stopping power*) between the fast particle and the medium is given by the Bethe–Bloch formula:

$$-\frac{1}{\rho}\frac{dE}{dX}=\frac{5.08\times10^{-31}z^2}{\beta^2}\Big[F(\beta)-\ln I\Big]\ \ \text{MeV cm}^{-1}$$

(5.1)

where ρ is the electron density of the medium traversed by the protons, z is the atomic number of the heavy particle, $\beta = v/c$ is the speed of the particle relative to c, I is the mean excitation energy of the medium, and

$$F(\beta)=\ln\frac{1.02\times10^6\beta^2}{1-\beta^2}-\beta^2$$ [values of $F(\beta)$ are tabulated in the work of Turner (1995)]. It should be noted that, in contrast to fast electrons, the radiative (i.e., bremsstrahlung) energy loss by heavier charged particles is entirely negligible.

The most prominent characteristic of Equation 5.1 that dictates the shape of the depth-dose curves shown in Figure 5.1 is that the energy transferred is inversely proportional to the square of the velocity of the particle (β). Thus, the greater the velocity of the particle, the smaller the energy deposited (per unit distance), and vice versa. Looking at Figure 5.1, this translates into low-energy deposition levels at small depths (where the particles have their highest energy and velocity) with a steep increase of energy deposition toward the end of the range, where little energy remains (due to the cumulative effect of collisions with the orbiting electrons of the atoms of the medium), and hence, the velocity is low.

An important feature of the PBPs shown in Figure 5.1 is that the carbon ion PBP seems much "sharper" than the proton PBP. The mass of the particle influences the sharpness of the dose falloff of the BP: the greater the mass the sharper the dose falloff. According to Table 5.1, the mass of a carbon ion is about 12 times that of a proton, which, in turn, is about 2000 times greater than the electron mass (m_o). The sharpness of the falloff in a particle beam is ultimately related to the range straggling of the particles, which, in turn, varies

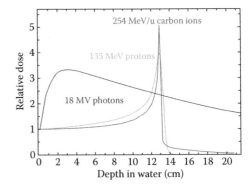

FIGURE 5.1 Comparison of a photon, proton, and carbon ion beam depth-dose profiles.

Table 5.1   Physical Properties of Electrons, Protons, and Carbon Ions

Particle	Charge	Mass
e	−1e	1 m_o
^{1}H(p)	+1e	1836 m_o = 1.008 u
^{12}C	+6e	12 u

approximately inversely to the square root of the mass of the particle.

Another effect that varies approximately inversely to the square root of the mass of the particle is the lateral dose falloff of the beam (called *apparent* penumbra). The distal dose falloff of the beam is related to inelastic interactions between the particle and the electrons and nuclei constituting the absorbing material, whereas the lateral dose depends on the multiple elastic Coulomb scattering (MCS) the particle suffers as it traverses the medium. In this elastic scattering process, the particles suffer numerous small angle deflections away from the central trajectory, resulting in a larger divergence of the beam. In the following sections, we will see in more detail how each of these types of interactions affects the shape of the depth-dose curve.

5.2.1.1   Range and Distal Falloff of the Depth-Dose Curve

The range of a particle is directly related to its initial energy. This is shown in Figure 5.2, where the range of a monoenergetic proton or carbon ion is plotted against particle energy. This range is the *median* or *nominal* range of the particle (R) and only applies to monoenergetic particle beams. In practice, at the depth of the BP, there is a spread of energies, because in each of the

inelastic collisions with bound atomic electrons along their track, different amounts of energy are transferred by the particle with a *maximum* possible energy proportional to m/M, where m is the mass of the electron, and M is the mass of the particle. The amount of energy transferred fluctuates between collisions, with those particles transferring larger total amounts of energy having a shorter range, and vice versa; this is known as *energy-loss straggling*, and it is the reason for the 'smearing' of the BP. Thus, a monoenergetic beam of particles traveling within the medium in the direction x with energy E and nominal range R will have associated a distribution of ranges, $s(x)$, given by a *Gaussian* distribution:

$$s(x) = \frac{1}{\sqrt{2\pi}\sigma_x} e^{-(x-R)^2/2\sigma_x^2} \tag{5.2}$$

In the region where this formula is valid ($2 < R < 40$ cm), σ_x is *almost* proportional to the (residual) range R and inversely proportional to the square root of the particle mass number A.

5.2.1.2   Lateral Width and Lateral Falloff of the Depth-Dose Curve

On entering the medium, the beam has a certain width (i.e., field size) and then broadens slowly as it penetrates the medium. This broadening is due to elastic interactions of the incident particles with the nuclei of the atoms constituting the medium, also known as MCS. The large number of small-angle deflections in an ion beam leads to lateral divergence of the beam, as can be observed in Figure 5.3.

According to Highland (Highland 1975), the distribution of angular deflections in an ion beam can be

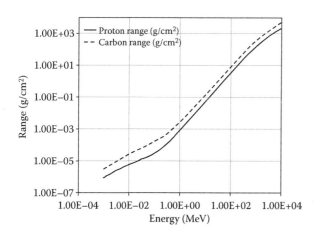

FIGURE 5.2   Residual range in water for different particles (PSTAR).

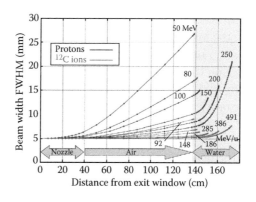

FIGURE 5.3   Lateral width of protons and carbon ions in air and water. (From Weber, U. and Kraft, G. 2009. *Cancer J.* 15: 325–332.)

Chapter 5

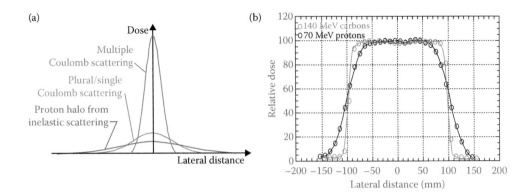

FIGURE 5.4 (a) Lateral profile of the dose contribution from the different interaction processes by which charged particles lose their energy. (From Goiten, M. 2008. *Radiation Onclogy: A Physicist's-Eye View.* Springer. p. 226.) (b) Lateral profile of a carbon and proton beam, showing the sharper penumbra of the carbon beam (IAEA 2007).

represented to the first order by a Gaussian, where the mean angle of multiple scattering is given by

$$\overline{\theta}_0 \approx \frac{z \cdot 21.2 \text{MeV}}{M_p v^2} \sqrt{\frac{1}{X_0}} \tag{5.3}$$

where z and M_p are the charge numbers (1 for protons and 6 for carbon ions) and mass of the particle, and X_0 is radiation length, characteristic of the scattering material. X_0 varies approximately as $A/NZ(Z + 1)$, where N is Avogadro's number, Z is the atomic number, and A the atomic weight of the target material. The dependence of the mean angle of multiple scattering on the inverse power of the ion (kinetic) energy explains the broadening of the BP near the end of the range. In addition, the inverse proportionality between $\overline{\theta}_0$ and the particle mass explains the difference on the magnitude of the effect for protons and carbon ions. The broadening of a proton pencil beam is approximately 3.5 times larger than for a carbon beam compared for the same range (Weber 2009). A factor of 2 comes from the different charge-to-mass ratio, and a factor of approximately 1.7 is due to the lower proton energies for the same range.

Inelastic nuclear interactions also have an important effect on the penumbra of charged particle beams, as shown in Figure 5.4, where secondary (or of any superior order) charged particles emerge from the collision at a small but not negligible angle to the direction of the incident particle and create a halo of dose around the beam that becomes larger at larger depths. This halo can also be approximated by a Gaussian distribution, which further contributes to the tails of the lateral dose distribution. The three quasi-Gaussian profiles shown in Figure 5.4a represent the source of

the lateral broadening toward the end of the range of the proton or carbon ion beams shown in Figure 5.3. It is clear that carbon ions have a small halo compared to that associated with protons, because the contribution from the three types of interaction to the overall Gaussian profile is larger for protons than for the carbon ions.

5.2.1.3 Dose Contribution from Secondary Particles and Heavier Fragments

Another effect produced by inelastic nuclear interactions is the fragmentation of the primary particles, a process by which the projectile nucleus, after suffering a nuclear collision with a target nucleus, is broken apart into several daughter particles, as shown in Figure 5.5.

Figure 5.6 shows the contribution to the dose profile of a 391-MeV/u carbon ion PBP from the stripping of neutrons from the target that convert the primary carbons into positron-emitting isotopes ^{11}C and ^{10}C. The dose distribution of ^{11}C and ^{10}C reaches its maximum a few millimeters before the BP, and their contribution to the total dose (TD) is ~14% and ~2% of the TD, respectively. A maximum of 40% of the TD deposited up to the BP corresponds to the sum of the absorbed

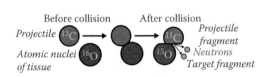

FIGURE 5.5 Sketch of a fragmentation process. (From Kraft, G. 2005. *Proceedings of the International Meeting of Particle Therapy Cooperative Group* (PTCOG) 43, Rinecker Proton Therapy Center, Munich, Germany.)

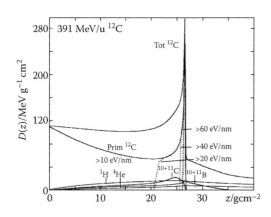

FIGURE 5.6   Dose contribution from primary and fragments of a 391-MeV/u carbon ion beam (Kempe et al. 2007).

doses from ^1H, ^4He, ^{10}B, ^{11}B, ^{10}C, and ^{11}C (Kempe et al. 2007).

5.2.2   Beam Modulation in Proton and Carbon Ions

Because the BP is rather narrow, it must be spread out (SOBP) in order to treat large tumors (see Figure 5.7). The SOBP is obtained by the superposition of many different BPs of different ranges. The depth of the BPs can be changed in steps by inserting material in the beam path (degrader) or by changing the energy of the beam produced in the accelerator. For clinical purposes, the former method requires the design of a degrader with variable thickness to model different beams with sufficient range spread-out (or modulation) to cover the spectrum of tumor sizes encountered in the clinic.

Whichever degrading system is used, the dose weighting carried by each individual BP (Figure 5.7b) must be such that a flat SOBP is produced. In Equation 5.4, the dose D at each depth z_i is a weighted (W_j) sum of the dose contributions d_i of n beams with different energies and, thus, different peak positions Z_i:

$$D(z_i) = \sum_{j=1}^{n} W_j d_j(z_i) \qquad (5.4)$$

5.2.3   LET Distributions in Proton and Carbon Ion Beams

Linear energy transfer (LET) is a well-established concept (Zirkle et al. 1952) for quantifying radiation quality. LET is defined by the International Commission on Radiation Units and Measurements (ICRU 1970) as the quotient of $d\overline{\varepsilon}$ by dl, where $d\overline{\varepsilon}$ is the average energy locally imparted to the medium by a charged particle of specified energy in traversing a distance of dl', i.e.,

$$L_\Delta = \left.\frac{d\overline{\varepsilon}}{dl}\right|_\Delta \qquad (5.5)$$

The subscript Δ refers to the word "locally" in the definition and indicates the cutoff limit of energy transfer. Any particle with energy greater than this cutoff limit will not be considered local and, therefore, not taken into account in the calculation of LET. L_∞ indicates that all possible energy transfers have been included, and this unrestricted LET is synonymous with collisional stopping power (see Figure 5.8). Common units are kiloelectron volts per micrometer in water of unit density for LET or linear stopping power and megaelectron volts per square centimeters per gram for mass stopping power.

(a)   (b)

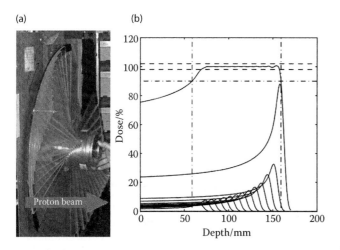

FIGURE 5.7   (a) Range modulator. (b) SOBP (Gottshalk 2004).

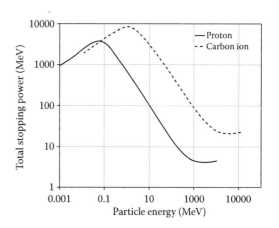

FIGURE 5.8   Stopping power of proton and carbon ions of different energy.

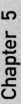
Chapter 5

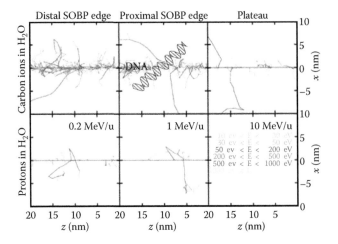

FIGURE 5.9 Structure of a proton (below) and carbon ion (above) track (Kramer and Kraft 1994).

The fact that clinical decisions in radiotherapy are based on dose distributions makes the dose-averaged LET distribution (LET_d) the preferred quantity to report results that relate radiation effectiveness with radiation quality. The questions at hand would be: How different are the track structure and LET_d distribution of protons and carbon ions? Is the LET distribution in a particle beam independent of its range (or initial energy)? What effect does the modulation of the beam have on its LET distribution?

In response to the first question, Figure 5.9 shows the difference in the pattern of energy deposition, even at different positions of a proton and carbon ion beam.

The larger density of energy transfer events to the medium per particle track in the case of carbon ions than in the case of protons translates into larger values of LET_d for carbon ions than for protons. Figure 5.10 provides an estimation of the difference in LET_d values that applies to proton and carbon ion beams where only the contribution from the primary particles is considered. It has recently been shown (Grassberger and Paganetti 2011; Kempe et al. 2007) that the inclusion of the energy deposited by secondary particles in the calculation of LET_d would moderately increase the profiles shown in Figure 5.10, mainly in the plateau region, because it is in this region where secondary particles mainly deposit their energy, as shown in Figure 5.6.

Figure 5.11 shows three different proton (left) and carbon ion (right) beams of different ranges but with the same modulation (1.8 and 3 cm for the proton and the carbon ion beams, respectively). In both figures, we can see how, due to the presence of more energetic particles in deeper beams and the fact that stopping power is inversely related to energy in therapeutic beams (Figure 5.8), the LET_d at any given depth of a shallow beam is always higher than the LET_d at the same depth of a deeper beam.

When the modulation of the beam increases, pencil beams carrying lower weight of the dose and with lower LET_d are introduced in the proximal portion of the SOBP, which will reduce the LET_d at the center of the SOBP, as observed in Figure 5.12.

5.3 RBE of Protons and Carbon Ions

Gammas (^{60}Co: LET_d ~0.4 keV/μm; ^{137}Cs: LET_d ~0.8 keV/μm) and X-rays (200 kV: LET_d ~3.5 keV/μm) [ICRP92] (Valentin 2003) have traditionally been considered the *reference* radiation to compare with the effectiveness

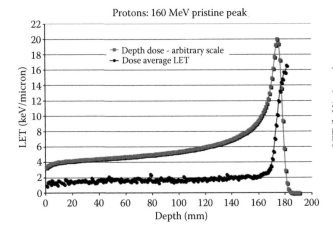

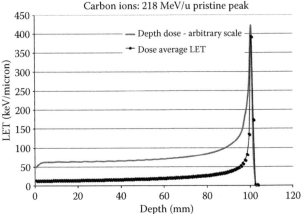

FIGURE 5.10 LET profiles of a proton and carbon ion pristine peak. The LET is only computed for the primary particles, so no fragmentation tail is observed on the carbon ion BP.

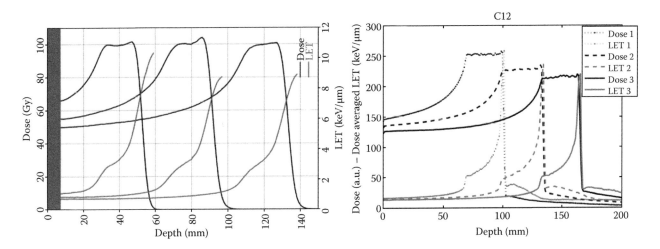

FIGURE 5.11 LET profiles of a proton (left) (Carabe et al. 2012a) and carbon ion (right) (Kantemiris et al. 2012) SOBP. Lower energy beams have larger values of LET across the beam profile.

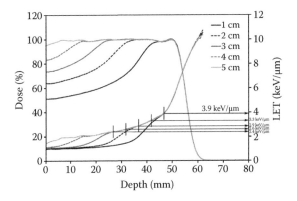

FIGURE 5.12 Change of LET at the center of the SOBP with the width of the beam SOBP (Carabe et al. 2012a).

of any other type of radiation (so-called *test radiation*). The ICRP (1990) and ICRU (1979) defined the RBE as "A ratio of the absorbed dose of a reference radiation to the absorbed dose of a test radiation to produce the same level of biological effect, other conditions being equal." More specifically, if *A* is the *test* (high-LET) radiation, *B* is the *reference* (X-rays) radiation, D_A and D_B are the doses necessary to produce the same effect of interest with radiation *A* and *B,* respectively, then the RBE of radiation *A* relative to radiation *B* is

$$RBE = \frac{D_A}{D_B}\bigg|_{Iso-effective} \qquad (5.6)$$

Special emphasis should be put on the words "same level of biological effect" and "other conditions being equal," because the ICRU specifies that "when two radiations produce an effect that is not of the same extent and/or nature, an RBE cannot be specified." Thus, D_A and D_B must be isoeffective, and they should

be measured in identical experimental conditions. It is precisely this last requirement that introduces some of the most important uncertainties in the measurement of RBE.

Among all variables on which the RBE depends, its variation has been best established for changes in dose per fraction, particle LET, and tissue or cellular $(\alpha/\beta)_{X\text{-rays}}$. The following sections will demonstrate these dependencies for proton ions and for carbon ions.

5.3.1 RBE as a Function Dose or Survival Level

The general rationale for using high-LET radiotherapy is the higher impact that this type of radiation has on certain radioresistant tumors, because cell killing is less affected by the position of the cells in the cell cycle or their oxygen status, and cells have less ability to repair high-LET radiation injury (Withers et al. 1982; Wambersie et al. 2004). These three aspects lead to the conclusion that fractionation effects are of diminished importance when using high-LET radiations, as illustrated in Figure 5.14. This figure shows how, for the neutron (high-LET) data, increasing the number of fractions (*n*) has little effect on the total dose (TD) required to achieve a given effect, except perhaps when *n* is quite small.

Figures 5.13 and 5.14 also demonstrate that high-LET radiations are more effective than low LET to produce a given effect: with any given number of fractions, the TD required to produce the same effect is always lower (Kramer et al. 2003) with high-LET$_d$ radiations. It is also important to notice in Figure 5.14 the continuous increase of the low-LET TD with the number of fractions, whereas the high-LET TD

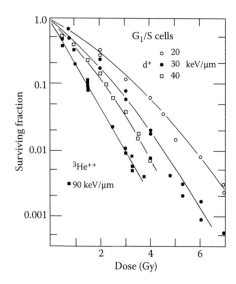

FIGURE 5.13 Survival curves for V79 cells exposed to different charged particles of different LET (Birds et al. 1980).

remains relatively insensitive to the changes of the fraction size. This implies that the ratio of the TD of low- and high-LET radiation required to produce the same effect increases with increasing the number of fractions or, equivalently, with decreasing the dose per fraction (Withers et al. 1982).

Based on this context, the difference in biological effectiveness of low- and high-LET radiations (RBE) can be formulated as the ratio of the TD required with both types of radiation to produce the same effect, $\text{RBE} = \dfrac{\text{TD}_{\text{low-LET}}}{\text{TD}_{\text{high-LET}}}$. This idea is represented in Figure 5.15a, where the solid line corresponds to the total isoeffective dose for low-LET radiations and the dashed line represents the isoeffective line for high-LET radiations. To understand the shape of these lines, we need to remember that, at very large doses per fraction (i.e., low number of fractions), lethality is mainly produced by the accumulation of sublethal damage, which is represented by the quadratic term in the linear quadratic formulation. This implies that the TD required when delivered in a small number of large fractions (high dose) is lower than in more fractionated regimens in order to produce the same effect (e.g., same number of lethal events). Figure 5.15 also shows how the TD ratio (RBE) increases with an increasing number of fractions, and vice versa. This implies that RBE must have a maximum (RBE_{max}) corresponding to the ratio of the asymptotic values of low- and high-LET TDs at an infinite number of fractions and a minimum (RBE_{min}) corresponding to the ratio of TDs

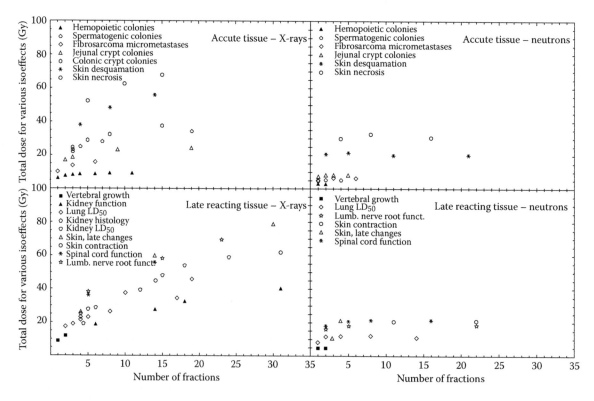

FIGURE 5.14 Total isoeffective doses versus the number of fractions for different early (top) and late (bottom) reacting tissues using low- (X-rays, left) and high-LET (neutrons, right) radiations.

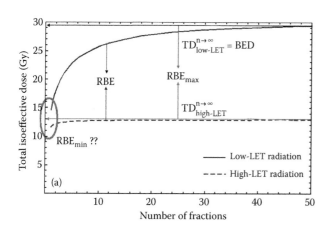
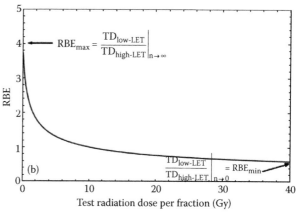

FIGURE 5.15 Variation of RBE with (a) total isoeffective dose and (b) dose per fraction.

obtained from the intersection of the solid and dashed lines with the y-axes, as shown in Figure 5.15b. Due to the asymptotic nature of RBE_{min}, as it corresponds to zero number of fractions, it is virtually impossible to know if it is equal to one (which would imply that both lines in Figure 5.15a intersect the y-axes at the same point) or if it has a value different than one.

The values of RBE_{max} and RBE_{min} can be easily obtained from the linear-quadratic (LQ) formulation as follows. If the selected end point is the production of a given number (N) of lethal events at both high- and low-LET, then

$$N\big|_{low\text{-}LET} = N\big|_{high\text{-}LET} \Rightarrow \alpha_L d_L + \beta_L d_L^2 = \alpha_H d_H + \beta_H d_H^2$$

$$(5.7)$$

At very low doses, the quadratic terms of Equation 5.7 are negligible compared to the linear components; hence, Equation 5.7 can be reduced to

$$\alpha_L d_L = \alpha_H d_H \Rightarrow RBE_{max} = \frac{d_L}{d_H}\bigg|_{d\approx 0Gy} = \frac{\alpha_H}{\alpha_L} \qquad (5.8)$$

In the opposite end, at very high doses, the linear terms can be neglected, and Equation 5.7 becomes

$$\beta_L d_L^2 = \beta_H d_H^2 \Rightarrow RBE_{min} = \frac{d_L}{d_H}\bigg|_{d\to\infty Gy} = \sqrt{\frac{\beta_H}{\beta_L}} \qquad (5.9)$$

Although there are substantial data for protons (Paganetti et al. 2002) and carbon ions (Ando and

Kase 2009), indicating that $\alpha_L \neq \alpha_H$ for both animal and human *in vitro* and *in vivo* experiments, no such body of data exist to substantiate that $\beta_L \neq \beta_H$. Recent efforts to prove the existence of $RBE_{min} \neq 1$ for neutrons (Jones 2010) and for carbon ions (Carabe-Fernandez et al. 2010) seem to suggest better prediction of the TD required in hypofractionation treatments with these particles when it is assumed that $RBE_{min} \neq 1$ than in the opposite case. The analysis for protons is inconclusive.

Although the use of the LQ formulation has been justified for doses up to 18 Gy (Brenner 2008), it is practically impossible to extrapolate the LQ formulation to dose levels of the order of those required to calculate RBE_{min}. However, it seems more feasible to make such extrapolation to zero number of fractions on fractionated data such as the one shown in Figure 5.16 (left). However, in this case, to obtain reliable values of RBE_{min} that could have an impact on the RBE relevant to hypofractionated regimens, a fractionation data set covering from a very low to a very large number of fractions is required. Very limited data of this type exist for high-LET radiations.

Figure 5.16 shows the difference in the predicted value of RBE at large doses per fraction when RBE_{min} is assumed to be equal to (gray curve) and different (black curve) from 1. In particular, changes have been reported (Carabe-Fernandez et al. 2010) on proton and carbon ion RBE between 7% and 9% at large doses per fraction (see Figure 5.17) when it is considered that $RBE_{min} \neq 1$. It is also suggested that these differences may be larger for tissues with low (α/β) than tissues with high (α/β). This may have important implications in hypofractionated treatments for hepatocellular cancer treated with 14 GyE/fraction carbon ions (Tsujii et al. 2007) and stage I lung cancer treated with proton

Chapter 5

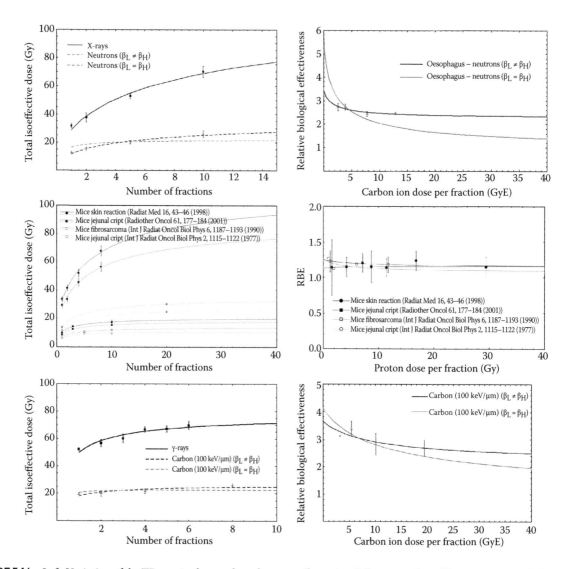

FIGURE 5.16   Left: Variation of the TD required to produce the same effect using different number of fractions to treat different tissues. Right: Variation of RBE with dose per fraction used to treat different tissues (top, esophagus (Hornsey and Field 1979); middle, various; bottom, skin (Ando et al. 1998)).

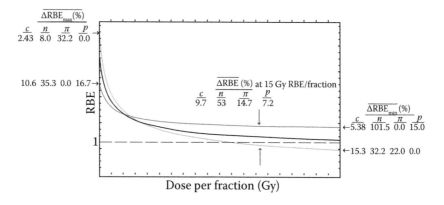

FIGURE 5.17   Uncertainty in the prediction of RBE due to the uncertainty of RBE_{min}. (Carabe-Fernandez, A., Dale, R. G., Hopewell, J. W., Jones, B., Paganetti, H. 2010. *Phys. Med. Biol.* 55:5685–5700.)

(14–18 GyE/fraction) or carbon ions (13–15 GyE/fraction). Not only did radiobiological experiments with neutrons (Laramore et al. 1997) and carbon ions (Ando et al. 2005) demonstrate a decrease in RBE with increasing doses per fraction but also that RBE for tumor tissue do not decrease as rapidly as for normal tissue, which substantiates the fact that the therapeutic ratio increases rather than decreases with an increasing fraction size. However, if RBE_{min} is substantially different from unity, the positive therapeutic ratio may be affected in either way. A larger amount of *in vivo fractionation* data for tissues of low and high (α/β) exposed to proton and carbon ions is required to study changes of RBE at large doses per fraction.

5.3.2 RBE as a Function of Particle Linear Energy Transfer

It was as early as 1935 when Zirkle (1935) reported enhanced biological effects produced by alpha particles as a function of ion concentration produced in their paths. However, it was not until the 1960s (Barendsen and Beusker 1960; Barendsen 1963) that changes of RBE with LET for different particles, biological systems, and end points were studied systematically.

Since Barendsen obtained, for the first time, a relationship between RBE and LET (Barendesen 1964) using deuterons and α particles of different energy levels, subsequent measurements with protons (Belli et al. 1989;

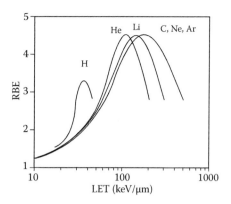

FIGURE 5.18 Schematic of the relationship between RBE and LET for cells exposed to different charged particles. (Raju, M. R. 1995. *Int. J. Radiat. Biol.* 67:237–259.)

Folkard et al. 1996; Perris et al. 1986) and heavy ions (Todd 1967; Cox et al. 1977; Blakely et al. 1979; Thacker 1979; Blakely 1992; Goodwin et al. 1994) have shown that the maximum RBE of charged particles shifts to lower LETs with a decreasing charge of the particles.

An immediate consequence of the shift of the maximum RBE with the particle charge is that particles of different charges but the same LET will have different RBE, which means that we cannot easily extrapolate the radiobiological data from one particle to another (see Figure 5.18). Different authors (Folkard et al. 1996; Belli et al. 1989, 1992; Goodhead et al. 1992) have shown that low-energy protons have higher effectiveness than α particles of comparable LET for V79 cell inactivation

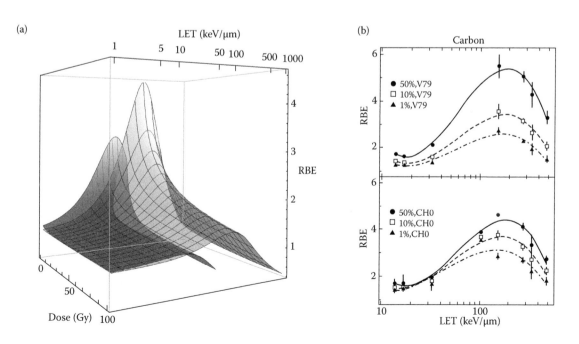

FIGURE 5.19 (a) Variation of proton (dark) and carbon ion (light) RBE with LET at different dose levels. (b) LET dependency of RBE measured at different survival levels. (Weyrather, W. K., Ritter, S., Scholz, M. and Kraft, G. 1999. *Int. J. Radiat. Biol.* 75:1357–1364.)

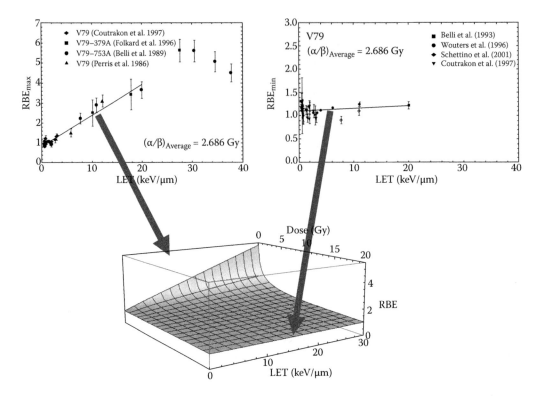

FIGURE 5.20   Variation of RBE with dose and LET for V79 cells exposed to protons.

and mutation induction. However, the yield of double-strand breaks (dsb) for protons was found not to be significantly higher than that for α particles, which suggests that different types of radiation may produce dsb with different biological consequences (Jenner et al. 1992). This difference in RBE associated with different particles of the same LET is related to the different pattern in which they deposit their energy microscopically along their track (also called track structure), as discussed earlier in reference to Figure 5.9. It is this

inadequacy of the concept of LET to differentiate the various effectiveness of particles with different charge, which justifies the use of track structure–based models such the local effect model (LEM; Scholz et al. 1997) and the microsimetric-kinetic model (MKM; Hawkins 1996; Kase et al. 2008) to characterize the RBE of the beam in high-LET (mainly carbon ions) radiotherapy clinics.

The values of RBE shown in Figure 5.18 correspond to a "static" picture of RBE for a given dose of the

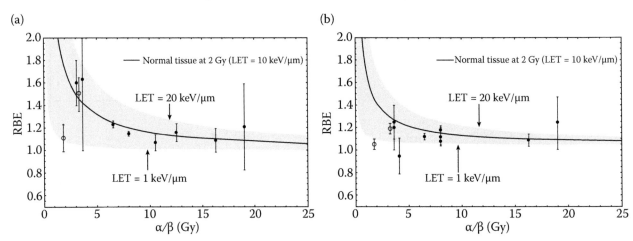

FIGURE 5.21   Variation of RBE with $(\alpha/\beta)_x$. The central line in these figures is obtained from Equation 5.12, assuming a LET_d value in RBE_{max} and RBE_{min} (Equations 5.10 and 5.11) of 10 keV/μm at the mid-SOBP position for all the beams. (Reported by Gerweck, L. E. and Kozin, S. V. 1999. *Radiother. Oncol.* 50:135–142.)

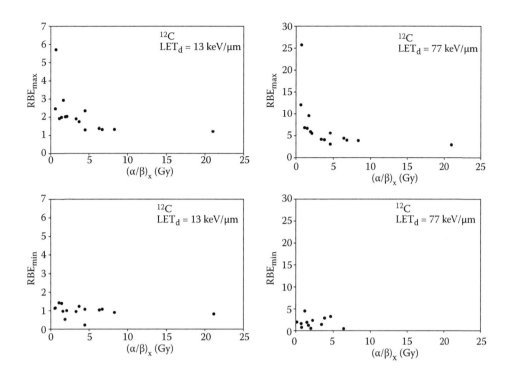

FIGURE 5.22   Variation of RBE with $(\alpha/\beta)_x$ for carbon ions of different LET_d. (Suzuki, M., Kase, Y., Yamaguchi, H., Kanai, T. and Ando, K. 2000. *Int. J. Radiat. Oncol. Biol. Phys.* 48:241–250.)

corresponding particle. However, we have to always keep in mind the dose dependency of RBE, which causes it to decrease from its maximum value (at zero dose or full survival), and therefore, the values presented will change for different dose levels, as shown in Figure 5.19a. This figure is a replot of the proton and carbon ion lines in Figure 5.18 and shows, in a more representative way, how RBE varies with LET at different dose levels (or, equivalently, at different survival levels, as indicated in Figure 5.19b, where high survival levels would correspond to lower doses).

Regardless of the complexity of the relationship between LET and RBE, Wilkens and Oelfke (2004)

have shown that, in the case of proton treatment plans, LET values no larger than 15 keV/µm are normally found, which, according to the animal data shown in Figure 5.20, could allow us to use a simple linear relationship between RBE and LET in the range between 1 and 20 keV/µm.

In the case of carbon ions, this simplification might not be applicable in all cases, as indicated in the work of Ando and Kase (2009) and Furusawa et al. (2000), which is another reason that more complex models such as LEM and MKM are required in order to calculate the RBE at different depths of the carbon ion BP. Nevertheless, Ando and Kase (2009) have tried to use simple linear

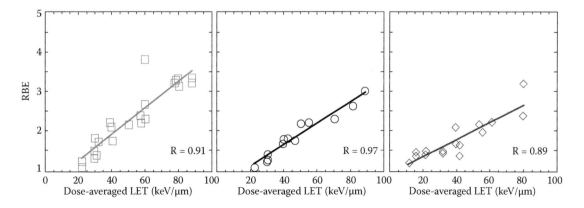

FIGURE 5.23   Linear regression analysis on carbon ions RBE versus LET for 10% survivals in colony formations of V79 (left), HSG (middle), and T-1 (right) cells. (Ando, K. and Kase, Y. 2009. *Int. J. Radiat. Biol.* 85:715–728.)

Chapter 5

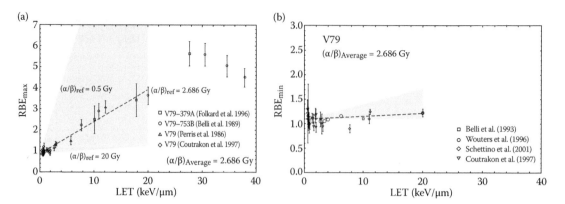

FIGURE 5.24 Variation of the RBE versus LET_d relationship with $(\alpha/\beta)_x$ for protons. (Carabe, A., Moteabbed, M., Depauw, N., Schuemann, J. and Paganetti, H. 2012a. *Phys. Med. Biol.* 57:1159–1172.)

regression analysis to predict RBE values for carbon ions of different LET and obtained an extremely good correlation for LET values up to 100 keV/μm.

More recently, a similar linear regression analysis has been used by Sorensen et al. (2011) for data on V79, HSG, and T1 exposed to different particle species. In their study, they proved that a high linear correlation between RBE and LET can be achieved not only for carbon ions but also for other types of particles with LET values ranging from 10 to 100 keV/μm. Therefore, because the values of LET distributions within the body for a proton or carbon ion beam are below 100 keV/μm, we could, in principle, use a simple linear formulation to predict values of RBE at different depths of the treatment beam.

5.3.3 RBE of Proton and Carbon Ions as a Function of the Target α/β Ratio

The question of predictability of the RBE of high-LET radiation from the $(\alpha/\beta)_{X\text{-ray}}$ was studied in detail by numerous authors for different particles. Stenerlow et al. (1995) showed that there was no clear relationship between RBE and $(\alpha/\beta)_{X\text{-ray}}$ for nitrogen and hydrogen. Similarly, for neutrons, Fertil et al. (1982) and Warenius et al. (1994) showed the same lack of correlation between neutron RBE and $(\alpha/\beta)_{X\text{-ray}}$. A correlation between the inherent radiosensitivity to photons and neutrons of 20 human cell lines was discovered by Britten (1992), but this correlation was found only in the low-dose region.

Gerweck and Kozin (1999) reported changes of proton RBE with $(\alpha/\beta)_{X\text{-ray}}$ ratios at two different dose levels, 2 and 6 Gy, and found that the RBE of mid-SOBP protons may be significantly higher in the 2-Gy dose range than at higher doses, for low $(\alpha/\beta)_{X\text{-ray}}$ tissues. However, because the RBE values were obtained from targets in the midposition of the SOBP (or, in

two cases, an average of several positions in the SOBP corresponding to the open symbols in Figure 5.21) of beams with different energy levels (i.e., range) and modulation, according to Figures 5.11 and 5.12, the LET_d at the mid-SOBP position could be different for each individual point in Figure 5.21. This uncertainty is evidenced by the gray area in Figure 5.21a and b for low (2 Gy) and large (>6 Gy) doses, respectively.

Suzuki et al. (2000) studied the change of RBE with radiosensitivity for 16 cell lines (normal and tumor tissues) measured for carbon ions with different LET. Single-dose cell survival curves were reported from where the variation of RBE with dose could be calculated, which allows the ability to assess the change of RBE with $(\alpha/\beta)_{X\text{-ray}}$ at different dose levels. Figure 5.22 shows that a similar relationship to that for protons can be established at low doses per fraction between the RBE of carbon ions and $(\alpha/\beta)_{X\text{-ray}}$, with a lower RBE for cell lines with larger $(\alpha/\beta)_{X\text{-ray}}$. However, at larger doses, the RBE for carbon ions seems to be independent of the cell $(\alpha/\beta)_{X\text{-ray}}$.

The immediate consequence of the inverse relationship between RBE and $(\alpha/\beta)_x$ is that the slope of the

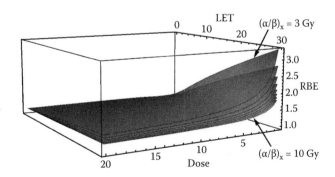

FIGURE 5.25 Change of RBE with dose (single exposure or fractionated), $(\alpha/\beta)_{X\text{-ray}}$ and LET. Observe how, at low doses, the RBE for low $(\alpha/\beta)_{X\text{-ray}}$ is higher than that for high $(\alpha/\beta)_{X\text{-ray}}$.

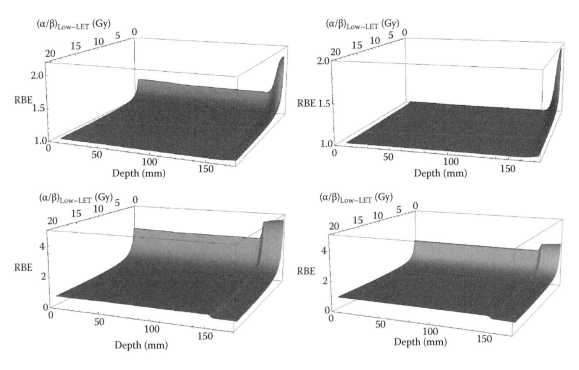

FIGURE 5.26 (See color insert.) Proton (top) and carbon (bottom) RBE versus depth (combined dose and LET$_d$ effect) and $(\alpha/\beta)_x$ for a SOBP normalized to 2 Gy (left) and 15 Gy (right).

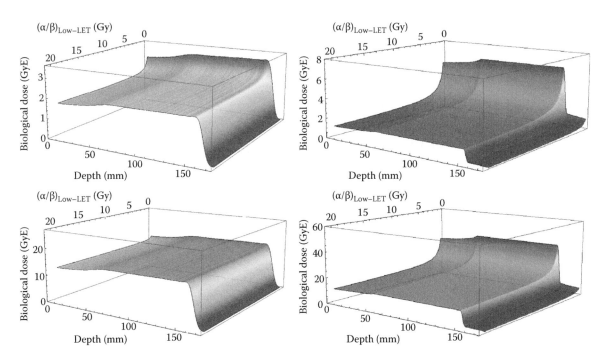

FIGURE 5.27 (See color insert.) Top left: Biological proton SOBP at 2 Gy for different $(\alpha/\beta)_{X\text{-ray}}$. Top right: Biological carbon ion SOBP at 2 Gy for different $(\alpha/\beta)_{X\text{-ray}}$. Bottom left: Biological proton SOBP at 15 Gy for different $(\alpha/\beta)_{X\text{-ray}}$. Bottom right: biological carbon ion SOBP at 15 Gy for different $(\alpha/\beta)_{X\text{-ray}}$.

Table 5.2 Values of α/β Ratio for a Variety of Early- and Late-Responding Normal as well as Tumor Human Tissues

Tissue/Organ	End Point	α/β (Gy)	95% CL (Gy)	Source
Early reactions				
Skin	Erythema	8.8	6.9; 11.6	Turesson and Thames (1989)
	Erythema	12.3	1.8; 22.8	Bentzen et al. (1988)
	Dry desquamation	~8	N/A	Chogule and Supe (1993)
	Desquamation	11.2	8.5; 17.6	Turesson and Thames (1989)
Oral mucosa	Mucositis	9.3	5.8; 17.9	Denham et al. (1995)
	Mucositis	15	−15; 45	Rezvani et al. (1991)
	Mucositis	~8	N/A	Chogule and Supe (1993)
Late Reactions				
Skin/vasculature	Telangiectasia	2.8	1.7; 3.8	Turesson and Thames (1989)
	Telangiectasia	2.6	2.2; 3.3	Bentzen et al. (1990)
	Telangiectasia	2.8	−0.1; 8.1	Bentzen and Overgaard (1991)
Subcutis	Fibrosis	1.7	0.6; 2.6	Bentzen and Overgaard (1991)
Breast	Cosmetic change in appearance	3.4	2.3; 4.5	START Trialists Group (2008)
	Induration (fibrosis)	3.1	1.8; 4.4	Yarnold et al. (2005)
Muscle/vasculature/ cartilage	Impaired shoulder movement	3.5	0.7; 6.2	Bentzen et al. (1989)
Nerve	Brachial plexopathy	<3.5*	N/A	Olsen et al. (1990)
	Brachial plexopathy	~2	N/A	Powell et al. (1990)
	Optic neuropathy	1.6	−7; 10	Jiang et al. (1994)
Spinal cord	Myelopathy	<3.3	N/A	Dische et al. (1981)
Eye	Corneal injury	2.9	−4; 10	Jiang et al. (1994)
Bowel	Stricture/perforation	3.9	2.5; 5.3	Deore et al. (1993)
Bowel	Various late effects	4.3	2.2; 9.6	Dische et al. (1999)
Lung	Pneumonitis	4.0	2.2; 5.8	Bentzen et al. (2000)
	Lung fibrosis (radiological)	3.1	−0.2; 8.5	Dubray et al. (1995)
Head and neck	Various late effects	3.5	1.1; 5.9	Rezvani et al. (1991)
Head and neck	Various late effects	4.0	3.3; 5.0	Stuschke and Thames (1999)
Supraglottic larynx	Various late effects	3.8	0.8; 14	Maciejewski et al. (1986)
Oral cavity + oropharynx	Various late effects	0.8	−0.6; 2.5	Maciejewski et al. (1990)
Tumors				
Head and neck				
Various		10.5	6.5; 29	Stuschke and Thames (1999)
Larynx		14.5*	4.9; 24	Rezvani et al. (1993)
Vocal cord		~13	'wide'	Robertson et al. (1993)
Buccal mucosa		6.6	2.9; ∞	Maciejewski et al. (1989)
Tonsil		7.2	3.6; ∞	Maciejewski et al. (1989)
Nasopharynx		16	−11; 43	Lee et al. (1995)
Skin		8.5[a]	4.5; 11.3	Trott et al. (1984)
Prostate[b]		1.1	−3.3; 5.6	Bentzen and Ritter (2005)

(continued)

Table 5.2 Values of α/β Ratio for a Variety of Early- and Late-Responding Normal as well as Tumor Human Tissues (Continued)

Tissue/Organ	End Point	α/β (Gy)	95% CL (Gy)	Source
Breast		4.6	1.1; 8.1	START Trialists Group (2008)
Oesophagus		4.9	1.5; 17	Geh et al. (2006)
Melanoma		0.6	−1.1; 2.5	Bentzen et al. (1989)
Liposarcoma		0.4	−1.4; 5.4	Thames and Suit (1986)

Source: Bentzen, S., and Joiner, M. 2009. In *Basic Clinical Radiobiology*. M. Joiner and A.V. Kogel (eds.). Edward Arnold.

Note: CL = Confidence limit. Reference details are available from Søren Bentzen. See also Thames et al. (1990) and Table 13.2.

[a] Reanalysis of original published data.

[b] Several more estimates are available from comparisons of outcome after brachytherapy versus external-beam therapy.

regression line correlating to RBE and LET$_d$ should be different for each individual cell line or tissue, as shown in Figure 5.23 for carbon ions. In the case of protons, most reported proton RBE values *in vitro* correspond to V79 data, and little data have been obtained from other cell lines. In order to generate tissue-dependent RBE versus LET$_d$ relationships for protons, the variation of RBE$_{max}$ and RBE$_{min}$ with LET$_d$ must be modeled in order to accommodate the dependence of RBE on (α/β)$_x$, and this is shown in Figure 5.24a and b. Equations 5.10 and 5.11 show such dependencies. The top and bottom boundaries in Figure 5.21a and b are obtained from Equation 5.12 for LET$_d$ values in RBE$_{max}$ and RBE$_{min}$ (Equations 5.10 and 5.11) of 20 and 1 keV/μm, respectively, whereas the central line corresponds to a LET$_d$ value of 10 keV/μm. The values of LET$_d$ = 1 and 20 keV/μm correspond to the range of LET$_d$ values where the linear fit provided by Equations 5.10 and 5.11 are valid in Figure 5.24a and b repectively.

Different mathematical models confirm the findings of an increased RBE at low (α/β)$_{X-ray}$ for protons and carbon ions. Paganetti et al. (2000) used the track structure model developed by Katz and coworkers (Butts and Katz 1967; Katz and Sharma 1973; Katz and Sharma 1974) to study the dependency of proton RBE with (α/β)$_{X-ray}$ and concluded that it is not possible to establish a one-to-one correlation between both parameters, with the consequence that there is an overall tendency for RBE to increase as (α/β)$_{X-ray}$ decreases. Similarly, the findings shown in Figures 5.21 and 5.22 are in accordance with the predicted relationship between RBE and (α/β)$_x$ by the MKM (Hawkins 1996), as shown in Equations 5.10 and 5.11.

A 3-D plot of Equation 5.12, bearing in mind the relationship between RBE$_{max}$ and RBE$_{min}$ with LET$_d$ and (α/β)$_x$ established in Equations 5.10 and 5.11 *for V79 cells,* would allow us to see how the RBE depends on the dose per fraction, LET, and (α/β)$_{X-ray}$ simultaneously. Figure 5.25 helps demonstrate how it is impossible to consider RBE dependent on "mainly" any one of the three parameters, as each one has an important impact (perhaps higher in normal tissues than in tumor) on the final clinical value of RBE.

According to this and bearing in mind that depth convolves two parameters (dose and LET) that affect the RBE in opposite directions, we could calculate the biological dose of a proton and carbon ion beam, taking a fair accountability of the impact that

$$\mathrm{RBE}_{max}\left[\mathrm{LET}_d,(\alpha/\beta)_x\right]=0.843+0.154\frac{2.686}{(\alpha/\beta)_x}\mathrm{LET}_d$$

(5.10)

$$\mathrm{RBE}_{min}\left[\mathrm{LET}_d,(\alpha/\beta)_x\right]=1.09+0.006\frac{2.686}{(\alpha/\beta)_x}\mathrm{LET}_d$$

(5.11)

$$\mathrm{RBE}=\frac{\sqrt{(\alpha/\beta)_x^2+4(\alpha/\beta)_x\,\mathrm{RBE}_{max}[\mathrm{LET}_d,(\alpha/\beta)_x]\cdot d_{particle}+4\mathrm{RBE}_{min}^2[\mathrm{LET}_d,(\alpha/\beta)_x]\cdot d_{particle}^2}}{2d_{particle}}-\frac{(\alpha/\beta)_x}{2d_{particle}}$$

(5.12)

Chapter 5

each of the three parameters so far discussed (dose, $(\alpha/\beta)_{X\text{-ray}}$, LET) has on the RBE. Indeed, looking at proton RBE versus depth and $(\alpha/\beta)_{X\text{-ray}}$ for an SOBP normalized to 2 Gy in Figure 5.26 (left) and the biological depth dose curve in Figure 5.27 (top left), we would argue that an increase of RBE of 10% or even 20% along the depth axis due to an increase of LET for normal and slow proliferating tissues could hardly compare with a 60% increase of RBE due to a decrease of $(\alpha/\beta)_x$. As the normalization dose is increased to 15 Gy (Figures 5.26 and 5.27, right), the influence on the RBE from $(\alpha/\beta)_x$ is less apparent. The LET effect is, however, much more pronounced, specially in the case of carbon ions.

For this reason, in the case of protons, it is *at least* equally important to know the LET distribution within the patient and to have human relevant data of α and β values for X-rays, protons, and carbon ions for a wide variety of human tissues. Only under these circumstances would it be possible to make proper use of proton and carbon ion radiobiology in order to do biological optimization of treatment planning. The inherent problem would be that human data would be

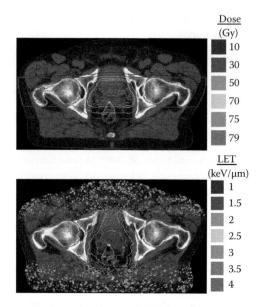

FIGURE 5.28 **(See color insert.)** Dose (top) and LET_d (bottom) distributions in a prostate plan with two parallel opposed beams.

restricted to *in vitro* experiments, which would introduce uncertainties of the type analyzed in the following section.

5.4 Biological Uncertainty in Particle Radiotherapy

The use of computational techniques such as Monte Carlo calculations allow us to obtain accurate dose and LET distributions within patients receiving protons or carbon ions. However, according to Figure 5.24, the uncertainty of the rate of the change of RBE with LET for cell lines with different $(\alpha/\beta)_x$ introduces equally large uncertainties in the values of RBE. To reduce this uncertainty, a wide spectrum of cell lines (from very radioresistant to very radiosensitive) would need to be exposed at different depths of a proton beam with known LET_d distributions. If the BP is spread out, according to Figures 5.11 and 5.12 and to avoid the uncertainty presented in Figure 5.21, the modulation of the beam must always be the same. Until this type of data is produced, it will be difficult to use RBE values with a *reasonable** level of uncertainty in clinical particle radiotherapy. Although these data exist to a certain extent for carbon ions (Ando and Kase 2009), in the case of protons, we can only assume a relationship between proton RBE and LET for different tissue types that changes with $(\alpha/\beta)_x$, as indicated in Equations 5.10 and 5.11.

However, even if we accept Equations 5.10 and 5.11 as valid, we still have to confront the problem of the intrinsic uncertainty of the values of $(\alpha/\beta)_x$ used clinically. Table 5.2 shows how, in some instances, the uncertainty associated with these values is too large to use the reported mean value with sufficient confidence. Nonetheless, although the case of the uncertainty presented in Figure 5.24 (and, thus, the uncertainty of the validity of Equations 5.10 and 5.11) can only be challenged by producing more experimental data, the uncertainty on the RBE introduced by

---

* Even if the experimental uncertainty is eliminated by precise and consistent measurements, there is still the question of how feasible it is to use animal *in vivo* or *in vitro* experimental data in clinical particle radiotherapy, or even human *in vivo* data.

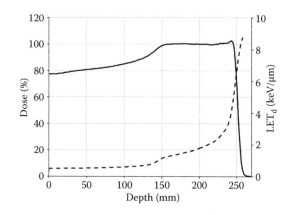

FIGURE 5.29 Dose and LET_d distribution of a proton beam of 25 cm in range and 10-cm SOBP width.

Table 5.3 Prescription and Biological Parameters Relevant to the Treatment of the Prostate

	Prescribed Dose and Critical Structure Dose Constrains	α/β (CI)
Prostate Gland (Standard Number of Fractions = 29)		
GTV + Seminal Vesicles	78 Gy (RBE)/2 Gy (RBE)	GTV: 1.5[a] (1.2[b], 5.6[c])
		SEM V.: 3.0 (2.5, 3.5)
Rectum	V75 < 10%, V70 < 70%	4.0[d] (2.5, 5.0)[e]
Bladder		4.0[d] (3.0, 7.0)[e]

Note: CI = confidence intervals, GTV = gross humor volume, SEM = standard error of mean.

[a] Brenner, D.J., Hall, E.J. Fractionation and protraction for radiotherapy of prostate carcinoma. *Int. J. Radiat. Oncol. Biol. Phys.* 1999; 43:1095–101.

[b] Ritter, M., Forman, J., et al. Hypofraction for prostate cancer. *Cancer J.* 2009; 15:1–6.

[c] Daşu, A. Is the alpha/beta value for prostate tumours low enough to be safely used in clinical trials? *Clin. Oncol. (R. Coll. Radiol.)* 2007; 19:289–301.

[d] Koukourakis, M.I., Abatzoglou, I., et al. Biological dose volume histograms during conformal hypofractionated accelerated radiotherapy for prostate cancer. *Med. Phys.* 2007; 34:76–80.

[e] Fowler, J.F. The radiobiology of prostate cancer including new aspects of fractionated radiotherapy. *Acta. Oncol.* 2005; 44:265–76.

the uncertainty of the values listed in Table 5.2 can be easily calculated, as described in Section 5.4.1. Table 5.2 reported values of α/β for different normal and tumor tissues (Bentzen and Joiner 2009).

There is also the effect that the uncertainty of RBE has on the "biological" range of a particle beam. According to different authors (Robertson et al. 1975; Paganetti et al. 2000), the increase of RBE at the distal falloff of the proton beam means that the depth corresponding to the distal 90% dose will be displaced 1–2 mm deeper. Similar results have been obtained recently (Carabe et al. 2012a) in a systematic study that aims at quantifying how much this displacement is for individual values of $(\alpha/\beta)_x$ and a quantification of the biological range of the particle beam due to the uncertainty of $(\alpha/\beta)_x$ (see Section 5.4.2).

5.4.1 Impact of the Uncertainty of Clinical Values of $(\alpha/\beta)_x$ on RBE

The main objective in this section is to calculate the difference in the biological dose* distributions produced when RBE is assumed constant (1.1) and when it is assumed to depend on $(\alpha/\beta)_x$ according to Equation 5.12. In both cases, the physical dose distributions are the same, and LET_d and $(\alpha/\beta)_x$ are the only variables to explain the differences between the two biological dose

distributions. In particular, the comparison is done as follows.

1. The biological dose distribution obtained with an RBE calculated using Equations 5.10 and 5.11 for the values of $(\alpha/\beta)_x$ reported in Table 5.2, third column, is compared with the biological dose distribution obtained with an RBE = 1.1.
2. The uncertainty of the biological dose distribution obtained for a variable RBE will be calculated from the 95% confidence interval of $(\alpha/\beta)_x$ (Table 5.2, fourth column).

This comparison has been done for prostate and brain patients (Carabe et al. 2012b).

5.4.1.1 Prostate

In the case of the prostate, the traditional technique for treatment planning consists of two lateral opposed fields (see Figure 5.28). Usually, the distal 98% percentage depth dose (PDD) matches the planning treatment volume (PTV), which requires ranges of approximately 25 cm and SOBP width of ~10 cm. According to Figure 5.29, the LET_d values normally found in a water phantom measurement in the middle of the SOBP of beams with these characteristics are on the order of 1.8 keV/μm, whereas in the falloff, the maximum LET_d is on the order of 8 keV/μm. However, the presence of bony structures such as the femur heads and pelvic structures alter and degrade the energy spectrum of the beam in such a way

---

* Biological dose = physical dose × RBE.

that reduces the maximum LET$_d$ to almost half of its value in water. This is shown in Figure 5.28 which, together with the noncritical nature of these tissues, shows typical 2-D dose (top) and LET (bottom) distributions within the patient in a prostate plan.

An important factor to observe in Figure 5.28 is that, in order to provide full dose coverage to the target, the elevated LET$_d$ values of the beam lay on the healthy tissue beyond the target. However, these values are only on the order of three to four times larger

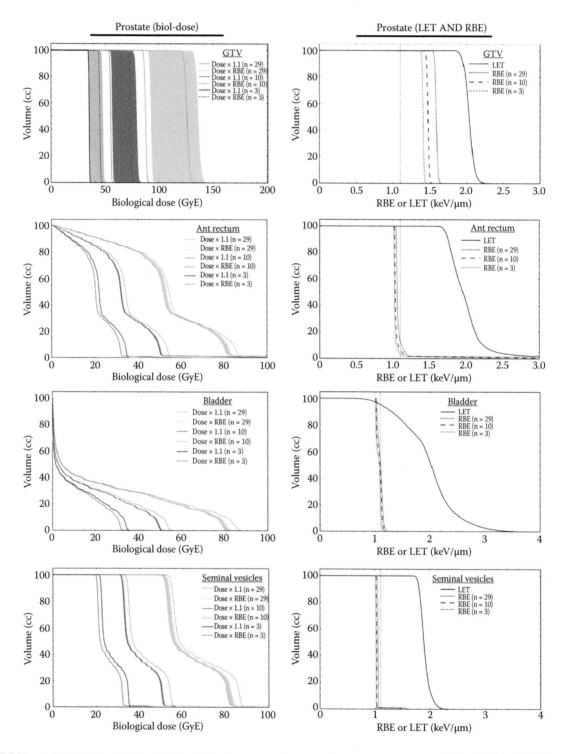

FIGURE 5.30 Left: BDVHs within the GTV and the dose constrain organs in a prostate treatment. Right: RBE and LET$_d$ histograms within the same volumes as on the left. Different colors are used for different fraction schemes (light → number of fractions (n) = 29; dark → n = 10; gray → n = 3).

Table 5.4 Prescription and Biological Parameters Relevant to the Treatment of Craniopharyngioma

	Prescribed Dose and Critical Structure Dose Constrains	α/β (CI)
Craniopharyngioma (Standard Number of Fractions = 29)		
GTV	72 GyRBE (X-rays + proton boost)	12[a] (10, 15)[a]
Brainstem	No hot spot	2.1[b] (1.5, 3.9)[b]
Chiasm	No hot spot	2.9[d] (1.5, 3.9)[b]
Optical Nerve	No hot spot (56–60 Gy(RBE))	1.6[c] (0.5, 10.3)[c]
Normal Brain		2.9[d] (1.5, 3.9)[b]

[a] Yuan, J., Wang, J.Z., et al. Hypofractionation regimens for stereotactic radiotherapy for large brain tumors. *Int. J. Radiat. Oncol. Biol. Phys.* 2008; 72:390–7.

[b] Meeks, S.L., Buatti, J.M., Foote, K.D., et al. Calculation of cranial nerve complication probability for acoustic neuroma radiosurgery. *Int. J. Radiat. Oncol. Biol. Phys.* 2000; 47:597–602.

[c] Jiang, G.L., Tucker, S.L., et al. Radiation-induced injury to the visual pathway. *Radio. Oncol.* 1994; 30:17–25.

[d] Lawrence, Y.R., Li, X.A., et al. Radiation dose-volume effects in the brain. *Int. J. Radiat. Oncol. Biol. Phys.* 76(3 Suppl):S20–7.

than those corresponding to megavoltage X-ray beams, which could explain the lack of post-treatment complications in the normal tissues adjacent to the PTV.

There has been a substantial revision of the $(\alpha/\beta)_x$ of the prostate (Fowler et al. 2001; Shridhar et al. 2009), and according to Table 5.3, it has associated a low value of 1.5 Gy with a wide uncertainty range (1.2, 5.0). The low $(\alpha/\beta)_x$ should therefore favor effectiveness in the target when using protons. We have two questions: The RBE should increase, but how much? What is the uncertainty in the RBE due to the uncertainty of the $(\alpha/\beta)_x$? To answer these questions, Figure 5.30 shows dose-volume histograms (DVHs) of the biological doses produced by assuming a constant RBE of 1.1 and an RBE derived from Equation 5.12, using as input data the dose and LET_d shown in Figure 5.24, and the $(\alpha/\beta)_x$ values shown in Table 5.3.

Figure 5.30 shows the resulting DVH of the biological doses [biological dose-volume histograms (BDVHs)] obtained from the fixed and variable RBE values across different critical organs involved in a prostate treatment. In this figure, the dashed lines correspond to the biological dose distribution obtained from an RBE derived from Equation 5.12, whereas the black solid line corresponds to the biological dose distribution obtained with an RBE of 1.1. The shadowed areas correspond to the uncertainty on the DVH produced by the uncertainty on the $(\alpha/\beta)_x$. Different colors are used for different number of isoeffective fractions. On the right panels, the RBE values (RBE-VH) obtained from Equation 5.12 (dashed lines) are compared with the 1.1

value, which is represented by the vertical dotted line. The solid black line corresponds to the LET_d volume histogram (LET_d-VH) within each individual region of interest contour.

The following conclusions can be extracted according to Figure 5.30.

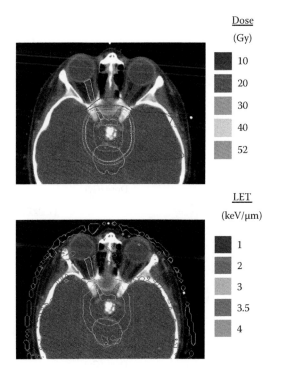

FIGURE 5.31 (See color insert.) Dose (XiO, solid line; Monte Carlo, dashed line) and LET_d distributions in the treatment of a craniopharyngioma.

Chapter 5

1. In the left panels (BDVH), the difference between the dashed and the black line is very pronounced for the very low $(\alpha/\beta)_x$ (i.e., GTV), whereas in the case of the nontarget organs, both lines are much closer.

2. In the left panels, the consideration of RBE = 1.1 underpredicts the biological dose in the prostate, whereas in the nontarget tissues, it seems to overpredict. This could potentially mean that less dose can be used in the target to induce the same tumor control at the same time the dose to the nontarget organs would be significantly lower. However, these results are based on the assumption that Equations 5.10 and 5.11 are valid.

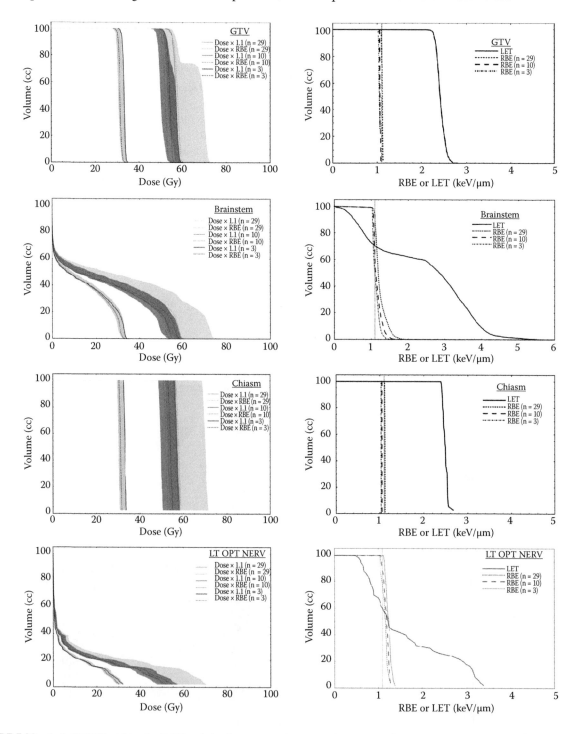

FIGURE 5.32 Left: BDVH within the GTV and the dose constrain organs in a craniopharyngioma treatment. Right: RBE and LET$_d$ histograms within the same volumes as on the left.

Table 5.5 Prescription and Biological Parameters Relevant to the Treatment of the Adenoid Cystic Carcinoma of the Lacrimal Gland

	Prescription [Gy (RBE)]	(α/β) [Gy]	$\Delta(\alpha/\beta)$ [Gy]
GTV	72 (X + P)	10.0	
CTV	56	3.0	1.5–3.9
Optic nerve	56–60	1.6	0.5–10.3
Optic chiasm	60	3.0	1.5–3.9
Macula	55Gy	2.9	1.5–4.1
Brainstem		2.1	1.5–3.9
Normal brain		2.9	1.5–3.9

It is difficult to separate the influence of the LET_d and the $(\alpha/\beta)_x$ on the RBE; thus, we have included the LET_d and RBE distributions within each volume for the nominal (central) values of $(\alpha/\beta)_x$ on the right panels of Figure 5.30. For example, for the prostate GTV and the seminal vesicles, the LET_d distributions are very similar, but because the physical dose received by the seminal vesicles is lower than the GTV, we might expect a larger RBE in the seminal vesicles. However, the RBE values in the prostate are higher than for the seminal vesicles, because prostate tissues have lower $(\alpha/\beta)_x$. In addition, Figure 5.30 seems to suggest that hypofractionating the dose decreases the RBE and the uncertainty of the biological dose.

5.4.1.2 Brain Craniopharyngioma

Due to the location of the malignancy, the suprasellar region, highly conformal radiation therapy is best indicated to treat craniopharyngiomas, which justifies the use of proton therapy. The critical structures involved for this treatment site are the brainstem, the optical chiasm, and optical nerves, as well as the normal brain.

According to Table 5.4, the critical tissues next to the target may see increased values of RBE due to their low value of $(\alpha/\beta)_x$, particularly in the case of the optical nerve. Figure 5.31 shows the proton boost dose and

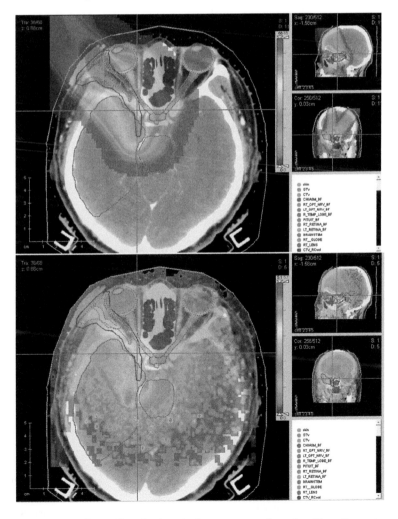

FIGURE 5.33 **(See color insert.)** Dose and LET_d distributions in the treatment of an adenoid cystic carcinoma of the lachrymal gland.

LET$_d$ distributions obtained in a treatment plan comprising three proton fields.

With the dose, LET$_d$, and $(\alpha/\beta)_x$ values, the biological dose distributions can be calculated using a constant and a variable RBE. Figure 5.32 shows the resulting BDVHs on the left plots, accompanied again by the RBE-VH and the LET$_d$-VH on the right. The chiasm is included within the GTV, so both volumes will have similar dose and LET$_d$ distributions but different $(\alpha/\beta)_x$ values, which, in the case of the chiasm, is about four times lower than the GTV. However, according to the data points in Figure 5.21, RBE becomes independent of $(\alpha/\beta)_x$ from 3 to 5 Gy onward, especially for hypofractionated regimes, which could explain the similarity on the BDVH of the GTV and the chiasm. The step in the BDVH curve of the GTV at the 80% volume level

in Figure 5.32 is due to an overlap of the brainstem contour with the GTV contour. Because of the low $(\alpha/\beta)_{X\text{-rays}}$ of the brainstem (1.5 Gy) compared to the GTV (10 Gy), we see an increase of the RBE weighted TD of the GTV. The higher $(\alpha/\beta)_{X\text{-rays}}$ boundary of the GTV BDVH is not equally affected, because the high $(\alpha/\beta)_{X\text{-rays}}$ cause a smaller variation in the already low RBE value.

5.4.1.3 Adenoid Cystic Carcinoma of the Lacrimal Gland

The GTV treatment of the lacrimal gland normally involves the eye and the optical nerve; however, the perineural extension of the treatment volume to create the CTV will imply dose to the sellar and suprasellar regions similar to the case of craniopharyngioma.

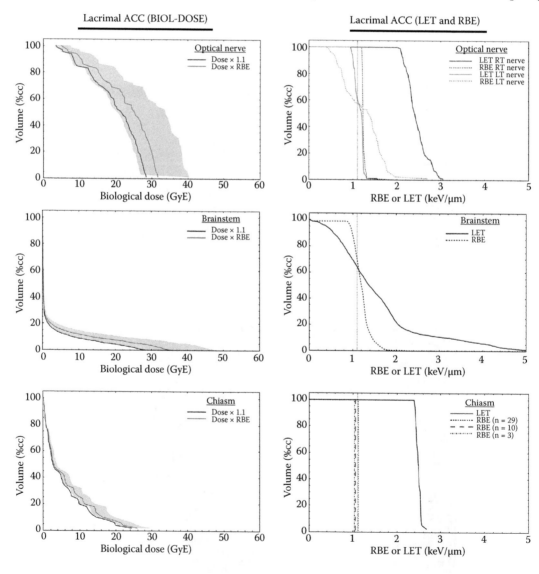

FIGURE 5.34 Left: BDVHs within the GTV and the dose constrain organs in an adenoid cystic carcinoma (ACC) treatment. Right: RBE and LET$_d$ histograms within the same volumes as on the left.

Therefore, the dose gradients that can be achieved with protons make this type of radiation indicated to spare healthy tissue. The typical treatment protocol used in this site is described in Table 5.5, along with the reported values of (α/β) of the tissues involved.

Figure 5.33 shows the usual dose and LET_d distributions of a proton treatment of the lacrimal gland. In order to ensure coverage of the entire CTV, part of the brainstem and the optical chiasm are exposed at the distal falloff of the beam. These organs receive doses much lower than the target but receive LET_d values larger than the target itself. A lower dose plus a larger LET_d implies enhanced RBE values. This increase could be larger in the case of the optical nerve due to low $(\alpha/\beta)_x$. In addition, due to the location of the disease, the beams involved in this type of treatment usually have short ranges and small modulation, which, according to Figures 5.11 and 5.12, could imply an increase of values of LET_d across the treatment volume. Figure 5.34 shows the RBE and biological dose distributions in the critical organs.

5.4.2 Characterization of the BER of a Particle Beam and Its Uncertainty as a Function of $(\alpha/\beta)_x$

Robertson et al. (1975) first indicated that the higher RBE at the most distal part of the BP extends the biologically effective range (BER) of the beam. In their measurements with hepatic cell lines (H4), they observed that this displacement was on the order of 1–2 mm; however, no other systematic analysis has been done in order to characterize the BER with the $(\alpha/\beta)_x$ of different tissues. A more recent study (Carabe et al. 2012a) has established this correlation for tissues with $(\alpha/\beta)_x$ between 0.5 and 20 Gy using beams with different ranges (4.8, 8.8, and 12.8 cm) and for doses between 2 and 15 Gy. The vertical axis in Figure 5.35 corresponds to the range difference between biological dose profiles obtained with a constant and a variable RBE. Each line in Figure 5.35 crosses the 0% range difference, corresponding to the situation where an

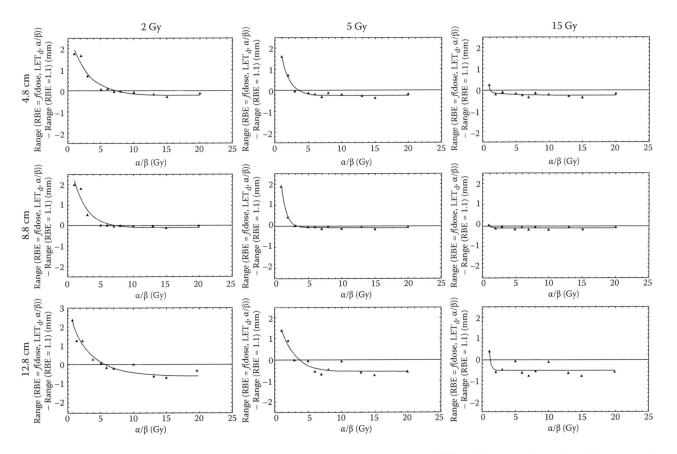

FIGURE 5.35 Range shift obtained at three different depths, 4.8 cm (top), 8.8 cm (middle), and 12.8 cm (bottom), and for 2, 5, and 15 Gy. For large $(\alpha/\beta)_x$, the range shift obtained from a variable RBE is shorter than for the constant RBE. At larger doses, lower values of $(\alpha/\beta)_x$ are required in order to counterbalance each other to produce no shift.

increase of the dose is counterbalanced by a decrease of the $(\alpha/\beta)_x$. Lower values of $(\alpha/\beta)_x$ would be required to compensate for larger doses, which is the reason for the 0% range difference at lower $(\alpha/\beta)_x$ when larger doses are delivered.

Clinical proton beams are always designed to avoid critical (dose-limiting) organs in the distal falloff of the beam, which is one of the reasons for the lack of toxicity in critical organs even if the RBE deviates from 1.1. There are situations, however, where conformality around the target cannot be achieved without compromising a critical organ, and it is in these cases where the incorporation of a varying RBE with $(\alpha/\beta)_x$ could

help in the design of the treatment plan. Therefore, the incorporation of the data reflected in Figure 5.35 could potentially help in guiding an oncologist while trying to avoid critical structures with beams pointing directly to the critical structures, as in the case of the lachrymal gland shown previously. This type of data can be used to calculate how much the biological effect of the beam is expanding beyond the distal falloff of the beam into the critical structures. The creation of curves similar to Figure 5.35 for specific tissues could help us better assess this effect and therefore reduce the normal tissue complication probability (NTCP) associated with specific treatments with difficult geometries.

5.5 Acknowledgments

The figures in Section 5.3 of this chapter were produced during the author's research fellowship at the Massachusetts General Hospital in Prof. Paganetti's lab. The author is indebted to Dr Paganetti for the many useful comments made to this chapter. The author would also like to thank Dr. Bassler from the Department of Physics and Astronomy at Aarhus University (Denmark) as well as Dr. Henkner from Heidelberg Ion-Beam Therapy Center (Germany) who provided useful carbon ion data. Finally, thanks are also extended to Dr. Alan Nahum for his comments in the physics section.

References

Ando, K. and Kase, Y. 2009. Biological characteristics of carbon-ion therapy. *Int. J. Radiat. Biol.* 85:715–728.

Ando, K., Koike, S., Nojima, K., Chen, Y.-J., Ohira, C., Ando, S., Kobayashi, N., Ohbuchi, T., Shimizu, W., and Kanai, T. 1998. Mouse skin reactions following fractionated irradiation with carbon ions. *Int. J. Radiat. Biol.* 74:129–138.

Ando, K., Koike, S., Uzawa, A., Takai, N., Fukawa, T., Furusawa, Y., Aoki, M., and Miyato, Y. 2005. Biological gain of carbon-ion radiotherapy for the early response of tumor growth delay and against early response of skin reaction in mice. *J. Radiat. Res. (Tokyo)* 46:51–57.

Barendsen, G. W. 1964. Impairment of the proliferative capacity of human cells in culture by α-particles with differing linear-energy *Transfer. Int. J. Radiat. Biol.* 8:453–466.

Barendsen, G. W., and Beusker T. L. J. 1960. Effects of different ionizing radiations on human cells in tissue culture—Part I: Irradiation techniques and dosimetry. *Radiat. Res.* 13:832–840.

Barendsen, G. W., Walter, H. M. D., Fowler, J. F., and Bewley, D. K. 1963. Effects of different ionization radiations on human cells in tissue culture—Part III: Experiments with cyclotron—Accelerated alpha-particles and deuterons. *Radiat. Res.* 18:106–119.

Belli, M., Cherubini, R., Finotto, S., Moschini, G., Sapora, O., Simone, G., and Tabocchini, M. A. 1989. RBE-LET relationship for the survival of V79 cells irradiated with low-energy protons. *Int. J. Radiat. Biol.* 55:93–104.

Belli, M., Goodhead, D. T., Ianzini, F., Simone, G., and Tabocchini, M. A. 1992. Direct comparison of biological effectiveness of protons and alpha-particles of the same LET—Part II: Mutation induction at the HPRT locus in V79 cells. *Int. J. Radiat. Biol.* 61:625–629.

Belli, M., Cera, F. et al. 1993. Inactivation and mutation induction in V79 cells by low energy protons: Re-evaluation of the results at the LNL facility. *Int. J. Radiat. Biol.* 3:331–337.

Bentzen, S., and Joiner, M. 2009. The linear-quadratic approach in clinical practice. In *Basic Clinical Radiobiology.* M. Joiner and A. V. Kogel (eds.). 2009. Edward Arnold.

Berger, M. J. 1985. Energy loss straggling of protons in water vapor. *Radiation Protection Dosimetry* 13(1):87–90.

Birds, R. P., Rohrig, N., Colvett, R. D., Geard, C. R., and Marino, S. A. 1980. Ianctivation of synchronized Chinese hamster V79 cells with charged-particle track segments. *Radiat. Res.* 82:277–289.

Blakely, E. A. 1992. Cell inactivation by heavy-charged particles. *Radiat. Environ. Biophys.* 31:181–196.

Blakely, E. A., Cornelius, A. T., Tracy, C. H. Y., Karen, C. S., and Lyman, J. T. 1979. Inactivation of human kidney cells by high-energy monoenergetic heavy-ion beams. *Radiat. Res.* 80:122–160.

Britten, R. A., Warenius, H. M., Parkins, C., and Peacock, J. H. 1992. The inherent cellular sensitivity to 62.5MeV$_{(p\rightarrow Be+)}$ neutrons of human cells differing in photon sensitivity. *Int. J. Radiat. Biol.* 61:805–812.

Butts, J. J., and Katz, R. 1967. Theory of RBE for heavy-ion bombardment of dry enzymes and viruses. *Radiat. Res.* 30:855–871.

Carabe, A., Moteabbed, M., Depauw, N., Schuemann, J., and Paganetti, H. 2012a. Range uncertainty in proton therapy due to variable biological effectiveness. *Phys. Med. Biol.* 57(5):1159–1172.

Carabe, A., España, S., Grassberger, C., and Paganetti, H. 2012b. Clinical consequences of relative biological effectiveness variations in proton radiotherapy of the prostate, brain and liver. *Phys. Med. Biol.* (submitted).

Carabe-Fernandez, A., Dale, R. G., Hopewell, J. W, Jones, B., and Paganetti, H. 2010. Fractionation effects in particle radiotherapy: Implications for hypofractionation regimes. *Phys. Med. Biol.* 55:5685–5700.

Carabe-Fernandez, A., Dale, R. G., and Jones, B. 2007. The incorporation of the concept of minimum RBE (RBE_{min}) into the linear-quadratic model and the potential for improved radiobiological analysis of high-LET treatments. *Int. J. Radiat. Biol.* 83:27–39.

Coutrakon, G., Cortese, J., Ghebremedhin, A., Hubbard, J., Johanning, J., Koss, P., Maudsley, G., Slater, C. R., and Zuccarelli, C. 1997. Microdosimetry spectra of the Loma Linda proton beam and relative biological effectiveness comparisons. *Med. Phys.* 24:1499–1506.

Cox, R., Thacker, J., Goodhead, D. T., Masson, W. K., and Wilinson, R. E. 1977. Inactivation and mutation of cultured mammalian cells by aluminum characteristic ultrasoft X-rays. *Int. J. Radiat. Biol.* 31:561–576.

Fertil, B., Deschavanne, P. J., Gueulette, J., Possoz, A., Wambersie, A., and Malaise, E. P. 1982. *In vitro* radiosensitivity of six human cells—Part II: Relation to the RBE of 50-MeV neutrons. *Radiat. Res.* 90:526–537.

Folkard, M., Prise, K. M., Vojnovic, B., Newman, H. C., Roper, M. J., and Michael, B. D. 1996. Inactivation of V79 cells by low-energy protons, deuterons and helium-3 ions. *Int. J. Radiat. Biol.* 69:729–738.

Fowler, J. F., Chappell, R., and Ritter, M. 2001. Is α/β for prostate tumors really low? *Int. J. Radiat. Biol.* 50:1021–1031.

Furusawa, Y., Fukutsu, K., Aoki, M., Itsukaichi, H., Eguchi-Kasai, K., Ohara, H., Yatagai, F., Kanai, T., and Ando, K. 2000. Inactivation of aerobic and hypoxic cells from three different cell lines by accelerated ^{3}He-, ^{12}C-, and ^{20}Ne-Ion beams. *Radiat. Res.* 154:485–496.

Gerweck, L. E., and Kozin, S. V. 1999. Relative biological effectiveness of proton beams in clinical therapy. *Radiother. Oncol.* 50:135–142.

Goitein, M. (ed.). 2008. *Radiation Oncology: A Physicist's-Eye View.* Springer. p. 226.

Goodhead, D. T. 1987. In *The Dosimetry of Ionizing Radiation* (Vol. 2). K. R. Kase, B. E. Bjarngard, and F. H. Attix (eds.), San Diego, CA: Academic Press, pp. 1–89.

Goodhead, D. T., Belli, M., Mill, A. J., Bance, D. A., Allen, L. A., Hall, S. C., Ianzani, F., Simone, G., Stevens, D. L., Stretch, A., Tabocchini, M. A., and Wilkinson, R. E. 1992. Direct comparison between protons and alpha-particles of the same LET. *Int. J. Radiat. Biol.* 61:611–624.

Goodwin, E. H., Blakelym E. A., and Tobias, C. A. 1994. Chromosomal damage and repair in G1-phase Chinese hamster ovary cells exposed to charged-particle beams. *Radiat. Res.* 138(3):343–351.

Gottschalk, B. 2004. Passive beam spreading in proton radiation therapy. http://huhepl.harvard.edu/~gottschalk.

Grassberger, C., and Paganetti, H. 2011. Elevated LET components in clinical protons beams *Phys. Med. Biol.* 56:6677–6691.

Hawkins, R. B. 1996. A microdosimetric-kinetic model of cell death from exposure to ionizing radiation of any LET, with experimental and clinical applications. *Int. J. Radiat. Biol.* 69:739–755.

Highland, V. L. 1975. Some practical remarks on multiple scattering. *Nucl. Instrum. Meth.* 129:497–499.

Hornsey, S., and Field, S. B. 1979. The effect of single and fractionated doses of x-rays and neutrons on the oesophagus. *Eur. J. Cancer* 15:491–498.

IAEA. 2007. Dose reporting in ion beam therapy, IAEA-TECDOC-1560, p. 10.

ICRP. 1990. RBE for deterministic effects. ICRP Publication 58. *Ann. ICRP* 20 (4).

ICRU-16 (International Commission on Radiation Units and Measurements). 1970. Linear energy transfer Report No. 16 (Washington: ICRU).

International Commission on Radiation Units and Measurements (ICRU). 1979. *Report No. 30.* Bethesda, MD. pp. 13–14.

Jenner, T. J., Belli, M., Goodhead, D. T., Ianzani, F., Simone, G., and Tabocchini, M. A. 1992. Direct comparison between protons and alpha-particles of the same LET. *Int. J. Radiat. Biol.* 61:631–637.

Jones, B. 2010. The apparent increase in the β parameter of the linear quadratic model with increased linear energy transfer during fast neutron irradiation. *Brit. J. Radiat.* 83:433–436.

Kantemiris, I., Karaiskos, P., Papagiannis, P., and Angelopoulos, A. 2011. Dose and dose average LET comparison of H, He, Li, Be, B, C, N, and O ion beams forming a spread-out Bragg peak. *Med. Phys.* 38:6585–6591.

Kase, Y., Kanai, T., Matsufuji, N., Furusawa, Y., Elsasser, T., and Scholz, M. 2008. Biophysical calculation of cell survival probabilities using amorphous track structure models for heavy-ion irradiation. *Phys. Med. Biol.* 53:37–59.

Katz, R., and Sharma, S. C. 1973. Response of cells to fast neutrons, stopped pions, and heavy ion beams. *Nucl. Instrum. Methods* 111:93–116.

Katz, R., and Sharma, S. C. 1974. Heavy particles in therapy: An application of track theory. *Phys. Med. Biol.* 19:413.

Kellerer, A. M., and Chmelevsky, D. 1975. Criteria of the applicability of LET. *Radiat. Res.* 63(2):226–234.

Kempe, J., Gudowska, I., and Brahme, A. 2007. Depth absorbed dose and LET distributions of therapeutic ^{1}H, ^{4}He, ^{7}Li and ^{12}C beams. *Med. Phys.* 34:183–192.

Kraft, G. 2005. *Proceedings of the International Meeting of Particle Therapy Cooperative Group* (PTCOG) 43, Rinecker Proton Therapy Center, Munich (Germany).

Kramer, M., and Kraft, G. 1994. Calculations of heavy ion track structure. *Radiat. Environm. Biophys.* 33:91–109.

Kramer, M., Weyrather, W. K., and Scholz, M. 2003. The increased biological effectiveness of heavy charged particles: From radiobiology to treatment planning. *Technol. Canc. Res. Treat.* 2:427–436.

Laramore, G. E., Krall, J. M., and Thomas, F. J. et al. 1993. Fast neutron radiotherapy for locally advanced prostate cancer: Final report of Radiation Therapy Oncology Group Randomized Clinical Trial. *Am. J. Clin. Oncol.* 16:164–167.

Mayo, C., Martel, M. K., Lawrance, B. M., Flickinger, J., Nam, J., and Kirkpatrick, J. 2010. Radiation dose-volume effects of optic nerves and chiasm. *Int. J. Radiat. Oncol. Biol. Phys.* 76:S28–S35.

Perris, A., Pialoglou, P., Katsanos, A. A., and Sideris, E. G. 1986. Biological effectiveness of low-energy protons—Part I: Survival of Chinese hamster cells *Int. J. Radiat. Biol.* 50:1093–1101.

Paganetti, H., Gerweck, L. E., and Goitein, M. 2000. The general relation between tissue response to X-radiation (α/β-values) and the relative biological effectiveness (RBE) of protons: Prediction by the Katz track-structure model. *Int. J. Radiat. Biol.* 76:985–998.

Paganetti, H., Niemierko, A., Ancukiewicz, M., Gerweck, L. E., Goitein, M., Loeffler, J. S., and Suit, H. D. 2002. Relative biological effectiveness (RBE) values for proton beam therapy. *Int. J. Radiat. Oncol. Biol. Phys.* 53:407–421.

PSTAR: http://physics.nist.gov/PhysRefData/Star/Text/PSTAR.html.

Raju, M. R. 1995. Proton radiobiology, radiosurgery and radiotherapy. *Int. J. Radiat. Biol.* 67:237–259.

Chapter 5

Robertson, J. B., Williams, J. R., Schmidt, R. A., Little, J. B., Flynn, D. F., and Suit, H. D. 1975. Radiobiological studies of a high-energy modulated proton beam utilizing cultured mammalian cells. *Cancer* 35:1664–1677.

Schettino, G., Folkard, M., Prise, K. M., Vojnovic, B., Bowey, A. G., and Michael, B. D. 2001. Low-dose hypersensitivity in Chinese hamster V79 cells targeted with counted protons using a charged-particle microbeam. *Radiat. Res.* 126: 526–534.

Scholz, M., Kellerer, A. M., Kraft-Weyrather, W., and Kraft, G. 1997. Computation of cell survival in heavy-ion beams for therapy. The model and its approximation. *Radiat. Environ. Biophys.* 36:59–66.

Shridhar, R., Bolton, S., Joiner, M. C., and Forman, J. D. 2009. Dose escalation using a hypofractionated, intensity-modulated radiation therapy boost for localized prostate cancer: Preliminary results addressing concerns of high or low alpha/beta ratio. *Clin. Genitourin. Canc.* 7:E52–E527.

Sorensen, B. S., Overgaard, J., and Basseler, N. 2011. *In vitro* RBE-LET dependence for multiple-particle types. *Acta Oncol.* 50:757–762.

Stenerlow, B., Petterson, O. A., Essand, M., Blomquist, E., and Carlsson, J. 1995. Irregular variations in radiation sensitivity when the linear energy transfer is increased. *Radiother. Oncol.* 36:133–142.

Suzuki, M., Kase, Y., Yamaguchi, H., Kanai, T., and Ando, K. 2000. Relative biological effectiveness for cell-killing effect on various human cell lines irradiated with heavy-ion medical accelerator in Chiba (HIMAC) carbon-ion beams. *Int. J. Radiat. Oncol. Biol. Phys.* 48:241–250.

Todd, P. 1967. Heavy-ion irradiation of cultured human cells. *Radiat. Res.* (Suppl.) 7:196–207.

Tsujii, H., Mizoe, J., Kamada, T., Baba, M., Tsuji, H., Kato, H., Kato, S., Yamada, S., Yasuda, S., Ohno, T., and Turner, J. E. 1995. *Atoms, Radiation, and Radiation Protection.* New York: Wiley-Interscience. Chapter 5.

Valentin, J. 2003. Relative biological effectiveness (RBE), quality factor (Q), and radiation weighting factor (wR): ICRP Publication 92. *Annals of the ICRP* 33(4):1–121.

Wambersie, A., Gahbauer, A. R., and Menzel, G. H. 2004. RBE and weighting of absorbed dose in ion-beam therapy. *Radiother. Oncol.* 73:S176–S182.

Warenius, H. M., Britten, R. A., and Peackok, J. H. 1994. The relative cellular radiosensitivity of 30 human in vitro cell lines of different histological type to high LET 62.5MeV (p◊Be+) fast neutrons and 4MeV photons. *Radiother. Oncol.* 30:83–89.

Weber, U., and Kraft, G. 2009. Comparison of carbon ions versus protons. *Cancer J.* 15:325–332.

Weyrather, W. K., Ritter, S., Scholz, M., and Kraft, G. 1999. RBE for carbon track-segment irradiation in cell lines differing repair capacity. *Int. J. Radiat. Biol.* 75:1357–1364.

Wilkens, J. J., and Oelfke, U. 2004. Three-dimensional LET calculations for treatment planning of proton therapy. *Z. Med. Phys.* 14:41–46.

Withers, H. R., Thames, H. D., and Peters, L. J. 1982. Biological bases for high RBE values for late effects of neutron irradiation. *Int. J. Radiat. Oncol. Biol. Phys.* 8:2071–2076.

Wouters, B. G., Lam, G. K. Y., Oelfke, U., Gardey, K., Durand, R. E., and Skarsgard, L. O. 1996. Measurements of relative biological effectiveness of the 70 MeV proton beam at TRIUMF using Chinese hamster V79 cells and the high-precision cell sorter assay. *Radiat. Res.* 146:159–170.

Yanagi, T., Imai, R., Kagei, K., Kato, H., Hara, R., Hasegawa, A., Nakajima, M., Sugane, N., Tamaki, N., Takagi, R., Kandatsu, S., Yoshikawa, K., Kishimoto, R., and Miyamoto, T. 2007. Clinical results of carbon ion radiotherapy at NIRS. *J. Radiat. Res.* 48 (Suppl A):A1–A13.

Zirkle, R. E. 1935. Biological effectiveness of alpha particles as a function of ion concentration produced in their paths. *Am. J. Cancer* 23: 558–567.

Zirkle, R. E., Marchbank, D. F., and Kuck, K. D. 1952. Exponential and sigmoid survival curves resulting from alpha and x irradiation of Aspergillus spores *J. Cell Physiol. Suppl.* 39:78–85.

6. Dosimetry for Proton and Carbon Ion Therapy

Jonathan B. Farr and Richard L. Maughan

6.1   Introduction

The role of dosimetry in radiation therapy is to provide quantitative radiation dose measurement to ensure that the radiation dose is delivered to the patient correctly. This may be viewed as primarily a quality assurance process. The overall dosimetry quality assurance process is many faceted, ranging from the measurements made during equipment acceptance testing ensuring that the radiation beams meet the vendor's specifications to commissioning the data planning system, beam calibration measurements, and, finally, routine clinical beam quality assurance (QA) procedures. Beam calibration requires absolute beam dose measurements,

Proton and Carbon Ion Therapy Edited by C.-M. Charlie Ma and Tony Lomax © 2013 Taylor & Francis Group, LLC. ISBN: 978-1-4398-1607-3.

Chapter 6

whereas many of the other QA functions require only knowledge of relative dose. To arrive at the prescribed dose, called the isoeffective dose, an additional biological factor must be included to represent the biological effect. According to the International Commission on Radiation Units and Measurements (ICRU), the product of the biological factor, relative biological effectiveness (RBE), and physical dose is called the RBE-weighted absorbed dose and may be identified in gray (RBE), with the appending RBE modifying the SI physical dose unit of gray (ICRU 1998). However, physical doses, absolute or relative, form the tangible, measureable quantities for sensors and detectors. This chapter describes those sensors and detectors and their use in determining the physical dose.

A wide variety of radiation dose detectors have been developed with varying sensitivities, energy-dependent response, and spatial resolution. Sensitivity and energy dependence may be a function of the radiation modality being measured (photons, electrons, neutrons, protons, or heavier charged particles). The spatial resolution depends on the structure and/or type of detector. To make meaningful dose measurements, it is important to understand the properties of the detector being employed.

Many dosimetry detectors and techniques originally developed for photon radiation therapy have been applied to the dosimetry of charge particle beams. It is therefore necessary to understand any limitations that are peculiar to the use of this instrumentation when applied to the measurement of absolute and relative dose in charged particle beams. This chapter discusses the types of detectors available for charged particle dosimetry and their use, suitability, and limitations in performing the QA functions outlined above. In charged particle therapy, the suitability of the detector may depend on the dose delivery modality employed (i.e., scattered or scanned beam). Scanned beams present some unique challenges, particularly with respect to measurement efficiency, and these challenges require unique and innovative solutions.

6.2 Sensors and Detectors

Many different detectors are used for the dosimetry of both conventional and particle beams. Among these are ionization chambers, radiographic film, thermoluminescent dosimeters (TLDs), diodes, and radiochromic film. Details on these "standard" radiologic detectors can be readily found throughout the radiological physics literature, including in several excellent textbooks (Knoll 1979; Attix 1986). This section focuses on particle therapy–specific detector properties. The detectors must satisfy several fundamental needs: traceable or absolute dosimetry, relative dosimetry of lateral and longitudinal beam profiles, and out-of-field dosimetry (primarily neutrons). A typical problem for detector selection and use in particle therapy is energy or linear energy transfer (LET) dependence. In this regard, air or tissue equivalent ionization chambers are taken as the gold standard, because their response exhibits minimal LET dependence. Detectors that exhibit a significant LET dependence such as TLD, film, solid-state, and gel will exhibit a response different from ionization chambers in the high-LET portions of ion beams.

6.2.1 Absolute and Traceable Detectors

Absolute dosimetry or dosimetry that can be determined from fundamental radiation-energy properties is important for developing national and international metrology standards and can also be used as a cross comparison from other dosimetric (possibly nonstandard) measurements. Two common absolute dosimetry methods, calorimetry and Faraday cup (FC) dosimetry, are discussed here. Details on rare types such as activation analysis can be found in the literature (Nichiporov 2003). Current national and international standards for traceable dosimetry are based on ionization chamber measurements (IAEA 2000; ICRU 1998, 2007). Therefore, ionization chambers are also discussed here. Their dosimetric protocol application is discussed in Section 6.4.

6.2.1.1 Calorimetry

Calorimetry is based on a direct measurement of the heat generated per unit mass of material as the radiation beam deposits energy in the material. In this respect, calorimetry represents a fundamental method of dose measurement and, therefore, the potential to serve as an absolute standard for particle therapy (Brede et al. 2006; Vynckier 2004; Palmans et al. 2004; Palmans et al. 1996; Vatnitsky, Siebers and Miller 1996; Siebers et al. 1995; Delacroix et al. 1997; Sassowsky and Pedroni 2005). However, there are a number of complications and uncertainties. First, in practice, as a differential temperature measurement, the thermal properties of the medium must be precisely known (i.e., the specific heat capacity). Because the heat capacity of graphite is lower than water but their radiological properties are similar, graphite calorimeters are sometimes preferred. Contrarily, due to the higher thermal conductivity, the graphite calorimeter requires more thermal isolation. Another disadvantage of this approach is that, because dose to water is the natural goal of reference dosimetry, an additional step of dose conversion is required. For ion beams, this implies accurate knowledge of the stopping power ratios, which for any material or material combination will depend on elastic and in-elastic processes. Second, because the quantity of heat produced is quite small, the measurements require elaborate thermally insulated equipment with associated painstaking setup and time-consuming analysis. The equipment is bulky, and the method is impractical for routine dose calibration. Usually, the measurement uses a thermistor within the insulated calorimeter connected to a Wheatstone bridge. Depending on whether the temperature is allowed to rise as a function of the absorbed dose or is actively controlled to be stable prior, during and subsequent to the irradiation, two measurement modes are defined, respectively: 1) quasi-adiabatic and 2) quasi-isothermal. In the quasi-isothermal mode, the energy measurement is differential from the energy required to maintain thermal stasis. Again, because the temperature change measured is small, on the order of milli-Kelvin (mK) thermal flow within the device can affect the measurement. As heat flow is a time-dependent behavior, the effect is coupled with radiation delivery temporal behavior or interplay (Sassowsky and Pedroni 2005). Third, not all of the absorbed energy is transferred as heat. A small proportion (a few percent) contributes to radiochemical reactions or lattice defects in the absorbing medium, and a correction for this thermal defect is required, which is a significant source of uncertainty in these measurements. This problem is exacerbated by varying LET in ion beams, because the defect depends on it (Medin 2010; Brede et al. 2006; Sassowsky and Pedroni 2005; Sarfehnia et al. 2010).

6.2.1.2 Faraday Cups

Historically, FCs have also been used for the dosimetry of charged particle beams (Grusell et al. 1995; Cambria et al. 1997; Lorin et al. 2008; Verhey et al. 1979). From a known fluence, the charge particles are completely stopped within the FC, and their charge is collected. As secondary charged particles are also produced when the beam strikes the detector, the cupped shape together with a guard electrode helps suppress the loss of these recoil particles that would otherwise affect the charge measurement. A thin window is called for so as to minimize electron production from it. To suppress the signal from ions produced in air, a vacuum is often applied up to 10^{-6} hPa. From the measured beam charge and with a precise knowledge of the beam cross-sectional area, it is possible to use FC measurements to calculate deposited radiation dose. The energy averaged stopping power in water is required to calculate the dose. In this sense, FC dosimetry is not "absolute" but can represent a practical secondary standard within a treatment facility. Because the FC measures collected charge and does not measure beam energy, the energy and spectrum must be independently known for a dose calculation. This can be problematic in mixed energy beams, for instance, within a spread-out Bragg peak (SOBP) in a scattering system. In practice, FC dosimetry can also be sensitive to effects from these beamlines, where distal collimation is required, probably due to increased angular confusion of the particles and low-energy particles scattered off the collimation. Therefore, although FC dosimetry was established early in the history of particle therapy (Raju 1978) but has fallen a bit out of use, it could experience a renaissance for use with scanned beams. This is because, in scanning systems, the beam profile is much smaller than the FC cross section, distal collimation is not used, the energy and spectrum are well defined, and stopping powers in water are well established for quasi-monoenergetic beams.

6.2.1.3 Ionization Chambers

Calorimeters and FCs are not available today commercially. Practically, as with radiation therapy, in general, ionization chambers have been adopted as the preferred instrumentation for performing calibration measurements at clinical facilities. They can be fabricated in a wide variety of shapes and sizes and are convenient and easy to use for routine measurements. The ionization chamber can be calibrated against a standard radiation source. Hence, current particle therapy system calibrations are based on ionization chamber dosimetry. Cylindrical chambers offer the advantages of accurate standards calibration and can be made in miniature formats. Hence, they are particularly well suited for absolute and small-field dosimetry. Parallel plate chambers offer a geometrical advantage, because their electrical collection plane is well defined, which is especially important for measurements in water at

Chapter 6

depths of less than 3 cm. Details of their protocol use for calibration are provided in the Section 6.4.

6.2.2 Absolute and Traceable Sensors

6.2.2.1 Alanine

Alanine is an amino acid that, when irradiated, generates radical unpaired electrons (CH_3–CH–$COOH$) that remain stably trapped by the surrounding structure and lend themselves to read out by electron spin resonance (ESR) techniques. The ESR signal depends on the dose delivered. Alanine is convenient to use due to its powder or pellet forms and wide dose dynamic range. Also attractive is its nearly water-equivalent density. LET dependence in particle beams has been studied but is inconclusive (Gall et al. 1996; Palmans 2003; Bassler et al. 2008).

6.2.3 Relative Detectors

In many cases, the goal of dosimetry measurements is to compare the measured dose spatially in two dimensions (2-D) or three dimensions (3-D) to the known dose under the calibration condition. In radiotherapy, this is part of the QA process where spatial information of how the dose is delivered to a patient is of importance. Traditionally, 2-D relative measurements have been made with an ionization chamber or film. However, several challenges in particle therapy dosimetry have led to the development of multichannel/multielement relative detectors. This evolution of relative dosimetry detectors is described in the following sections.

6.2.3.1 Film

The use of film for relative dosimetry, using either conventional silver-based emulsions or radiochromic film, has many advantages: high spatial resolution up to hundreds of line pairs per centimeter; long-term record; availability in many sizes; and ease of use. The primary disadvantage of film dosimetry in particle therapy is the well-known quenching effect observed in high-LET regions (Vatnitsky 1997; Spielberger et al. 2001; Kirby et al. 2010). Because the effect is LET dependent, it is even more problematic with heavy ions.

Recently, with the reduction in cost and concurrent increase in sensitivity, the use of the latest Gafchromic films [International Specialty Products (ISP), Wayne, New Jersey] in proton therapy relative dosimetry has become widespread. Three products from ISP are prominent: types MD-V2-55, EBT, and the newest, EBT2. The type MD-V2-55 represents older technology. It comprises two sensitive layers of pentacosa-10,12-dyinoic acid (PCDA) suspended in gelatin and constrained by outer clear polyester layers. The EBT offers 10 times higher sensitivity, giving adequate darkening with standard radiotherapy single fraction doses and faster optical density stabilization (Cheung, Butson, and Yu 2005). ISP also claims better sensitive layer uniformity for EBT compared to MD-V2-55 from its production process. The EBT formulation includes lithium in the PCDA makeup, LiPCDA. The LiPCDA crystals are more linear in dose response than those of PCDA (Rink et al. 2008). For both formulations, exposure to ionizing radiation causes polymerization through the free radical process, resulting in chains of the characteristically blue-colored polydiacetylene. The absorption spectrum for both types shows two sensitivity peaks, 618/676 and 583/635 nm for MD-V2-55 and EBT, respectively (Niroomand-Rad et al. 1998; Devic et al. 2007). By including a yellow dye in the newest EBT2 type, sensitivity to nonionizing radiation [visible and ultraviolet (UV) light] is decreased an order of magnitude. The "background" signal from the yellow dye can also be used to control the film uniformity on a piece-by-piece basis. A synthetic polymer in EBT2 replaces the natural gelatin used in EBT, providing a potential improvement in consistency between lots. EBT and EBT2 also differ in their structural layer arrangement. EBT is formed as a symmetric sandwich of 97-μm polyester, 17-μm active layer, 8-μm surface layer, 17-μm active layer, and 97-μm polyester, whereas EBT2 is not symmetric in cross section, being composed of the following layers: 50-μm polyester, 25-μm adhesive layer, 8-μm surface layer, 30-μm active layer, and 175-μm polyester. The change to a synthetic polymer from gelatin enabled the fabrication of the PCDA into a single layer in EBT2. In total, EBT2 is slightly thicker at 285 μm compared to 236 μm for EBT.

Due to its stated advantages, the use of film in ion therapy dosimetry has an extensive experience (Kirby et al. 2010; Mumot et al. 2009; Spielberger et al. 2001; Vatnitsky 1997; Vatnitsky et al. 1999; Piermattei et al. 2000; Daftari et al. 1999). Again, a quenching effect in the Bragg peak (BP) and distal portion of the SOBP has been repeatedly reported. The effect depends on the LET, so it is even more pronounced in the dosimetry of heavier ion beams. Two theories from gel dosimetry are suggested to explain the observations: the quenching may be related to recombination-free radicals in the active layer (Gustavsson et al. 2004) or, alternatively, to the density of activation sites within the layer (Jirasek and Duzenli 2002). Regardless of the underlying cause, the known quenching effect suggests limitations for the use of film in particle therapy dosimetry. That is, it should not be used in regions of high LET. There remains, however, a relevant valid indication for use within the proximal and middle portions of uniform proton fields where the high LET components contributing to the dose are relatively minor in comparison (Zhao and Das 2010). Uniformity concerns from the new type EBT2 have been preliminarily reported and needs further investigation (Hartmann, Martisikova and Jakel 2010).

6.2.3.2 Solid-State Detectors

Types of solid-state detectors used in particle therapy include diodes, metal–oxide–semiconductor field-effect transistors (MOSFETs), and diamond detectors. Their principle advantages are to provide high spatial resolution and sensitivity. Both n- and p-type MOSFETs have been considered for use in particle therapy (Verhey et al. 1979; Vatnitsky et al. 1999; Raju 1966; Onori et al. 2000). The detectors are usually used without applied bias voltage. Studies have indicated that the n-type MOSFET is not suitable for use due to an observed overresponse in the BP region. The increased sensitivity to LET has been explained from a differential increase in cross section to radiation-induced damage in comparison to electron capture (Grusell and Medin 2000). All types of MOSFETs show a decrease in sensitivity with irradiated dose due to induced defects in the lattice structure. For p-type diodes, the sensitivity becomes relatively constant after a significant preirradiation. Therefore, preirradiation is necessary prior to use; otherwise, dose-dependent effects can be observed (Kaiser, Bassler and Jakel 2010). An interesting p-type MOSFET developed for use in proton therapy is the Hi-pSi (IBA Dosimetry, Schwarzenbruck, Germany), consisting of the p substrate doped with silicon. It has been tested in low- and high-energy proton beams and found to perform within 4% accuracy with respect to a Markus ionization chamber within the BP region (Grusell and Medin 2000). However, some uncertainties with preirradiation and dose rate dependence have been reported (De Angelis et al. 2002).

Diamonds are also used in particle therapy dosimetry. Due to the amount of impurities and the variation of those impurities in naturally occurring diamonds, manmade chemical types are preferred. They can be formed by high-pressure, high-temperature (HPHT), or chemical vapor deposition (CVD) processes. Detectors can be simply fabricated by sandwiching the diamond between two conductive plates with leads. The high density of strong bonding in the diamond lattice structure giving high radiation hardness is attractive. Because of high charge mobility, its dosimetry response is also rapid. The low-Z-carbon base of diamond is another attractive property. Even with HPHT/CVD diamonds, residual impurities can present a problem as they modify the response of the detector, causing intersample variability. Diamond detectors also have the need for preirradiation. Although they do not exhibit a significant LET dependence, they have strong energy dependence (Fidanzio et al. 2002; Pacilio et al. 2002; van Luijk et al. 2001; Onori et al. 2000; Vatnitsky et al. 1999; Vatnitsky et al. 1995). In use, they can be first cross calibrated at the measurement energy with an ionization chamber, for instance, and then used for high-resolution small-field or penumbra measurements.

6.2.3.3 TLDs

The use of TLDs, especially LiF:Mg,Ti-type TLD-100, in proton therapy has been well studied (Vatnitsky et al. 1995; Carlsson and Carlsson 1970; Richmond et al. 1987). Although TLDs exhibit a relatively thick water-equivalent thickness and LET dependence, their use is interesting, especially with regard to *in vivo* or surface patient dosimetry. Their use in modulated proton beams tends to smooth out the effect of LET dependence. Recent studies suggest that the use of TLD-100 is suitable for use in this manner, with dose accuracy and reproducibility similar to applications in photon and electron beam TLD dosimetry (Sabini et al. 2002; Zullo et al. 2010). Another convenient use of TLD in proton therapy is to facilitate mailable remote dose auditing (Section 6.3.1.2). Studies of TLD use in heavy ion therapy have also been performed. Although TLD LET dependence is usually considered to be a disadvantage, in this case, it can also be used to quantify the LET distribution in beams exhibiting changing LET (e.g., proton in the Bragg region, neutron, carbon, and other heavy ion species). This is performed using TLD glow curve deconvolution methods most commonly using types LiF:Mg,Cu,P (TLD 700) and CaF2:Tm (TLD-300) (Angelone et al. 1998; Berger et al. 2006; Yudelev, Hunter and Farr 2004; Patrick et al. 1976; Czopyk et al. 2007; Skopec et al. 2006; Olko et al. 2004; Olko et al. 2002; Sabini et al. 2002).

6.2.3.4 OSLDs

A recent development has been the introduction of detectors based on optically stimulated luminescence (OSL). Most OSL detectors (OSLDs) are based on the doped aluminum oxide Al_2O_3:C (Botter-Jensen et al. 2003). OSLDs can operate in active or passive arrangements. In the most common passive situation, OSLDs function as an analog to TLDs; the carbon doping provides deep electron traps that become filled due to ionization processes. A readout process excites the material, releasing the electrons back to the conduction band and emitting a visible photon in the process. The visible light is then collected as a related function of the ionization. TLDs require heat as the excitation for readout, and OSLDs require light. When connected to a light pipe, the OSLD has the additional capability of being polled for readout in a quasi-real-time active mode. This provides a possibility for use as an *in vivo* dosimeter as well.

Having their origin in health physics, OSLDs are already used as space flight vehicle and personnel dosimeters. The mixed LET nature of cosmic radiation and the desire to use the devices in higher LET beams or beam regions has driven investigations into their LET dependence. Initial findings indicate a proportional decrease in response with increasing LET (Reft 2009; Sawakuchi et al. 2008). This property can be useful in determining differential LETs (Yukihara et al. 2004; Sawakuchi et al. 2010). A greater understanding of the fundamental solid-state theory may allow the future use of these detectors in mixed LET beams. This is because the OSL reacts to ionization through two channels: 1) an ultraviolet (UV) emission band centered at 335 nm and 2) a primary F-center emission band centered at 420 nm (Yukihara and McKeever 2006). The UV emissions increase in proportion to the F-center emissions with increasing LET. This property may be used in the future to correct or suppress the overall OSLD LET dependency. Considering the problem from the single-hit/multihit track theory, it has been reported that OSLD performance can remain LET independent at proton energies higher than 10 MeV and doses less than 0.3 Gy (Edmund et al. 2007).

6.2.3.5 Multielement Detectors

Multielement detectors based on the ionization chamber technology are becoming increasingly important in charged particle therapy. There are both commercial and custom detectors being developed. The detectors are constructed from discrete elements, usually ionization chambers, using parallel electronics with software. Usually, the design goal is to provide attractive ionization chamber attributes such as excellent reproducibility and LET insensitivity while also providing high spatial resolution at multiple sampling positions. For multielement detectors, the challenge of building low-noise multichannel electrometers and electronics interfaced with control and analysis software is significant. Generally, the detectors have either a lateral (Arjomandy et al. 2008; Farr et al. 2008) or a transverse beam measurement capability (Nichiporov et al. 2007). In addition, quasi 3-D ionization chamber–based detectors have also been developed (Cirio et al. 2004; Karger, Jakel and Hartmann 1999). Two examples of ionization chamber–based multielement detectors are depicted in Figure 6.1.

Longitudinal multielement detectors are constructed from two different geometries. The common idea is to collect charge discretely in layers defined by the sensitive volume of two facing electrodes. If the layer-sensitive area is significantly smaller than the

FIGURE 6.1 Discrete ionization chamber array developed at the PSI (left) and a commercial array (IBA Dosimetry, Schwarzenbruck, Germany Model: IM'RT MatriXX) with custom waterproof sleeve and 1-D motor drive in a water phantom (right). The arrows indicate the incident beam direction and motion directions. The PSI detector provides a water-adjustable depth with a reservoir and bellows system. In addition it can rotate to any gantry angle. The MatriXX apparatus allows variable water depth, but is restricted to horizontal use.

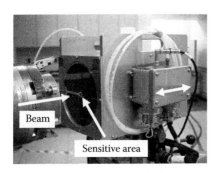

FIGURE 6.2 Multilayer ionization chamber constructed of 122 small-volume ionization chambers stacked at approximately 2-mm water-equivalent uniform separations. The detector provides 1-mm-range measurement accuracy.

beam profile, a multilayer ionization chamber (MLIC) results (Figure 6.2). If the area is large with respect to the beam transverse profile, a multilayer Faraday cup (MLFC) results. MLIC designs can be constructed from water equivalent materials or slightly higher density materials selected to give a water-equivalent range loss when the separating gas thickness, usually ambient air, is accounted for. MLFCs can also be constructed this way but may alternatively be fabricated from alternating layers of conductive metals and dielectric, giving a more compact and efficient detector. An MLFC, which is a form of a range telescope, works on the same principle as an FC, but instead of collecting the charge from stopped protons in a single layer, multiple conductive layers separated by thin dielectrics are used to collect charge in a stacked manner. A differential stopping power calibration/scaling is then required to result in water-equivalent range readings. Together, the measured points represent discrete points on the depth profile (Burrage et al. 2009; Hsi et al. 2009c; Paganetti and Gottschalk 2002).

6.2.3.6 Scintillating Detectors

Light emission–type detectors can also be used for high-definition transverse beam measurements. They are based either on material scintillation or gas electron multiplier (GEM) technology. In a scintillator, usually fabricated from Lanex Gd2O2S:Tb, the resulting final output is visible light, which may be reflected by a beam splitter and detected in to charge-coupled device camera (Ma, Geis and Boyer 1997). Unfortunately, common scintillation systems are known to suffer quenching (Archambault et al. 2008; Seravalli et al. 2009). A successful attempt was tested to compensate for this effect by mixing two "counteracting"

scintillator materials, Gd2O2S:Tb and (Zn, Cd)S:Ag, to achieve a combined quasi-LET independence in proton beams (Safai, Lin and Pedroni 2004). Another development path has been in the direction of GEM detectors, which suffer less quenching than regular scintillators, especially for use with heavier ions than proton (Seravalli et al. 2008). The gas used in a GEM is usually an Ar and CF_4 mixture, which, as a gas, is not as limited as solid Lanex in terms of ionization density for high-LET types/regions. GEM detectors are, however, not yet available commercially. Their routine use is also questionable from findings of reproducibility uncertainty (Seravalli et al. 2009). Today, the development and use of GEM detectors remains limited to nuclear physics laboratories.

6.2.3.7 Gel Detectors

Gel dosimetry comes with the attraction of three-dimensional (3-D) dosimetry possibilities. Gel dosimetry in particle therapy has been studied but less extensively than other types. Historically, its study has been likely hindered by special handling conditions and readout complexity. The use of gels for particle therapy dosimetry has also been limited from an observed quenching effect from varying LET (Back et al. 1999; Heufelder et al. 2003; Stiefel et al. 2004). In addition to the observations, the quenching effect has been correlated to LET based on computer modeling (Gustavsson et al. 2004). An additional complication in carbon ion therapy is the variety of charge states including those of fragmentation products. This is because, in addition to LET dependence, gels also exhibit a charge dependence that was observed in carbon ion testing (Ramm et al. 2000). Recent developments have attempted to ameliorate some of these conditions.

The advent of products such as solid polyurethane polymer dosimeters (e.g., PRESAGE) containing a radiochromic dye has made handling easier (Brown et al. 2008; Guo, Adamovics and Oldham 2006). An additional simplification is provided by automated optical readout systems that are obviating the need for CT or MR gel scans (Oldham et al. 2001). Quenching in PRESAGE was, however, observed in the BP region from proton irradiation (Al-Nowais et al. 2009). This remaining problem in gel dosimetry for proton therapy has received attention by the active firm MGS Research, Madison, CT, in the development of a series of new gel formulations for proton therapy, the latest of which—type BANG3-Pro2—indicates reported improved response (Zeiden et al. 2010).

Chapter 6

6.2.4 Nuclear Interaction Detectors

6.2.4.1 Neutron Detectors

Neutron detection is important in particle therapy for two reasons. First, from a patient safety perspective, it is necessary to estimate the neutron dose to the patient outside the treatment volume. Second, from a radiation shielding and personnel protection perspective, the radiation shielding must be designed to attenuate the neutrons produced inside the treatment room, beam transport room, and the accelerator vault when the particle beam interacts with materials within those areas. Neutron detectors are needed to validate the constructed facility shielding.

When energetic charged particles interact with matter, there are a significant number of nuclear interactions that may produce secondary particles such as neutrons. Interactions occurring in the particle beam nozzle (which may include an energy modulator, monitor ionization chambers, scattering system, aperture, and compensator) and interactions occurring in the patient may result in the production of neutrons. When a scattered beam is used the potential for secondary neutron production is greater than for a scanned beam, the scanned beam does not usually require multiple beam-modifying devices (scatterers, modulators, compensators, and apertures).

The magnitude of the neutron dose produced by scattering in the patient compared to that penetrating a shielding wall may vary by many orders of magnitude, and therefore, the detection methods used must be chosen appropriately. Radiation protection measurements involve neutron instantaneous dose equivalent rates of <20 μSv hr^{-1} (or a physical dose rate of <2 μGy hr^{-1}). Most radiation shielding walls have areas outside the walls where dose-equivalent rates may be even 10–100 times lower. Such dose levels require the use of sensitive detectors such as ^{3}He or BF$_3$ counters, scintillators, or bubble detectors, and even the sensitivity of these detectors, which essentially count individual particles, may not allow measurement of the lowest calculated levels outside a shield. The neutron physical dose delivered to the patient outside of the treatment field is typically 10^{-3} to 10^{-4} times lower than the prescribed proton dose for proton therapy. A typical therapeutic dose rate for proton therapy is 2 Gy min^{-1}, resulting in a corresponding neutron dose rate of approximately 0.1–1 mGy min^{-1} or 6–60 mGy hr^{-1}. Assuming a radiation weighting factor of 10 for carcinogenesis, this gives neutron dose equivalent rates of 60–600 mSv hr^{-1}.

Hence, the neutron dose rates detected in shielding measurements are typically 3 to 6 orders of magnitude lower than those seen in the patient outside of the treatment field during proton radiation therapy. The sensitive devices used for radiation protection measurements generally exhibit considerable neutron energy sensitivity and, therefore, accurate calibration is not possible. Although Bonner spheres, bubble detectors and track etch detectors (Mesoloras et al. 2006; Wang et al. 2010; Yan et al. 2002) have been used for assessing neutron doses in the patient the uncertainties in these measurements are quite large. Bonner spheres have the disadvantage that they are large, typically 30 cm in diameter and, therefore, are unsuitable for spot measurements close to the edge of a proton therapy treatment field.

Various detectors may be used for detecting neutrons in radiation protection situations. The most sensitive detectors are used in this application and these detectors rely on the detection of individual particles. Until recently the preferred detector for neutrons in this application has been a Bonner sphere or long counter (Knoll 1979) which utilizes a boron trifluoride (BF$_3$) filled proportional counter to detect thermal neutrons. The principle of these detectors is based on thermalizing the fast neutrons in a large polyethylene moderator which surrounds the BF$_3$ counter. Typically the moderators are spherical or cylindrical in shape with a diameter/length of approximately 250 to 300 mm. The moderator thermalizes the neutrons which interact with the BF$_3$ counter positioned at the center of the moderator. The BF$_3$ counter is cylindrical in geometry with an outer diameter of about 25 mm and a length of 100 to 150 cm. A high voltage is applied between a fine wire axial conductor and the outer casing and the device can be operated as a proportional counter using appropriate pulse detection circuits. The thermal neutrons have a large interaction cross-section with ^{10}B and the products are recoil alpha and ^{7}Li nuclei through the reaction:

$$n_{\text{th}} + {}^{10}\text{B} = {}^4\text{He} + {}^7\text{Li}.$$

These ions are detected in the proportional counter. The detector electronics integrates the pulse counter rate to display a "constant" reading on an analog or digital scale. The detectors must be calibrated against a known neutron fluence. An Am-Be source is usually used; therefore, the calibration is at a low neutron energy (average energy ~4 MeV). The range of neutron energies that can be efficiently detected is determined by the size of the moderator. For a handheld detector, the size of the moderator is ~30 cm in diameter to maintain a manageable weight and limits the upper detectable fast neutron energy to about 10 to 15 MeV.

The energy response is a limitation of Bonner sphere detectors when used for radiation protection measurements around high-energy accelerators, as there may be a significant number of high-energy neutrons penetrating the shielding walls. It is certainly a limitation if these devices are used for off-axis neutron dose measurements inside the proton treatment room. The need for neutron detectors with a wider energy response has led to the development of new detectors. Modified moderators for long counters allow the energy range to be extended considerably. The Wendi-2 detector developed at Los Alamos National Laboratory (Olsher et al. 2000) uses a polyethylene moderator that incorporates a tungsten powder shell embedded in the polyethylene. This detector uses a ^{3}He proportional counter rather than BF_3. ^{3}He has replaced BF_3 as the preferred proportional counter gas, as BF_3 is now classified as a hazardous material, which makes the use of ^{3}He cost-effective in spite of its high cost. The thermal neutron reaction with ^{3}He yields hydrogen and tritium nuclei:

$$n_{th} + {}^3He = {}^1H + {}^3H$$

and has the advantage that it has a higher cross section (5330 mbarn) than the corresponding boron reaction with thermal neutrons (3840 mbarn), thus producing a more efficient detector. The modified moderator allows for detection of neutrons in the energy range from 25 meV to 5 GeV. Neutron dose rates in the range of 0.01 μSv h^{-1} to 100 mSv h^{-1} may be detected. The disadvantage of this detector is that the moderator is very bulky.

An alternative detector with an extended energy response from 25 meV to 100 MeV uses a scintillation counter to detect recoil protons from elastic scattering (Olsher et al. 2004). Although this detector has a much smaller energy range than the Wendi-2 detector, it has the advantage of being much lighter and readily portable. The detector is marketed as the Priscilla detector. Neutron dose rates from 0.01 μSv h^{-1} to 10 mSv h^{-1} may be detected.

Another type of neutron detector is the "bubble" detector. This depends on the formation of small bubbles in a superheated gel, a technique first suggested by Apfel (Apfel 1979). The technology was also studied at the Atomic Energy of Canada Limited by Ing (1986) and has been commercialized (BTI, Chalk River). In the commercial system, the gel is held in a test tube–shaped

container with a special twist cap that allows the contents of the tube to be pressurized. The detector is placed in the neutron field for a known time. After exposure, the gel is depressurized by releasing the cap, and small bubbles form in the gel, where neutron interactions have occurred. The gels are manufactured to be suitable for reading integrated neutron doses over a limited range specified by the user. Multiple detectors are necessary to cover an extended dose range. The detectors are precalibrated to read in bubbles per micro-Sievert. The useable dose range is 1 μSv–5 mSv, and the user may specify the sensitivity to be in the range 0.31–30 bubbles μSv^{-1}. Typically, the aim is to create 40–100 bubbles in the detector, which can be countered by eye or using a special automatic counting unit. The accuracy of the measurements depend on the number of bubbles created; 100 bubbles will give a standard deviation of 10%, whereas 40 bubbles increases the standard deviation to 16%. Various types of bubble detector are available; the energy response for the detector recommended for personal neutron dosimetry is approximately 200 keV–15 MeV. A repressurizing device is used to regenerate the clear gel, and the detectors can be reused hundreds of times. The detectors are precalibrated to read in micro-Sieverts; therefore, the assumption of a neutron radiation weighting factor is implicit in the calibration process.

The assumption of a radiation weighting factor is a problem with the calibration of all these detectors, which are designed primarily as instruments for radiation protection measurements. Of course, this is not a problem when the detectors are being used for this purpose, because large uncertainties are acceptable in these circumstances, and dose levels are very small. However, when used to assess the neutron dose to the patient from scattered neutrons, the choice of radiation weighting factor may not be appropriate. The reason for this is that the off-axis scattered neutron doses are considerably higher than typical neutron doses outside a personnel radiation shield, and the RBE of neutrons increases significantly as the dose decreases from normal therapeutic levels through scattered therapy dose levels to radiation protection levels. In addition, the RBE of neutrons depends on the neutron energy, and the spectra of scattered neutrons in a patient may be considerably different from that penetrating the treatment room shielding walls.

For this reason, neutron detectors other than those used for radiation protection measurements may be more appropriate for assessing the scattered dose in the patient. The ideal detector may be a tissue-equivalent proportional counter to provide microdosimetry data

(Binns and Hough 1997). Such data, when used carefully, may provide reasonably accurate dose measurements for the independent components of the dose from the neutron interactions (photons, recoil protons, alphas, and heavy recoils) and, in addition, provides LET information for each of these components. The LET information can be used to assess the appropriate radiation weighting factor or RBE. The disadvantages of these detectors are that the data collection and analyses systems are relatively complex and the measurements are time consuming. Another technique for measuring scattered neutron dose outside the radiation field was recently proposed by (Diffenderfer et al. 2009; Diffenderfer et al. 2011). This is based on the dual detector technique that was developed for the neutron dosimetry of therapeutic fast neutron beams. The technique uses two ionization chambers of similar physical dimensions but constructed from different materials, which results in the chambers having different responses to the neutron and gamma ray components of the radiation dose. Details of the dual detector and microdosimetry measurement techniques may be found in Sections 6.3.4.1 and 6.3.4.2, respectively.

6.2.4.2 Other Nuclear Interaction Detectors

Apart from using neutron detectors for assessing the hazards of the neutron dose resulting from nuclear interactions, other nuclear interaction detectors are under consideration for assessing both delivered dose distributions and determining proton beam range.

When the proton beam interacts with the biological tissue (particularly, C, O, and N), a variety of nuclear interactions occur, which result in the formation of the positron emitters ^{11}C, ^{15}O, and ^{13}N. It was recognized as early as 1970 in pion therapy that, if the distribution of induced positron emitters in the patient's body could be imaged using PET, then it may be possible to retrieve information on the pion dose distribution in the patient (Taylor, Phillips and Young 1970). Further studies by Litzenberg et al. (Litzenberg et al. 1999) established the feasibility of this technique for proton therapy. Nuclear interactions of the proton beam with biological tissue also result in many of the nuclei being left in excited states from which they decay by gamma-ray emission. Most of these excited states have very short half-lives (less than a few nanoseconds), and the decay gammas are referred to as prompt gamma emission. These gamma rays can, in principle, be imaged using single photon emission imaging techniques and may also provide dose distribution and proton beam range information. Both these techniques and their potential for future development are discussed in greater detail in Sections 6.3.5.2 and 6.3.5.3, respectively.

6.3 Measurement Techniques

The evolution of the multidimensional detectors described above in Section 6.2.3.5 has been prompted not only by advances in detector technology but also, and more significantly, by advances in beam delivery technology. Therefore, their indicated use is related to the beam delivery type. Passive lateral and longitudinal particle beam scattering systems represent the most basic beam delivery system and are also the easiest to characterize using the simplest detectors. The development of the multidimensional detectors has been necessitated by the introduction of beam scanning; this applies to both uniform scanning and modulated scanning absolute and relative dosimetry.

6.3.1 Absolute Dosimetry

6.3.1.1 System Calibration

Although there are well-established protocols for the measurement of absolute dose with both scattering and scanning systems, the use of scanning systems is not yet common. Therefore, additional absolute dosimetry measurements are usually indicated in comparison to scattering systems. The additional measurements may include the use of TLD, alanine, and calorimetry, as mentioned in Section 6.2.1.

6.3.1.2 Dosimetry Cross Comparisons and Intercomparisons

After facility system calibration, recalibration, or as an assurance measure at any time, it may be desirable to compare the dose calibration with other centers and/or standards organizations. This can be performed on-site using ionization chambers and the measurement protocols from other centers (Vatnitsky et al. 1996). For distance-based comparisons, the use of other detectors is indicated. Although films can be used in this role, they are typically not. Most common are TLDs, as they are readily mailed and can be irradiated without special preparation (Kirby et al. 1986; Kirby, Hanson and Johnston 1992; Ibbott 2008). Alanine dosimeters can be used in this application (Onori et al. 1997; Nichiporov et al. 1995).

6.3.2 Beam Monitoring

Another application for FCs is beam monitoring, for instance, to monitor the beam fluence rate at the end of a trunkline. Simple wire chambers are also used for beam monitoring as are secondary emission monitors. In general, the trend in beam monitors is to develop low-noise, minimally perturbing detectors so as to maintain optimal beam quality for scanning applications.

6.3.3 Relative Dosimetry

6.3.3.1 Scattering

For scattering systems where the radiation field delivery is approximately invariant with time, a scanning ionization chamber can efficiently perform continuous lateral and longitudinal measurements.

6.3.3.2 Small Field

Small-field dosimetry and penumbra measurements are best performed with high-resolution sensors and detectors such as film or diamond. To adjust for diamond energy dependence or receive an accurate film calibration, the field can be initially delivered with a larger cross section at the same energy to an ionization chamber. Then, with this calibration, the field can be closed down to size and a relative output measured.

6.3.3.3 Scanning

The next step in beam delivery complexity is to vary the dose delivery longitudinally in time, called "dose stacking." Because the beam is not always "seen" by a discrete detector such as an ionization chamber, other than time-consuming point-by-point measurements, triggering of the detector would be needed for efficient use. Although such a triggering scheme could be employed, it is not typically done in practice. Instead, multielement detectors or an integrating sensor such as film are used. Due to particle streaming through physically small gaps, the use of film is contraindicated for longitudinal measurements with dose stacking, leaving the MLIC or MLFC as reasonable choices, depending on particular application, physical dose measurements, or beam monitoring, respectively. Modulated scanning beam systems (also known as pencil beam scanning) where the dose deposition varies with time both laterally and longitudinally require these detectors with additional requirements on dose rate linearity, geometric accuracy, and beam capture profile.

The use of multichannel detectors is associated with acquiring a large number of highly accurate lateral and longitudinal data sets. This is due to the many options of energy, energy modulation, spot size, and field size for such systems. It is not efficient to achieve this with point-by-point measurements. However, a reduced set of individual point measurements with a calibrated small-volume ionization chamber can serve as a valuable cross-check.

6.3.4 Neutron Dosimetry and Microdosimetry

6.3.4.1 Dual Ionization Chamber Neutron Dosimetry

The dual ionization chamber measurement technique may be used to evaluate the separate absorbed doses of neutrons and photons in a mixed-radiation field (ICRU 1977). The application to measuring scattered neutron dose outside a primary proton beam has been suggested by Diffenderfer et al. (2009, 2011). One ionization chamber (T) is built to have approximately the same sensitivity to neutrons and photons, whereas the wall material used for the second chamber (U) results in a lower sensitivity to neutrons than to photons. The dual chambers used here are an A150 tissue-equivalent plastic (TEP) ionization chamber filled with methane-based tissue equivalent gas (T) and a magnesium ionization chamber filled with argon (U).

For a given mixed field, the quotients of the responses of the dosimeters by their sensitivities to the gamma rays used for calibration R'_T and R'_U, respectively, may be written as

$$R'_T = k_T \cdot D_N + h_T \cdot D_G \tag{6.1}$$

$$R'_U = k_U \cdot D_N + h_U \cdot D_G \tag{6.2}$$

where D_N an D_G are the absorbed doses in tissue of neutrons and of photons in the mixed field, k_T and k_U are the ratios of the sensitivities of each dosimeter to neutrons to its sensitivity to the gamma rays used for calibration, and h_T and h_U are the ratios of the sensitivities of each dosimeter to the photons in the mixed field to its sensitivity to the gamma rays used for calibration, respectively.

Solving (6.1) and (6.2) simultaneously gives

$$D_N = (h_U \cdot R'_T - h_T \cdot R'_U)/(h_U \cdot k_T - h_T \cdot k_U)$$

and

$$D_G = (k_T \cdot R'_U - k_U \cdot R'_T)/(h_U \cdot k_T - h_T \cdot k_U).$$

Chapter 6

It is generally assumed that, in most cases, $h_T = h_U = k_T = 1$ and, therefore, these two equations simplify to

$$D_N = (R'_T - R'_U)/(1 - k_U)$$

and

$$D_G = (R'_U - k_U \cdot R'_T)/(1 - k_U).$$

The application of this method to measuring the neutron dose outside a proton field is not without problems. Foremost among these is the magnitude of the neutron dose rate in these situations is such that the current measured by commercial TEP and Mg ion chambers with measuring volumes of 1 cm³ is small in comparison to chamber leakage currents. This limitation may adversely affect the accuracy of the measurements. Diffenderfer overcame this limitation by manufacturing custom-made chambers of 10-cm³ volume without adversely affecting spatial resolution. The second limitation is the uncertainty in the k_u value, which is neutron energy dependent. Although k_u values have been well determined for the neutron energies and spectra associated with neutron therapy beams, they are not well known for the higher energy neutron spectra associated with neutrons from scattered proton beams. Fortunately, photon-neutron fields resulting from scattered proton beams are dominated by the neutron dose, and hence, the measurements are relatively insensitive to the exact k_u value. It is estimated that neutron dose can be measured with an uncertainty of better than 2%–3% in many situations. Once the physical neutron dose has been measured, a radiation weighting factor or RBE must be applied appropriately to the biological end point under consideration. As discussed above, uncertainties associated with that factor may be considerable, since it will vary as a function of neutron energy. However, the advantage of the dual chamber method is that it gives an accurate measure of the physical neutron and photon dose components in the beam. Measured information on the LET distribution in a mixed field beam may be obtained using microdosimetry techniques, which are discussed in the next section.

6.3.4.2 Microdosimetry of Proton Beams and Scattered Neutrons

Microdosimetry can be defined as the systematic study and quantification of the spatial and temporal distribution of absorbed energy in irradiated matter (Rossi and Zaider 1996). It encompasses a system of concepts and of physical quantities and their measurements. In the current context, we are most concerned with the measurement of LET spectra and their interpretation to yield data on both physical dose and radiation weighting factors or RBE. The measurement of microdosimetric or LET spectra was pioneered by Rossi (Rossi and Rosenzweig 1955) using tissue-equivalent proportional counters (TEPC). More recently, semiconductor microdosimetry detectors have been successfully used in this application (Rosenfeld et al. 2002; Rosenfeld et al. 2000).

The TEPC is typically a spherical or cylindrical proportional counter with an A150 wall and a central wire as a charge collector. A150 is a conductive plastic, and a high voltage is applied between the wall and the central conductor. The counter is enclosed in a gas-tight container and filled with a TEPC gas to a pressure that mimics the typical dimensions of a mammalian cell (10 μm in diameter); that is, the mass of gas in the actual geometrical volume of the counter is equal to the mass of a mammalian cell. Primary particles (protons in a proton beam) on secondary particles (protons, alphas, O, C, and N heavy recoils) set in motion by scattered neutrons traverse the counter volume, producing ionizing tracks that can be detected as single pulses with suitable amplifiers. The magnitude of an individual pulse is proportional to the LET, and using a multichannel analyzer, it is possible to measure an LET spectrum. The features of the spectrum can be measured to obtain information on the LET of the incident beam and radiation weighting factor or RBEs may inferred from the spectra.

To study the RBE of proton beams, the counter may be placed directly in the proton beam. The main feature of a proton beam LET spectrum are a single well-defined peak. As the proton beam energy decreases (e.g., if the counter is moved to deeper depth in a water phantom), the peak moves to higher LET, reaching a maximum in the distal edge of the BP or (SOBP) value, where the proton stopping power reaches its maximum value as the protons slow down and stop. The physical dimensions of a TEPC is typically 2–3 cm, so they are generally unsuitable for accurately resolving radiation weighting factor variations in the distal edge of the BP. Miniature, small-volume TEPCs may be built for this purpose, but the semiconductor microdosimetric counters, which may be fabricated with dimensions of a few microns, may be more suitable for this application (Rosenfeld et al. 2002).

The structure of the LET spectra observed outside of the proton beam resulting from scattered neutrons is more complex than that measured in the primary proton beam. Neutron LET spectra have multiple peaks. When neutrons are produced in nuclear interactions, there is always some accompanying gamma ray dose that will appear as broad low LET peaks in the spectrum with LET values between about 5×10^{-2} (typically the lower level of detection for a TEPC) and 5 keV μm^{-1}. The largest peak resulting from secondary recoil protons occurs between LETs of 1 and 150 keV μm^{-1}. There may be structure within this peak, but the high LET limit of the peak is characterized by a steep falling edge, known as the proton edge, and represents the highest attainable proton LET as the proton slows down and stops. Beyond this peak, between about 100 and 400 keV μm^{-1}, is an alpha peak and a corresponding alpha edge. The alpha particles are produced by neutron-induced nonelastic nuclear interactions predominantly in ^{12}C and ^{16}O. Finally, the heavy recoils (^{12}C, ^{14}N, and ^{16}O) have very similar stopping powers and form one indistinguishable peak extending to a maximum LET value of about 2×10^3 keV μm^{-1}.

If the detector LET axis (kiloelectron volts per micrometer) is carefully calibrated against an internal alpha source or a well-defined proton edge, it is possible to determine the absorbed dose in the detector with reasonable accuracy by determining the total number of events of each LET region and assuming the average chord length in the detecting volume. Furthermore, the dose components due to each of the types of secondary particle may be separately determined as can the average LET of each of the peaks. With knowledge of this average LET, it is possible to determine a separate radiation weighing factor (RBE) for the dose due to each of the secondary particle types or use these data to determine an LET weighted average radiation weighting factor (Pihet et al. 1988).

6.3.5 *In Vivo* Dosimetry and Range Determination

6.3.5.1 Positron Emission Tomography

Another interesting and emerging area of particle therapy dosimetry is applied *in vivo* measurements for dose distribution and range verification. These are typically performed during or shortly after treatment. These measurements are made by detecting the annihilation photons from positron-electron interactions. During particle therapy, positrons emitters are generated: ^{11}C (20.4 min half-life); ^{13}N (9.97 min half-life) and ^{15}O (2.04 min half-life) from inelastic processes during proton irradiation; and ^{11}C as a breakup product from ^{12}C therapy. The magnitudes of the half-lives are such that it is possible either to detect the positron signal in the treatment room using an in-room PET scanner (Enghardt et al. 1999; Nishio et al. 2010; Pawelke et

al. 1996; Parodi, Enghardt and Haberer 2002; Nishio et al. 2006; Parodi and Enghardt 2000) or to quickly transfer the patient to a prepared positron emission tomography–computed tomography (PET-CT) scanner (Hishikawa et al. 2002; Hsi et al. 2009c; Oelfke, Lam and Atkins 1996). A PET-CT scan indicates the positron concentration correlated geometrically with patient anatomy. The particle range cannot be directly observed, because the reactions producing the positron emitters all have negative Q-values. However, events can be detected close enough to the end of the proton range (within 3–5 mm) that an extrapolation can be made with acceptable accuracy. The application of this technique to the reconstruction of detailed in vivo dose distributions is more problematic. To achieve this aim, it is necessary to perform a Monte Carlo calculation to predict the planned particle beam dose distribution and the corresponding distribution of positron emitter isotopes and compare these to the measured isotope distribution generated by the delivered beam. There are uncertainties associated with the MC program database of cross sections for the PET isotope producing nuclear reactions. Additional uncertainties result from biological processes, particularly washout effects related to circulatory processes (Knopf et al. 2009; Parodi et al. 2007; Tomitani et al. 2003). As muscle tissue is composed predominantly of oxygen (~70% by weight) and the cross sections for the ^{15}O producing nuclear reactions are large, ^{15}O gives the largest signal for imaging in a proton beam. A disadvantage of ^{15}O is its short half-life (120 s), which requires that imaging must be performed in the treatment room immediately after delivery of the dose if the full advantage is to be achieved. An advantage of the short half-life is that the imaging time can be short, since 75% of the total counts are obtained in 4 min; this is also an advantage with respect to biological washout problems. However, imaging in the room takes up valuable room time and reduces the number of patients that can be treated in the room each day; an important consideration when the capital costs of proton and ^{12}C facilities are so high. In proton beams, ^{11}C imaging is also a possibility. A typical muscle tissue is composed of approximately 15% carbon; therefore, the ^{11}C signal is not as strong as ^{15}O. However, ^{11}C has a half-life of 20 min so that imaging outside the treatment room is possible for this isotope, although imaging times will be longer than for ^{15}O. Both of these imaging techniques are under consideration for proton beams. It should be noted that the ratio of carbon to oxygen nuclei in adipose tissue is different from that in muscle tissue and may

Chapter 6

vary considerably; typical percentages by the weight of oxygen and carbon in fat are ~27% and ~59%, respectively, and these may vary considerably. For carbon ion beams, the problem is well defined, since ^{11}C is the predominate isotope produced. A major disadvantage of PET *in vivo* dosimetry is the length of time spent inside or outside the treatment room for detection.

6.3.5.2 Prompt Gamma

A technique that is potentially faster than PET imaging under investigation is the detection of prompt gamma ray emission during particle therapy. The most detailed study on the feasibility of this technique for clinical proton beam imaging has been published by Polf et al. (Polf et al. 2009a), who have made a theoretical study with Monte Carlo techniques using the MCNPX code. They investigated the excited states produced by proton interactions with the most abundant nuclei in mammalian tissues (i.e., O, C, H, N, and Ca). Prominent gamma ray emissions occur at 6.13, 6.92, and 7.12 MeV. For calcium, less prominent emissions occur at 3.74 and 3.91 MeV, and for nitrogen, emissions occur at 2.33 MeV.

Although the techniques of activation analysis have been used for many years in medical science for determining body tissue composition (Chamberlain et al. 1968), it is only recently that tomographic techniques have been applied in this field (Floyd et al. 2006). In proton therapy applications, the interpretation of results will be difficult, since the cross sections for the reactions of interest vary with incident proton energy; typically, the cross sections peak at incident proton energies in the range 20–30 MeV. Measurements of prompt gamma emission spectra in a clinical proton beam have been reported by several authors (Min et al. 2007; Polf et al. 2009b). The challenge is that the gamma ray energies of the most prominent peaks are in the 4–7 MeV range, where the efficiency of the detectors most commonly used in nuclear medicine imaging are low. However, it has been demonstrated that, with carefully designed collimation and shielding, gamma rays originating in a phantom can be detected above background levels and measured spectra agree well with Monte Carlo calculations (Polf et al. 2009b). The limited results obtained to date seem promising, at least as a potential method for in vivo range verification.

6.3.5.3 Other Methods for Dose Range Detection and Verification

Implantable dosimeters using MOSFET technology with wireless readout have been available for several years (Black et al. 2005) and have been applied in conventional radiation therapy for verifying delivered dose in prostate cancer patients (Beyer et al. 2007). The devices are glass encapsulated and are available commercially in a capsule as small as 2.1 mm in diameter and 20 mm in length. This capsule is still large enough to produce significant perturbations in (Lu 2008b, 2008a) proton range and may be difficult to place at the interface between a tumor and critical organ close to the end of the proton range. However, some interesting strategies have been proposed for overcoming these problems and using these devices for in vivo proton range verification.

The first method can be used for passively scattered range modulated beams that use a moving modulator such as a rotating stepped wheel. The detector is positioned at some point in the treatment volume. The time dependence of the dose rate at this point is measured, and these data may be used to reconstruct the residual range at the point of measurement with millimeter accuracy (Lu 2008a). Unfortunately, at present, the implanted dosimeters do not have the necessary time resolution for this technique to be feasible; however, it should be possible to make these measurements with an intracavity detector such as a diode.

The second measurement has wider applicability and can be applied to all proton dosimetry delivery techniques, both passive scattering and scanning (Lu 2008b). In this method, an implantable dosimeter is positioned in the target volume, and instead of delivering a single SOBP, the dose is delivered as two separate sloping spread-out peaks, sloped in opposite directions. The doses delivered from each field are measured, and the ratio of these doses is calculated. From this ratio, it is possible to determine the water-equivalent path length to the position of the dosimeter. By comparing this to the treatment plan, it is possible to infer the residual range and, thus, the position of the distal edge of the BP. If the dosimeter is positioned reasonably close to the end of the range, it should be possible to infer the total range with reasonable accuracy.

6.4 Dosimetry Protocols and Uncertainty

Charged particle therapy began in the 1950s, and initially, dosimetry protocols were developed by each institution for their own use (Verhey et al. 1979; Astrakhan et al. 1970). As the field expanded, it was recognized that there was a need for nationally and internationally accepted dosimetry protocols. The first significant dosimetry recommendations were published by the American Association of Physicists in Medicine (AAPM) in 1986 (AAPM 1986). The document published was AAPM Report No. 16, "Protocol for Heavy Charged-Particle Therapy Beam Dosimetry," which presented the state of charged particle absolute and relative dosimetry at the time and made recommendations for its practice. It dealt with a wide range of heavy-charged particles, including negative pions, protons, and helium, and ions of heavier elements such as carbon, neon, and argon, reflecting the charged particles then under investigation in facilities in the United States. For absolute dosimetry, the AAPM report suggested three methods—calorimetry, FC dosimetry, and ionization chamber dosimetry—and states the uncertainties and major contributing sources for each method as ±2.5% (thermal defect), ±5% (beam area, fluence, secondary particle effects), and ±5%–10% (mean energy required to produce an ion pair w), respectively. Although, at that time, FC dosimetry was practiced for facility calibration, a subsequent interfacility dosimetry comparison found a larger variation in measured calibration factors among institutions using the FC method than among those using an ionization chamber as their standard (Vatnitsky et al. 1996). Hence, the use of FCs for facility calibration stopped, and ionization chamber dosimetry was adopted as the standard calibration technique for charged particle beam calibrations. Other recommendations of AAPM Report No. 16 consistent with its era were to specify dose in tissue and to use a w of 34.3 J C^{-1} for ionization chamber dosimetry.

The next initiative in defining a charged particle dosimetry protocol came from the European Charged Heavy Particle Dosimetry Group (ECHED), which published its own code of practice for clinical proton dosimetry in 1991 (Vynckier, Bonnett and Jones 1991). A supplement to this code followed in 1994 (Vynckier, Bonnett and Jones 1994). The original 1991 code called for using an air-filled tissue-equivalent ionization chamber as the field instrument for beam calibration, the beam calibrations being performed at the center of the SOBP in a water phantom. The preferred method for absolute calibration of the ionization chamber was against a TEP or graphite calorimeter. If a calorimeter was not available then calibration against an FC was the next best option; calibration against a ^{60}Co source was considered the least preferred option. The protocol provided recommendations on the values of the physical parameters required for the dose calibration together with a worksheet. The estimated error in the calibration was less than ±5% in all cases. The final dose

calibration was stated as proton dose to the tissue, with the tissue defined as ICRU muscle. At the time that the original ECHED protocol was being finalized, new stopping power data were in preparation, which were subsequently published in 1993 as ICRU Report 49 (ICRU 1993). The availability of these data and reconsideration by ECHED of the best medium in which to specify the absorbed dose led to the publication of a supplement to the ECHED protocol in 1994 (Vynckier, Bonnett and Jones 1994). In this supplement, new stopping power data from ICRU Report 49 were included for dose specification in water, which was now adopted as the preferred material for absorbed dose specification in order to provide uniformity with photon and electron beam calibration practices.

The next published protocol for absolute dosimetry appeared in 1998 as a part of ICRU Report 59 (ICRU 1998), "Clinical proton dosimetry—Part I: Beam production, beam delivery and measurement of absorbed dose." This report recommended that air-filled cylindrical ionization chambers with A150 or graphite walls should be used for the reference absorbed dose measurements with a ^{60}Co calibration traceable to an appropriate standards laboratory. In this protocol, it was further recommended that calibration coefficients should be stated as air kerma, exposure, or absorbed dose to water. The ICRU report recommends a value for the energy to produce an ion pair in a proton beam $w_p = (34.8 \pm 0.7)$ J C^{-1} in humid air. The estimated relative uncertainty in the absorbed dose measurement is 2.6% (1-σ). This report dealt only with proton beams.

In 2000, the International Atomic Energy Agency (IAEA) published a report in its Technical Report Series, TRS 398 (IAEA 2000), that was a code of practice for radiotherapy dosimetry and included recommendations for photon, electron, proton, and heavier charged-particle dosimetry. In this report, the use of ionization chambers having calibration factors specified in terms of dose to water is recommended. The ionizations chambers may have a cylindrical or plane parallel geometry. TRS 398 uses a w_p value of (34.23 ± 0.13) J C^{-1} in dry air. The stated overall relative uncertainties in the absorbed dose at the 1-σ level are 2.0% and 2.3% for cylindrical and plane-parallel chambers, respectively.

The absorbed dose calibrations obtained using the ICRU and IAEA protocols differ by as much as 3.1%, depending on the type of chamber used and other factors. In ICRU Report 78, which was jointly prepared by the ICRU and the IAEA, these discrepancies are resolved. This report makes the following recommendation (ICRU 2007):

Chapter 6

"It is recommended that the TRS 398 code of practice be adopted as the standard proton dosimetry protocol: it is simple to use; it provides tabulated beam correction factors for a wide range of common cylindrical and plane-parallel ionization chambers; it provides a formula for calculating $s_{w,air}$ (the water to air stopping power ratio) for proton beams in terms of the beam quality parameter (residual range); it harmonizes with the protocols for conventional radiotherapy and heavy-ion beams (also given in TRS 398), which are being adopted in many institutions; the uncertainties in dose determinations are less; more recent and accurate physical constants are used; and the formalism is more robust and rigorous than that of ICRU 59."

TRS 398 also provides a code of practice for the determination of absorbed dose for heavy ion beams. Heavy-ion therapy is complicated by the fact that it is a high-LET radiation and that the relative biological effectiveness of the heavy-ion beam increases with depth. Therefore, in treatment planning for clinical applications, it is necessary to use a biological effective dose, which is defined as the product of the physical dose and the relative biological effectiveness at any point. The use of biological effective dose allows the clinical results obtained with heavy-ion beams to be compared with those from conventional therapy in a meaningful manner. It is not within the scope of TRS 398 or of this chapter to discuss how biological effective dose should be determined but, rather, to concentrate on the accurate determination of absolute or relative physical dose. Recently, heavy-ion therapy has only been delivered at relatively few centers, dominated by the National Institute of Radiological Science's group in Chiba, Japan, and Gesellschaft für Schwerionenforschung (GSI)/German Cancer Research Center (DKFZ)/Heidelberg University collaboration in Germany. In the next few years, a significant number of new heavy-ion centers are scheduled to open, particularly in Germany and Japan. It is likely that TRS 398 will be adopted as the physical absorbed-dose measurement protocol, although there are several physical parameters in heavy-ion dosimetry that need reevaluation. The largest source of uncertainty in heavy-ion-beam absorbed-dose determination is the beam quality correction factor k_Q, which is largely dependent on the ratios of the stopping power of ions in water and air ($s_{w,air}$), and the energy to produce an ion pair in air (w_{air}). TRS 398 uses single values for these parameters, which actually vary as a function of the residual range (i.e., energy) of the heavy ion beam. However, a larger issue for the stopping power data is that of its absolute accuracy. Stopping power calculations depend critically on the mean excitation energy (I) of the atoms of the material through which the beam is traveling. Recently, some uncertainty has arisen about the correct I value for water, with published values varying between 67.2 and 80.8 eV. The impact of these variations on the calculated stopping power in water and, hence, on the water-to-air stopping power ratio is discussed in detail by Henkner et al. (Henkner et al. 2009). Uncertainties in the stopping power of ions in water also affect the measurement of depth-dose distributions. The ICRU is presently preparing a report on heavy-ion therapy to be entitled, "Prescribing, Recording, and Reporting Ion-Beam Therapy," which may provide new recommendations on the values of the physical parameters used in absorbed-dose calculation.

6.5 Acceptance Testing

The applications of the various described detectors to the system acceptance procedures is outlined in this section. The goal of acceptance testing is to receive a therapy system ready for the next step, commissioning. Acceptance testing has the following three goals (Nath et al. 1994).

1. It assures the safety of the patients and machine operators.
2. It provides the mechanism by which the institution determines that the system meets its performance specifications (i.e., determines that it received what it intended to purchase).
3. It provides critical baseline data for future quality assurance reviews.

6.5.1 Safety

The first point may seem surprising, as the system must have already gone through an approval process by the governing state body. However, although particle therapy systems move along the path of generalized, universal commercial systems such as medical linear accelerators, this goal has not yet been achieved in practice. Many recent advances have been made in their specification, notably by the Digital Imaging and Communication in Medicine (DICOM) Working Group and the International Electrotechnical Commission Subcommittee, which is preparing recommendations on requirements for the basic safety and essential performance of light ion accelerators.

However, particle therapy systems remain extremely complex, and installations usually contain significant new "features" requiring private, vendor-specific information to be used as part of the treatment process for each new installation. Hence, it is crucial to validate the safe performance of the installation prior to clinical use. This is most appropriately done together with the vendor. As part of safety testing with the particle beam, it is crucial to check a number of conditions that could potentially result in an unintended irradiation of patients and/or staff. As a guide, some possible tests that we used are listed in Table 6.1. However, it should be noted that this is likely not a complete list and the list can naturally vary with the type of delivery system. Therefore, until standard tests become available, we should perform a detailed failure mode and effects analysis (FMEA) and base safety tests on it. Acceptable results to pass the safety tests are that at no time can an out-of-tolerance condition in the irradiation system change the delivered patient dose within a certain tolerance. The tolerance needs to be determined by 1) regulatory requirements and 2) the professional experience of the staff at the center. Usually, the tolerance is stated in the percentage of fractional prescribed dose, although, sometimes, also as the percentage of the total prescribed dose cumulated over all fractions. Nozzle

Table 6.1 Possible Irradiation Safety Acceptance Tests

Failure Mode	Test Description
All Delivery Modes	
Excess dose rate	For each nominal field size/energy, request the maximum fluence rate from the accelerator. Confirm irradiation stop within dose tolerance.
Symmetry X,Y	At several energies, purposefully and rapidly set the distal quadrupole magnet X and, thereafter, Y to cause severe unsymmetric beam delivery. Confirm irradiation stop within dose tolerance.
Symmetry X,Y	At several energies, if possible, insert a physical one-half beam block proximal to the nozzle symmetry detector. Take care not to damage the nozzle. Confirm irradiation stop within dose tolerance.
MU overrun	Temporarily calibrate the dose monitor system to stop on the second channel first. Confirm irradiation stop within dose tolerance.
MU overrun	Temporarily calibrate both dose monitor channels to underrespond by more than the time/counter limit. Confirm irradiation stop within dose tolerance.
Scanned Delivery Modes	
Incorrect scan size	Force the scan size larger and smaller than the planned field delivery. Confirm irradiation stop within dose tolerance.
Incorrect beam profile	Defocus (enlarge) the beam profile magnetically or physically. Confirm detection and irradiation stop within dose tolerance.
Incorrect beam profile	Reduce the beam profile less than the nominal minimum by physical pinhole aperture. Confirm detection and irradiation stop within dose tolerance.
Incorrect beam profile	Defocus (enlarge) the beam profile magnetically or physically in one dimension and then diagonally. Confirm detection and irradiation stop within dose tolerance.
Beam position	a. Deliver the grid of beam spots at multiple energies/multiple gantry angles. Confirm the radiologic alignment of beam centroid at ±0.5 mm or better, depending on the application. Offset the grid delivery by 1 mm at the isocenter. Confirm detection and stop by the beam delivery system. b. Align the beam for scanning delivery. Shift the magnet to cause 1-mm off position. The system should detect and set the interlock.
Interrupt/restart	Using lateral and longitudinal multielement detectors: a. Set up a field delivery and deliver the entire field to the detector. b. Reset the field, deliver partially, and record detector data. c. Restart the delivery of partial field, deliver to completion, and record detector data. d. Sum the detector data from (b) and (c) and confirm that the result from (a) was obtained within tolerance.

Chapter 6

leakage (film) and out-of-field dose measurements (Section 6.2.4) should also be performed during safety acceptance testing.

6.5.2 Performance

Acceptance performance testing must demonstrate the functionality of the system. This does not require testing at each machine setting but, rather, at the extremes of performance and at the nominal reference condition.

6.5.2.1 Uniform Fields

Practically, the performance testing can be divided into three conditions: 1) nominal reference condition; 2) minimum condition; and 3) maximum condition. The reference irradiation condition can vary between centers but usually consists of approximately a 1-L volume centered at a preferred reference depth. The reference depth chosen by the center can be at the middle or high energy. Prior to testing, the system should be calibrated at this condition by protocol (Section 6.2.4). This reference condition is then extended to minimum and maximum test conditions with energy, field size, and energy modulation. In some cases, the nominal condition may also be the maximum condition, depending on how the system is calibrated (in this case, an additional

medium condition is also desirable for testing). Within these divisions, it is necessary to confirm that the desired dose is delivered. This depends on the correct absolute and relative doses being delivered. Absolute dose performance needs to be confirmed to be invariant with MU quantity and fluence rate. Representative tests for a uniform scanning dose delivery system are shown in Figure 6.3.

Typical performance criteria for uniform field quality are specified laterally and longitudinally as field size, flatness, symmetry, penumbra, and duration of delivery. This type of characterization is quite straightforward and well represented in the literature (Farr et al. 2008; Hsi et al. 2009b; Kanai et al. 1983; Koehler, Schneider and Sisterson 1975; Koehler, Schneider and Sisterson 1977; Lu and Kooy 2006; Oozeer et al. 1997).

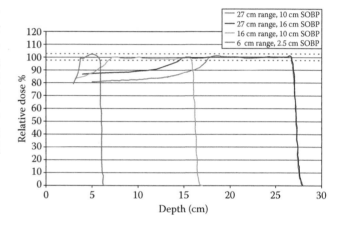

FIGURE 6.4 Example of a percent depth-dose acceptance test showing the limits of the SOBP delivery from the narrowest to the widest at the minimum, nominal, and maximum ranges in water.

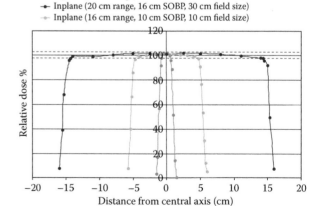

FIGURE 6.5 Example of a beam lateral profile acceptance test showing the limits of field size delivery from the narrowest to nominal to the widest at the reference range in water (16 cm in this case).

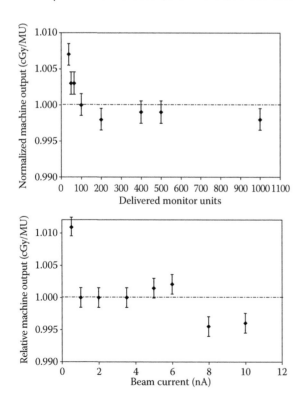

FIGURE 6.3 Monitor unit (MU) linearity test (above) and fluence rate linearity test (below.)

FIGURE 6.6 Star shot showing radiologic alignment at five gantry angles. The limit of convergence is measured to be better than 2 mm. Physical alignment measurements of the beam defining aperture were better than 1 mm.

Examples of acceptance test result excerpts for SOBP and field size constraints are given in Figures 6.4 and 6.5, respectively. The figures show only a subset of the required testing that must validate the dose quality throughout a characterization volume or irradiated volume. This volume definition, being empirical, is not standardized and has been proposed in several ways (Chu 1993; Arduini et al. 1996; Farr et al. 2008; Chu et al. 1993).

The dose delivery quality to the irradiated volume must also be shown not to depend on gantry angle more than the tolerance of 1 mm. Scattering and uniform dose delivery systems use a beam-defining

aperture. Therefore, mechanical alignment tests give an excellent indication of beam alignment. However, it is possible to demonstrate the radiological alignment with gantry angle by performing a star shot or with a Miller–Lutz test. With radiographic or radiochromic film placed edge on to the incident beam direction, multiple unmodulated irradiations are done at different (five to seven) gantry angles. An example of a star shot is presented in Figure 6.6. Similar testing can be performed with the patient positioning systems. Some therapy systems integrate the positioning system with gantry setting to counteract known gantry translations as a function of angle. In this case, an integrated star shot can also be performed.

For uniform field deliveries from modulated scanning systems, also called spot scanning, without apertures, the relative physical dose delivery quality must be demonstrated through the irradiated volume. Such an irradiated volume from modulated scanning can be planned to result in single-field uniform dose (SFUD; Lomax, Pedroni et al. 2004). SFUD deliveries should be verified for energy, modulation, penumbra, homogeneity, spot size, field size and gantry, and angle. Practically, the initial goal is to demonstrate a series of beam spot pattern deliveries based on these criteria. Figure 6.7 shows one such possible measurement setup using a scintillation detector coupled with a mirror and camera (Ma, Geis and Boyer 1997). Because the individual beam spot quality influences the irradiated field quality, it should be measured and quantified. Although a 3-D characterization is required for dose calculation, the specification for testing is usually two-dimensional (2-D).

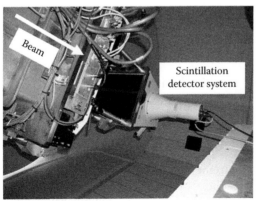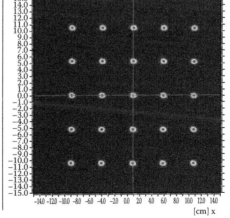

FIGURE 6.7 Experimental setup for demonstrating beam spot placement accuracy. In the left photo, a proton-modulated scanning nozzle delivers a scan of 25 spots on 5-cm centers at 164 MeV. A scintillator detector system including a CCD camera detects the proton dose at high resolution.

Chapter 6

Table 6.2 Isocentric Planar Beam Spot Quality Properties and Tolerances

Property	Description	Tolerance
Size	Specified as FWHM (in millimeters) in air at the isocenter or statistical standard deviation sigma (σ) (in millimeters) of the probability density function probability density function (PDF). For Gaussian distributions, the equivalence is $1-\sigma = 2.35$ FWHM.	For highly conformal fields delivered by pure PBS (no final aperture) and 7–14 mm FWHM (3–6 mm σ).
Shape	Described by a PDF, usually one, or the weighted sum of two 2-D Gaussians.	Invariance to energy, centroid position in the field, and gantry angle; overall, ±5% of σ.
Tilt	The rotation of the beam shape about the beam axis.	Invariance to energy, centroid position in the field, and gantry angle; overall, 10°.
Divergence	Beam parallel	

Table 6.2 lists common 2-D attributes and suggested tolerances. After the scanned beam quality has been demonstrated across the range of operational parameters, uniform fields can be scanned and characterized, as already described.

6.5.2.2 Nonuniform Fields

Nonuniform field delivery must be tested for full intensity-modulated proton therapy (IMPT), where the weighting of each individual spot is allowed to vary independently within multiple fields as part of an optimization process. Because IMPT optimization usually results in high-dose gradients from regions of a few highly weighted spots to low- or zero-weighted spots, the delivery ability of the therapy system must be validated in acceptance testing. This can be done initially with challenging 2-D fluence maps and, then, with 3-D geometries representative of high-dose gradient targets. The 2-D test pattern can be analogous to X-ray test tools used to evaluate diagnostic kilovoltage systems. In addition to the properties listed in Table 6.2, the test pattern should be capable of testing:

- Scan resolution in line pairs per centimeter
- Spot position accuracy in ordinate directions and diagonally
- Scan speed in ordinate directions
- Dynamic range of primary dose monitoring system
- Line scans in ordinate directions and diagonally
- Areas of high dose surrounding low dose (i.e., annuli)

The gamma index (Low et al. 1998) methodology can be applied to the results.

Following the 2-D tests, highly conformal 3-D test targets can be used, such as extruded polygons or spheres with dose "holes" cut out. Finally, anthropomorphic tests can be performed to test the entire treatment chain.

6.6 Beam Commissioning

After successful acceptance testing, additional data are needed to enable proper planning and delivery of radiation treatment. The detection methods used to collect the data will depend on the dose delivery mode. For a double scattering beam delivery system, percentage depth dose data should be measured by scanning in a water phantom using ionization chambers. Beam profile measurements are often made by scanning in air using a diode detector to provide better spatial resolution. For uniform scanning and modulated scanning systems, detector scanning, in air or in a water phantom, is problematic, and point measurements

combined with a full scan are impractical. The solution to these problems is to use a multilayer ionization chamber for percentage depth dose measurements of a single spot while using film measurement in a plastic or solid water phantom for beam profile measurements. A comprehensive set of beam data must be acquired and entered into the radiotherapy treatment planning system (TPS). "Commissioning" refers to the process whereby the needed machine-specific beam data are acquired and operational procedures are defined. It includes but is not limited to (Nath et al. 1994)

1. Beam data acquisition.
2. Entry of beam data into a TPS system and testing of its accuracy.
3. Benchmarking of the TPS system by measurements in some common and extreme operating conditions, which should be designed to provide an understanding of any limitations of the TPS
4. Development of operational procedures
5. Training of all personnel concerned with the operation.

The input data must be tested "closed loop" to verify that the patient radiation fields planned with acquired data are delivered within the accuracy requirements. In its final stage, this testing requires end-to-end tests of near-clinical plans and treatments on phantoms. Procedures based on facility workflow must be developed prior to human use. Sufficient staff training must be broad based from theory to the use of individual systems and, then, the whole system integrated workflow.

The same methods and materials are required; it is a matter of filling in with more detailed measurements to acquire the total beam data for the TPS necessary for clinical operation. Each TPS has different data requirements for input provided by the manufacturer. Typically, however, measured data are required at all energies, field sizes, SOBPs, spot sizes, and so forth. In addition, treatment devices such as immobilization and couch should be characterized for particle beam range loss. While the data are being collected, the TPS should be checked in parallel for the validity of the dose calculation engine, including for inhomogeneous fields (Fraass et al. 1998; Technical Reports Series No. 430: Commissioning and Quality Assurance of Computerized Planning Systems for Radiation Treatment of Cancer 2004; IAEA 2004). Two remaining steps in beam commissioning are suggested: 1) several mock-patient treatments should be prepared and verified dosmetrically and 2) an independent check of absolute dosimetry is performed and compares within 5%.

6.7 Clinical QA Procedures

6.7.1 Machine

Much of the preparation for machine QA is done during acceptance testing. Based on the acceptance testing, FMEA indications should be taken for daily, monthly, and annual QA. Annual QA may employ significant portions of the acceptance test, whereas daily QA must by nature be more limited. On a daily basis, it is generally considered to be prudent to measure at a minimum the machine output (dose calibration), particle beam range, transverse and longitudinal uniform field quality, and for scanning spot delivery accuracy. Some of these test conditions can be varied from day to day so that the range of machine operation is covered within a week (Kohno et al. 2006). For monthly QA, an independent monthly dose calibration with a second calibrated ionization chamber is in order, along with dose delivery quality checks at several gantry angles. Additional emphasis of monthly QA is given to patient alignment and positioning systems (Arjomandy et al. 2009).

The goals of radiological patient-specific field QA are to validate that

1. The field is deliverable.
2. The output factor measures correctly, as predicted by the TPS and/or an independent calculation.

3. The field-relative dosimetry corresponds to the expectation.
4. The uncertainty of the field delivery is understood and is acceptable.

In the beginning of a new center's operation, these properties are likely to be checked for every patient field. Using existing literature on predictive output factors and building on experience at the center, the quantity of patient specific may cautiously be reduced (Engelsman et al. 2009; Hsi et al. 2009a; Sahoo et al. 2008). Eventually, only spot checks may be required for "standard field types" such as for prostate whereas measured QA may be reserved for unusual fields such as those with low energy, small field size, or significant in-homogeneities. Uniform scanned patient fields can be verified in the same way. Full-intensity-modulated fields can be checked in an analogous manner to intensity-modulated photon therapy fields by mapping the patient plan to a suitable physics phantom/detector and comparing the expected delivery to the planned delivery (Smith 2009). For particle therapy, this is usually performed at a minimum of two depths in the test medium (Lomax, Bohringer et al. 2004).

Chapter 6

6.8 Summary

Many—most probably all—of the detectors and detection techniques described here are used in conventional photon and electron radiation therapy beam measurement applications. Ionization chambers are the accepted standard instrumentation for absolute light ion beam calibration. Film and diode measurements are also widely used with ion beams, although as the beam energy varies with depth, care must be taken to account for energy-dependent effects if accurate quantitative measurements are required. Some aspects of charged particle therapy with light ions place unique requirements on detection systems. One such requirement is the need to accurately assess particle range because of the uncertainties in range calculations, which arise from the inadequate knowledge of the stopping power. Another is the need for measuring the 3-D dose distributions created using modulated scanning in a quick and efficient manner. To date, the most efficient solution to this problem has been to use some form of 2-D detector or detector array and to scan through the third dimension in a water phantom. Solutions using bulk 3-D detectors such as gels have met with limited success, and a major shortcoming is that such systems are inconvenient and time consuming to use. The major need in light-ion beam dose measurement is a true 3-D detector that could obtain the necessary dose distribution information in a single scan of the irradiated volume, with adequate resolution (2–3 mm) and supported by software that would be capable of analyzing the data in 1 or 2 min. The availability of such a device would considerably reduce the time required for the acceptance and commissioning of modulated scanning beams and would also allow for comprehensive dose distribution measurements to be made for patient-specific quality assurance of IMPT.

Another requirement of light ion dosimetry is the need to make scattered neutron dose measurements to understand the potential risks to both the patient and staff. The scattered neutron doses associated with passively scattered neutron beams are generally higher than those measured in conventional photon therapy. This is particularly important for the patient when passive scattering systems are used for beam delivery. There is scope for improvement of the measurement techniques and more comprehensive measurements, including microdosimetry measurements, to better understand these risks.

A wide variety of detectors are available today, and many new techniques and technologies are under investigation. The topic of light-ion beam dose detection will remain an interesting and vibrant research field for many years to come.

References

American Association of Physicists in Medicine (AAPM). 1986. Report No. 16, Protocol for Heavy Charged-Particle Therapy Beam Dosimetry. New York: American Association of Physicists in Medicine.

Apfel, R. E. 1979. The superheated drop detector. *Nucl. Instrum. Methods* 162:603–608.

Archambault, L., Polf, J. C., Beaulieu, L., and Beddar, S. 2008. Characterizing the response of miniature scintillation detectors when irradiated with proton beams. *Phys. Med. Biol.* 53 (7):1865–1876.

Arduini, G., Cambria, R., Canzi, C., Gerardi, F., Gottschalk, B., Leone, R., Sangaletti, L., and Silari M. 1996. Physical specifications of clinical proton beams from a synchrotron. *Med. Phys.* 23 (6):939–951.

Arjomandy, B., N. Sahoo, Ding, X., and Gillin M. 2008. Use of a two-dimensional ionization chamber array for proton therapy beam quality assurance. *Med. Phys.* 35 (9):3889–3894.

Arjomandy, B., Sahoo, N., Zhu, X. R., Zullo, J. R., Wu, R. Y., Zhu, M., Ding, X., Martin, C., Ciangaru, G., and Gillin, M. T. 2009. An overview of the comprehensive proton therapy machine quality assurance procedures implemented at The University of Texas M. D. Anderson Cancer Center Proton Therapy Center-Houston. *Med. Phys.* 36 (6):2269–2282.

Astrakhan, B. V., Boreiko, V. F., Bugarchev, B. B., Vainberg, M. S., Valuev, Iu, and Kalinin, A. I. 1970. Dosimetry of the proton beam. *Med. Radiol. (Mosk)* 15 (7):55–62.

Attix, F. H. 1986. *Introduction to Radiological Physics and Radiation Dosimetry.* New York: Wiley.

Back, S. A., Medin, J., Magnusson, P., Olsson, P., Grusell, E., and Olsson, L. E. 1999. Ferrous sulphate gel dosimetry and MRI for proton beam dose measurements. *Phys. Med. Biol.* 44 (8):1983–1996.

Bassler, N., Hansen, J. W., Palmans, H., Holzscheiter, M. H., Kovacevic, S., and AD 4 ACE Collaboration. 2008. The antiproton depth-dose curve measured with alanine detectors. *Nucl. Instrum. Meth. B* 266 (6):929–936.

Berger, T., Hajek, M., Fugger, M., and Vana, N. 2006. Efficiency-corrected dose verification with thermoluminescence dosemeters in heavy-ion beams. *Radiat. Prot. Dosim.* 120 (1–4):361–364.

Beyer, G. P., Scarantino, C. W., Prestidge, B. R., Sadeghi, A. G., Anscher, M. S., Miften, M., Carrea, T. B., Sims, M., and Black, R. D. 2007. Technical evaluation of radiation dose delivered in prostate cancer patients as measured by an implantable MOSFET dosimeter. *Int. J. Radiat. Oncol. Biol. Phys.* 69 (3):925–935.

Binns, P. J., and Hough, J. H.1997. Secondary dose exposure during 200 MeV proton therapy. *Radiat. Prot. Dosim.* 70:441–444.

Black, R. D., Scarantino, C. W., Mann, G. G., Anscher, M. S., Ornitz, R. D., and Nelms, B. E. 2005. An analysis of an implantable dosimeter system for external beam therapy. *Int. J. Radiat. Oncol. Biol. Phys.* 63 (1):290–300.

Botter-Jensen, L., Andersen, C. E., Duller, G. A. T., and Murray, A. S. 2003. Developments in radiation, stimulation and observation facilities in luminescence measurements. *Radiat. Meas.* 37 (4–5):535–541.

Brede, H. J., Greif, K. D., Hecker, O., Heeg, P., Heesc, J., Jones, D. T., Kluge, H., and Schardt, D. 2006. Absorbed dose to water determination with ionization chamber dosimetry and calorimetry in restricted neutron, photon, proton and heavy-ion radiation fields. *Phys. Med. Biol.* 51 (15):3667–3682.

Brown, S., Venning, A., De Deene, Y., Vial, P., Oliver, L., Adamovics, J., and Baldock, C. 2008. Radiological properties of the PRESAGE and PAGAT polymer dosimeters. *Appl. Radiat. Isot.* 66 (12):1970–1974.

Burrage, J. W., Asad, A. H., Fox, R. A., Price, R. I., Campbell, A. M., and Siddiqui, S. 2009. A simple method to measure proton beam energy in a standard medical cyclotron. *Australas Phys. Eng. Sci. Med.* 32 (2):92–97.

Cambria, R., Herault, J., Brassart, N., Silari, M., and Chauvel, P. 1997. Proton beam dosimetry: A comparison between the Faraday cup and an ionization chamber. *Phys. Med. Biol.* 42 (6):1185–1196.

Carlsson, C. A., and Carlsson, G. A. 1970. Proton dosimetry: measurement of depth doses from 185-MeV protons by means of thermoluminescent LiF. *Radiat. Res.* 42 (2):207–219.

Chamberlain, M. J., Fremlin, J. H., Peters, D. K., and Philip, H. 1968. Total body calcium by whole body neutron activation: New technique for study of bone disease. *Br. Med. J.* 2 (5605):581–583.

Cheung, T., Butson, M. J., and Yu, P. K. 2005. Post-irradiation colouration of Gafchromic EBT radiochromic film. *Phys. Med. Biol.* 50 (20):N281–N285.

Chu, W. T., Staples, J. W., Ludewigt, B. A., Renner, T. A., Singn, R. P., Nyman, M. A., Collier, J. M., Daftari, I. K., Kubo, H., Petti, P. L., Verhey, L. J., Castro, J. R., and Alonso, J. R. 1993. LBL Report No. 33749, Performance Specifications for Proton Medical Facility. Berkeley: Lawrence Berkeley Laboratory, University of California.

Cirio, R., Garelli, E., Schulte, R., Amerio, S., Boriano, A., Bourhaleb, F., Coutrakon, G., Donetti, M., Giordanengo, S., Koss, P., Madon, E., Marchetto, F., Nastasi, U., Peroni, C., Santuari, D., Sardo, A., Scielzo, G., Stasi, M., and Trevisiol, E. 2004. Two-dimensional and quasi-three-dimensional dosimetry of hadron and photon beams with the Magic Cube and the Pixel Ionization Chamber. *Phys. Med. Biol.* 49 (16):3713–3724.

Czopyk, L., Klosowski, M., Olko, P., Swakon, J., Waligorski, M. P. R., Kajdrowicz, T., Cuttone, G., Cirrone, G. A. P., and Di Rosa, F. 2007. Two-dimensional dosimetry of radiotherapeutical proton beams using thermoluminescence foils. *Radiat. Prot. Dosim.* 126 (1–4):185–189.

Daftari, I., Castenadas, C., Petti, P. L., Singh, R. P., and Verhey, L. J. 1999. An application of GafChromic MD-55 film for 67.5-MeV clinical proton beam dosimetry. *Phys. Med. Biol.* 44 (11):2735–2745.

De Angelis, C., Onori, S., Pacili, M., Cirrone, G. A., Cuttone, G., Raffaele, L., and Sabini, M. G. 2002. Preliminary results on a dedicated silicon diode detector for proton dosimetry. *Radiat. Prot. Dosim.* 101 (1–4):461–464.

Delacroix, S., Bridier, A., Mazal, A., Daures, J., Ostrowsky, A., Nauraye, C., Kacperek, A., Vynkier, S., Brassard, N., and Habrand, J. L. 1997. Proton dosimetry comparison involving ionometry and calorimetry. *Int. J. Radiat. Oncol. Biol. Phys.* 37 (3):711–718.

Devic, S., Tomic, N., Pang, Z., Seuntjens, J., Podgorsak, E. B., and Soares, C. G. 2007. Absorption spectroscopy of EBT model GAFCHROMIC film. *Med. Phys.* 34 (1):112–118.

Diffenderfer, E., Maughan, R., Ainsley, C., Avery, S., and McDonough, J. 2009. Determination of neutron dose due to a therapeutic proton beam incident on a closed tungsten MLC using the dual hydrogenous/nonhydrogenous ionization chamber method. *Med. Phys.* 36:2733.

Diffenderfer, E. S., Ainsley, C. G., Kirk, M. L., McDonough, J. E., and Maughan, R. L. 2011. Comparison of secondary neutron dose in proton therapy resulting from the use of a tungsten alloy MLC or a brass collimator system. *Med. Phys.* 38:6248–6256.

Edmund, J. M., Andersen, C. E., Greilich, S., Sawakuchi, G. O., Yukihara, E. G., Jain, M., Hajdas, W., and Mattsson, S. 2007. Optically stimulated luminescence from Al2O3: C irradiated with 10-60 MeV protons. *Nucl. Instrum. Meth. A* 580 (1):210–213.

Engelsman, M., Lu, H. M., Herrup, D., Bussiere, M., and Kooy, H. M. 2009. Commissioning a passive-scattering proton therapy nozzle for accurate SOBP delivery. *Med. Phys.* 36 (6):2172–2180.

Enghardt, W., Debus, J., Haberer, T., Hasch, B. G., Hinz, R., Jakel, O., Kramer, M., Lauckner, K., and Pawelke, J. 1999. The application of PET to quality assurance of heavy-ion tumor therapy. *Strahlenther Onkol.* 175 (Suppl) 2:33–36.

Farr, J. B., Mascia, A. E., Hsi, W. C., Allgower, C. E., Jesseph, F., Schreuder, A. N., Wolanski, M., Nichiporov, D. F., and Anferov, V. 2008. Clinical characterization of a proton beam continuous uniform scanning system with dose layer stacking. *Med. Phys.* 35 (11):4945–4954.

Fidanzio, A., Azario, L., De Angelis, C., Pacilio, M., Onori, S., Kacperek, A., and Piermattei, A. 2002. A correction method for diamond detector signal dependence with proton energy. *Med. Phys.* 29 (5):669–675.

Floyd, C. E., Jr., Bender, J. E., Sharma, A. C., Kapadia, A., Xia, J., Harrawood, B., Tourassi, G. D., Lo, J. Y., Crowell, A., and Howell, C. 2006. Introduction to neutron stimulated emission computed tomography. *Phys. Med. Biol.* 51 (14):3375–3390.

Fraass, B., Doppke, K., Hunt, M., Kutcher, G., Starkschall, G., Stern, R., and Van Dyke, J. 1998. American Association of Physicists in Medicine Radiation Therapy Committee Task Group 53: Quality assurance for clinical radiotherapy treatment planning. *Med. Phys.* 25 (10):1773–1829.

Gall, K., Desrosiers, M., Bensen, D., and Serago, C. 1996. Alanine EPR dosimeter response in proton therapy beams. *Appl. Radiat. Isot.* 47 (11–12):1197–1199.

Grusell, E., Isacsson, U., Montelius, A., and Medin, J. 1995. Faraday cup dosimetry in a proton therapy beam without collimation. *Phys. Med. Biol.* 40 (11):1831–1840.

Grusell, E., and Medin, J. 2000. General characteristics of the use of silicon diode detectors for clinical dosimetry in proton beams. *Phys. Med. Biol.* 45 (9):2573–2582.

Guo, P. Y., Adamovics, J. A., and Oldham, M. 2006. Characterization of a new radiochromic three-dimensional dosimeter. *Med. Phys.* 33 (5):1338–1345.

Gustavsson, H., Back, S. A., Medin, J., Grusell, E., and Olsson, L. E. 2004. Linear energy transfer dependence of a normoxic polymer gel dosimeter investigated using proton beam absorbed dose measurements. *Phys. Med. Biol.* 49 (17):3847–3855.

Chapter 6

Hartmann, B., Martisikova, M., and Jakel, O. 2010. Technical Note: Homogeneity of Gafchromic (R) EBT2 film. *Med. Phys.* 37 (4):1753–1756.

Henkner, K., Bassler, N., Sobolevsky, N., and Jakel, O. 2009. Monte Carlo simulations on the water-to-air stopping power ratio for carbon ion dosimetry. *Med. Phys.* 36 (4):1230–125.

Heufelder, J., Stiefel, S., Pfaender, M., Ludemann, L., Grebe, G., and Heese, J. 2003. Use of BANG polymer gel for dose measurements in a 68 MeV proton beam. *Med. Phys.* 30 (6):1235–1240.

Hishikawa, Y., Kagawa, K., Murakami, M., Sakai, H., Akagi, T., and Abe, M. 2002. Usefulness of positron-emission tomographic images after proton therapy. *Int. J. Radiat. Oncol. Biol. Phys.* 53 (5):1388–1391.

Hsi, W. C., Indelicato, D. J., Vargas, C., Duvvuri, S., Li, Z., and Palta, J. 2009c. *in vivo* verification of proton beam path by using posttreatment PET/CT imaging. *Med. Phys.* 36 (9):4136–4146.

Hsi, W. C., Moyers, M. F., Nichiporov, D., Anferov, V., Wolanski, M., Allgower, C. E., Farr, J. B., Mascia, A. E., and Schreuder, A. N. 2009b. Energy spectrum control for modulated proton beams. *Med. Phys.* 36 (6):2297–2308.

Hsi, W. C., Schreuder, A. N., Moyers, M. F., Allgower, C. E., Farr, J. B., and Mascia, A. E. 2009a. Range and modulation dependencies for proton beam dose per monitor unit calculations. *Med. Phys.* 36 (2):634–641.

International Atomic Energy Agency (IAEA). 2000. Absorbed dose determination in external beam radiotherapy: Technical Report 398. Vienna: IAEA.

IAEA. 2000. TRS-398: Absorbed dose determination in external beam radiotherapy: An international code of practice for dosimetry based on standards of absorbed dose to water. Vienna, Austria. http://wwwnaweb.iaea.org/nahu/dmrp/pdf_files/COPV11b.pdf.

IAEA. 2004. TRS No. 430, Commissioning and quality assurance of computerized planning systems for radiation treatment of cancer. In *Technical Reports Series*. Vienna: International Atomic Energy Agency.

Ibbott, G. 2008. The Radiological Physics Center TLD proton dosimetry credentialling program.

International Commission on Radiation Units and Measurements (ICRU). 1977. Report No. 26, Neutron dosimetry for biology and medicine. Bethesda, MD: International Commission on Radiation Units and Measurements.

ICRU. 1993. Report No 49, Stopping powers and ranges of protons and alpha particles. Bethesda, MD: International Commission on radiation Units and Measurements.

ICRU. 1998. Report No 59, Clinical Proton dosimetry—Part I: Beam production, beam delivery and measurement of absorbed dose. Bethesda, MD: International Commission on Radiation Units and Measurements.

ICRU. 2007. ICRU Report 78, Prescribing, recording and reporting proton-beam therapy. *J. ICRU* 7 (2).

Ing, H. 1986. The status of bubble damage polymer detectors. *Nucl. Tracks Rad. Meas.* 12:49–54.

Jirasek, A., and Duzenli, C. 2002. Relative effectiveness of polyacrylamide gel dosimeters applied to proton beams: Fourier transform Raman observations and track structure calculations. *Med. Phys.* 29 (4):569–577.

Kaiser, F. J., Bassler, N., and Jakel, O. 2010. COTS Silicon diodes as radiation detectors in proton and heavy charged particle radiotherapy—Part 1. *Radiat. Environ. Bioph.* 49 (3):365–371.

Kanai, T., Kawachi, K., Matsuzawa, H. and Inada, T. 1983. Broad beam three-dimensional irradiation for proton radiotherapy. *Med. Phys.* 10 (3):344–346.

Karger, C. P., Jakel, O., and Hartmann, G. H. 1999. A system for three-dimensional dosimetric verification of treatment plans in intensity-modulated radiotherapy with heavy ions. *Med. Phys.* 26 (10):2125–2132.

Kirby, D., Green, S., Palmans, H., Hugtenburg, R., Wojnecki, C., and Parker, D. 2010. LET dependence of GafChromic films and an ion chamber in low-energy proton dosimetry. *Phys. Med. Biol.* 55 (2):417–433.

Kirby, T. H., Hanson, W. F., Gastorf, R. J., Chu, C. H. and Shalek, R. J. 1986. Mailable TLD system for photon and electron therapy beams. *Int. J. Radiat. Oncol. Biol. Phys.* 12 (2):261–265.

Kirby, T. H., Hanson, W. F. and Johnston, D. A. 1992. Uncertainty analysis of absorbed dose calculations from thermoluminescence dosimeters. *Med. Phys.* 19 (6):1427–1433.

Knoll, G. F. 1979. *Radiation Detection and Methods*. New York: Wiley.

Knopf, A., Parodi, K., Bortfeld, T., Shih, H. A., and Paganetti, H. 2009. Systematic analysis of biological and physical limitations of proton beam range verification with offline PET/CT scans. *Phys. Med. Biol.* 54 (14):4477–4495.

Koehler, A. M., Schneider, R. J., and Sisterson, J. M. 1977. Flattening of proton dose distributions for large-field radiotherapy. *Med. Phys.* 4 (4):297–301.

Koehler, A. M., Schneider, R. J., and Sisterson, J. M. 1975. Range Modulators for protons and heavy ions. *Nucl. Instrum. Meth.* 131:437–440.

Kohno, R., Nishio, T., Miyagishi, T., Matsumura, K., Saito, H., Uzawa, N., Sasano, T., Nakamura, T., and Ogino, T. 2006. Evaluation of daily quality assurance for proton therapy at National Cancer Center Hospital East. *Igaku Butsuri* 26 (4):153–162.

Litzenberg, D. W., Roberts, D. A., Lee, M. Y., Pham, K., Vander Molen, A. M., Ronningen, R., and Becchetti, F. D. 1999. Online monitoring of radiotherapy beams: Experimental results with proton beams. *Med. Phys.* 26 (6):992–1006.

Lomax, A. J., Bohringer, T., Bolsi, A., Coray, D., Emert, F., Goitein, G., Jermann, M., Lin, S., Pedroni, E., Rutz, H., Stadelmann, O., Timmermann, B., Verwey, J., and Weber, D. C. 2004. Treatment planning and verification of proton therapy using spot scanning: Initial experiences. *Med. Phys.* 31 (11):3150–3157.

Lomax, A. J., Pedroni, E., Rutz, H., and Gotien, G. 2004. The clinical potential of intensity modulated proton therapy. *Med. Phys.* 14:147–152.

Lorin, S., Grusell, E., Tilly, N., Medin, J., Kimstrand, P., and Glimelius, B. 2008. Reference dosimetry in a scanned pulsed proton beam using ionisation chambers and a Faraday cup. *Phys. Med. Biol.* 53 (13):3519–3529.

Low, D. A., Harms, W. B., Mutic, S., and Purdy, J. A. 1998. A technique for the quantitative evaluation of dose distributions. *Med. Phys.* 25 (5):656–661.

Lu, H. M. 2008a. A potential method for *in vivo* range verification in proton therapy treatment. *Phys. Med. Biol.* 53 (5):1413–1424.

Lu, H. M. 2008b. A point dose method for *in vivo* range verification in proton therapy. *Phys. Med. Biol.* 53 (23):N415–N422.

Lu, H. M., and Kooy, H. 2006. Optimization of current modulation function for proton spread-out Bragg peak fields. *Med. Phys.* 33 (5):1281–1287.

Ma, L. J., Geis, P. B., and Boyer, A. L. 1997. Quality assurance for dynamic multileaf collimator modulated fields using a fast beam imaging system. *Med. Phys.* 24 (8):1213–1220.

Medin, J. 2010. Implementation of water calorimetry in a 180-MeV scanned pulsed proton beam including an experimental determination of k(Q) for a Farmer chamber. *Phys. Med. Biol.* 55 (12):3287–3298.

Mesoloras, G., Sandison, G. A., Stewart, R. D., Farr, J. B., and Hsi, W. C. 2006. Neutron scattered dose equivalent to a fetus from proton radiotherapy of the mother. *Med. Phys.* 33 (7):2479–2490.

Min, C. H., Kim, J. W., Youn, M. Y., and Kim, C. H. 2007. Determination of distal dose edge location by measuring right-angled prompt-gamma rays from a 38-MeV proton beam. *Nucl. Instrum. Meth.* A 580:562–565.

Mumot, M., Mytsin, G. V., Molokanov, A. G., and Malicki, J. 2009. The comparison of doses measured by radiochromic films and semiconductor detector in a 175-MeV proton beam. *Phys. Med.* 25 (3):105–110.

Nath, R., Biggs, P. J., Bova, F. J., Ling, C. C., Purdy, J. A., van de Geijn, J., and Weinhous, M. S. 1994. AAPM code of practice for radiotherapy accelerators: Report of AAPM Radiation Therapy Task Group No. 45. *Med. Phys.* 21 (7):1093–1121.

Nichiporov, D. 2003. Verification of absolute ionization chamber dosimetry in a proton beam using carbon activation measurements. *Med. Phys.* 30 (5):972–978.

Nichiporov, D., Kostjuchenko, V., Puhl, J. M., Bensen, D. L., Desrosiers, M. F., Dick, C. E., McLaughlin, W. L., Kojima, T., Coursey, B. M., and Zink, S. 1995. Investigation of applicability of alanine and radiochromic detectors to dosimetry of proton clinical beams. *Appl. Radiat. Isot.* 46 (12):1355–1362.

Nichiporov, D., Solberg, K., Hsi, W., Wolanski, M., Mascia, A., Farr, J., and Schreuder, A. 2007. Multichannel detectors for profile measurements in clinical proton fields. *Med. Phys.* 34 (7):2683–2690.

Niroomand-Rad, A., Blackwell, C. R., Coursey, B. M., Gall, K. P., Galvin, J. M., McLaughlin, W. L., Meigooni, A. S., Nath, R., Rodgers, J. E., and Soares, C. G. 1998. Radiochromic film dosimetry: Recommendations of AAPM Radiation Therapy Committee Task Group 55. American Association of Physicists in Medicine. *Med. Phys.* 25 (11):2093–2115.

Nishio, T., Miyatake, A., Ogino, T., Nakagawa, K., Saijo, N., and Esumi, H. 2010. The development and clinical use of a beam online PET system mounted on a rotating gantry port in proton therapy. *Int. J. Radiat. Oncol. Biol. Phys.* 76 (1):277–286.

Nishio, T., Ogino, T., Nomura, K., and Uchida, H. 2006. Dose-volume delivery guided proton therapy using beam on-line PET system. *Med. Phys.* 33 (11):4190–4197.

Oelfke, U., Lam, G. K., and Atkins, M. S. 1996. Proton dose monitoring with PET: Quantitative studies in Lucite. *Phys. Med. Biol.* 41 (1):177–196.

Oldham, M., Siewerdsen, J. H., Shetty, A., and Jaffray, D. A. 2001. High resolution gel-dosimetry by optical-CT and MR scanning. *Med. Phys.* 28 (7):1436–1445.

Olko, P., Bilski, P., Budzanowski, M., and Molokanov, A. 2004. Dosimetry of heavy charged particles with thermoluminescence detectors—Models and applications. *Radiat. Prot. Dosim.* 110 (1–4):315–318.

Olko, P., Bilski, P., Budzanowski, M. Waligorski, M. P. R., and Reitz, G. 2002. Modeling the response of thermoluminescence detectors exposed to low- and high-LET radiation fields. *J. Radiat. Res.* 43:S59–S62.

Olsher, R. H., Hsu, H. H., Beverding, A., Kleck, J. H., Casson, W. H., Vasilik, D. G., and Devine, R. T. 2000. WENDI: An improved neutron rem meter. *Health Phys.* 79 (2):170–181.

Olsher, R. H., Seagraves, D. T., Eisele, S. L., Bjork, C. W., Martinez, W. A., Romero, L. L, Mallett, M. W., Duran, M. A., and Hurlbut, C. R. 2004. PRESCILA: A new, lightweight neutron rem meter. *Health Phys.* 86 (6):603–612.

Onori, S., d'Errico, F., De Angelis, C., Egger, E., Fattibene, P., and Janovsky, I. 1997. Alanine dosimetry of proton therapy beams. *Med. Phys.* 24 (3):447–453.

Onori, S., De Angelis, C., Fattibene, P., Pacilio, M., Petetti, E., Azario, L., Miceli, R., Piermattei, A., Tonghi, L. B., Cuttone, G., and Lo Nigro, S. 2000. Dosimetric characterization of silicon and diamond detectors in low-energy proton beams. *Phys. Med. Biol.* 45 (10):3045–3058.

Oozeer, R., Mazal, A., Rosenwald, J. C., Belshi, R., Nauraye, C., Ferrand, R., and Biensan, S. 1997. A model for the lateral penumbra in water of a 200-MeV proton beam devoted to clinical applications. *Med. Phys.* 24 (10):1599–604.

Pacilio, M., De Angelis, C., Onori, S., Azario, L., Fidanzio, A., Miceli, R., Piermattei, A., and Kacperek, A. 2002. Characteristics of silicon and diamond detectors in a 60 MeV proton beam. *Phys. Med. Biol.* 47 (8):N107–N112.

Paganetti, H., and Gottschalk, B. 2002. Test of Monte Carlo nuclear interactions model for polyethylene (CH2) using a multi-layer faraday cup. *Med. Phys.* 29 (6):1328.

Palmans, H. 2003. Effect of alanine energy response and phantom material on depth dose measurements in ocular proton beams. *Technol. Cancer Res. T.* 2 (6):579–586.

Palmans, H., Seuntjens, J., Verhaegen, F., Denis, J. M., Vynckier, S., and Thierens, H. 1996. Water calorimetry and ionization chamber dosimetry in an 85-MeV clinical proton beam. *Med. Phys.* 23 (5):643–650.

Palmans, H., Thomas, R., Simon, M., Duane, S., Kacperek, A., DuSautoy, A., and Verhaegen, F. 2004. A small-body portable graphite calorimeter for dosimetry in low-energy clinical proton beams. *Phys. Med. Biol.* 49 (16):3737–3749.

Parodi, K., and Enghardt, W. 2000. Potential application of PET in quality assurance of proton therapy. *Phys. Med. Biol.* 45 (11):N151–N156.

Parodi, K., Enghardt, W. and Haberer, T. 2002. In-beam PET measurements of beta+ radioactivity induced by proton beams. *Phys. Med. Biol.* 47 (1):21–36.

Parodi, K., Paganetti, H., Shih, H. A., Michaud, S., Loeffler, J. S., DeLaney, T. F., Liebsch, N. J., Munzenrider, J. E., Fischman, A. J., Knopf, A. and Bortfeld, T. 2007. Patient study of *in vivo* verification of beam delivery and range, using positron emission tomography and computed tomography imaging after proton therapy. *Int. J. Radiat. Oncol. Biol. Phys.* 68 (3):920–934.

Patrick, J. W., Stephens, L. D., Thomas, R. H., and Kelly, L. S. 1976. The efficiency of 7LiF thermoluminescent dosimeters to high LET-particles, relative to 63Co gamma-rays. *Health Phys.* 30 (3):295–296.

Pawelke, J., Byars, L., Enghardt, W., Fromm, W. D., Geissel, H., Hasch, B. G., Lauckner, K., Manfrass, P., Schardt, D., and Sobiella, M. 1996. The investigation of different cameras for in-beam PET imaging. *Phys. Med. Biol.* 41 (2):279–296.

Piermattei, A., Miceli, R., Azario, L., Fidanzio, A., delle Canne, S., De Angelis, C., Onori, S., Pacilio, M., Petetti, E., Raffaele, L., and Sabini, M. G. 2000. Radiochromic film dosimetry of a low-energy proton beam. *Med. Phys.* 27 (7):1655–1660.

Pihet, P., Gueulette, J., Menzel, H. G., Grillmaier, R. E., and Wambersie, A. 1988. Use of microdosimetric data of clinical relevance in neutron therapy planning. *Radiat. Prot. Dosim.* 23:471–474.

Polf, J. C., Peterson, S., Ciangaru, G., Gillin, M., and Beddar, S. 2009a. Prompt gamma-ray emission from biological tissues during proton irradiation: A preliminary study. *Phys. Med. Biol.* 54 (3):731–743.

Chapter 6

Polf, J. C., Peterson, S., McCleskey, M., Roeder, B. T., Spiridon, A., Beddar, S., and Trache, L. 2009b. Measurement and calculation of characteristic prompt gamma ray spectra emitted during proton irradiation. *Phys. Med. Biol.* 54 (22):N519–N527.

Raju, M. R. 1966. The use of the miniature silicon diode as a radiation dosemeter. *Phys. Med. Biol.* 11 (3):371–6.

Raju, M. R. 1978. *Heavy Particle Radiotherapy*. New York: Academic Press.

Ramm, U., Weber, U., Bock, M., Kramer, M., Bankamp, A., Damrau, M., Thilmann, C., Bottcher, H. D., Schad, L. R., and Kraft, G. 2000. Three-dimensional BANG gel dosimetry in conformal carbon ion radiotherapy. *Phys. Med. Biol.* 45 (9):N95–N102.

Reft, C. S. 2009. The energy dependence and dose response of a commercial optically stimulated luminescent detector for kilovoltage photon, megavoltage photon, and electron, proton, and carbon beams. *Med. Phys.* 36 (5):1690–1699.

Richmond, R. G., Badhwar, G. D., Cash, B., and Atwell, W. 1987. Measurement of differential proton spectra onboard the Space Shuttle using a thermoluminescent dosimetry system. *Nucl Instrum. Meth. A* A256:393–397.

Rink, A., Lewis, D. F., Varma, S., Vitkin, I. A., and Jaffray, D. A. 2008. Temperature and hydration effects on absorbance spectra and radiation sensitivity of a radiochromic medium. *Med. Phys.* 35 (10):4545–4555.

Rosenfeld, A. B., Bradley, P. D., Cornelius, I., Allen, B. J., Zaider, M., Maughan, R. L., Yanch, J. C., Coderre, J., Flanz, J. B., and Kobayashi, T. 2002. Solid-state microdosimetry in hadron therapy. *Radiat. Prot. Dosim.* 101 (1–4):431–434.

Rosenfeld, A. B., Bradley, P. D., Cornelius, I., Kaplan, G. I., Allen, B. J., Flanz, J. B., Goitein, M., Van Meerbeec, Y., and Hayakawa, Y. 2000. A new silicon detector for microdosimetry applications in proton therapy. *IEEE Trans. Nucl. Sci.* 47 (4):1386–1394.

Rossi, H. H., and Rosenzweig, W. 1955. A device for the measurement of dose as a function of specific ionization. *Radiology* 64:404–411.

Rossi, H. H., and Zaider, M. 1996. *Microdosimetry and Its Applications*. Berlin, Germany: Springer-Verlag.

Sabini, M. G., Raffael, L., Bucciolini, M., Cirrone, G. A., Cuttone, G., Lo Nigro, S., Mazzocchi, S., Salamone, V. and Valastro, L. M. 2002. The use of thermoluminescent detectors for measurements of proton dose distribution. *Radiat. Prot. Dosim.* 101 (1–4):453–456.

Safai, S., Lin, S., and Pedroni, E. 2004. Development of an inorganic scintillating mixture for proton beam verification dosimetry. *Phys Med. Biol* 49 (19):4637–4655.

Sahoo, N., Zhu, X. R., Arjomandy, B., Ciangaru, G., Lii, M., Amos, R., Wu, R., and Gillin, M. T. 2008. A procedure for calculation of monitor units for passively scattered proton radiotherapy beams. *Med. Phys.* 35 (11):5088–5097.

Sarfehnia, A., Clasie, B., Chung, E., Lu, H. M., Flanz, J., Cascio, E., Engelsman, M., Paganetti, H. and Seuntjens, J. 2010. Direct absorbed dose to water determination based on water calorimetry in scanning proton beam delivery. *Med. Phys.* 37 (7):3541–3550.

Sassowsky, M., and Pedroni, E. 2005. On the feasibility of water calorimetry with scanned proton radiation. *Phys. Med. Biol.* 50 (22):5381–5400.

Sawakuchi, G. O., Sahoo, N., Gasparian, P. B. R., Rodriguez, M. G., Archambault, L., Titt, U., and Yukihara, E. G. 2010. Determination of average LET of therapeutic proton beams using Al2O3:C optically stimulated luminescence (OSL) detectors. *Phys. Med. Biol.* 55 (17):4963–4976.

Sawakuchi, G. O., Yukihara, E. G., McKeever, S. W. S., Benton, E. R., Gaza, R., Uchihori, Y., Yasuda, N., and Kitamura, H. 2008. Relative optically stimulated luminescence and thermoluminescence efficiencies of Al2O3:C dosimeters to heavy charged particles with energies relevant to space and radiotherapy dosimetry. *J. Appl. Phys.* 104 (12).

Seravalli, E., de Boer, M., Geurink, F., Huizenga, J., Kreuger, R., Schippers, J. M., van Eijk, C. W., and Voss, B. 2008. A scintillating gas detector for 2D dose measurements in clinical carbon beams. *Phys. Med. Biol.* 53 (17):4651–4665.

Seravalli, E., de Boer, M. R., Geurink, F., Huizenga, J., Kreuger, R., Schippers, J. M., and van Eijk, C. W. 2009. 2D dosimetry in a proton beam with a scintillating GEM detector. *Phys. Med. Biol.* 54 (12):3755–3771.

Siebers, J. V., Vatnitsky, S. M., Miller, D. W., and Moyers, M. F. 1995. Deduction of the air W-value in a therapeutic proton beam. *Phys. Med. Biol.* 40 (8):1339–1356.

Skopec, M., Loew, M., Price, J. L., Guardala, N., and Moscovitch, M. 2006. Discrimination of photon from proton irradiation using glow curve feature extraction and vector analysis. *Radiat. Prot. Dosim.* 120 (1–4):268–272.

Smith, A. R. 2009. Vision 20/20: Proton therapy. *Med. Phys.* 36 (2):556–568.

Spielberger, B., Scholz, M., Kramer, M., and Kraft, G. 2001. Experimental investigations of the response of films to heavy-ion irradiation. *Phys. Med. Biol.* 46 (11):2889–2897.

Stiefel, S., Heufelder, J., Pfaender, M., Ludemann, L., Grebe, G., and Heese, J. 2004. BANG-polymer dosimetry in the proton therapy of eye neoplasms. *Z. Med. Phys.* 14 (1):48–54 (in German).

Taylor, M. C., Phillips, G. C., and Young, R. C. 1970. Pion cancer therapy: Positron activity as an indicator of depth-dose. *Science* 169 (943):377–378.

Technical Reports Series No. 430: Commissioning and Quality Assurance of Computerized Planning Systems for Radiation Treatment of Cancer. 2004. Vienna.

Tomitani, T., Pawelke, J., Kanazawa, M., Yoshikawa, K., Yoshida, K., Sato, M., Takami, A., Koga, M., Futami, Y., Kitagawa, A., Urakabe, E., Suda, M., Mizuno, H., Kanai, T., Matsuura, H., Shinoda, I., and Takizawa, S. 2003. Washout studies of ¹¹C in rabbit thigh muscle implanted by secondary beams of HIMAC. *Phys. Med. Biol.* 48 (7):875–889.

van Luijk, P., van t' Veld, A. A. Zelle, H. D., and Schippers, J. M. 2001. Collimator scatter and 2-D dosimetry in small proton beams. *Phys. Med. Biol.* 46 (3):653–670.

Vatnitsky, S. M. 1997. Radiochromic film dosimetry for clinical proton beams. *Appl Radiat Isot* 48 (5):643–651.

Vatnitsky, S. M., Miller, D. W., Moyers, M. F., Levy, R. P., Schulte, R. W., Slater, J. D., and Slater, J. M. 1999. Dosimetry techniques for narrow proton beam radiosurgery. *Phys. Med. Biol.* 44 (11):2789–2801.

Vatnitsky, S. M., Siebers, J. V., and Miller, D. W. 1996. k(Q) factors for ionization chamber dosimetry in clinical proton beams. *Med. Phys.* 23 (1):25–31.

Vatnitsky, S., Miller, D., Siebers, J., and Moyers, M. 1995. Application of solid-state detectors for dosimetry of therapeutic proton beams. *Med. Phys.* 22 (4):469–473.

Vatnitsky, S., Siebers, J., Miller, D., Moyers, M., Schaefer, M., Jones, D., Vynckier, S., Hayakawa, Y., Delacroix, S., Isacsson, U., Medin, J., Kacperek, A., Lomax, A., Coray, A., Kluge, H., Heese, J., Verhey, L., Daftari, I., Gall, K., Lam, G., Beck, T., and Hartmann, G. 1996. Proton dosimetry intercomparison. *Radiother. Oncol.* 41 (2):169–177.

Verhey, L. J., Koehler, A. M., McDonald, J. C., Goitein, M., Ma, I. C., Schneider, R. J., and Wagner, M. 1979. The determination of absorbed dose in a proton beam for purposes of charged-particle radiation therapy. *Radiat. Res.* 79 (1):34–54.

Vynckier, S. 2004. Dosimetry of clinical neutron and proton beams: An overview of recommendations. *Radiat. Prot. Dosim.* 110 (1–4):565–572.

Vynckier, S., Bonnett, D. E., and Jones, D. T. 1991. Code of practice for clinical proton dosimetry. *Radiother. Oncol.* 20 (1):53–63.

Vynckier, S., Bonnett, D. E., and Jones, D. T. 1994. Supplement to the code of practice for clinical proton dosimetry. ECHED (European Clinical Heavy Particle Dosimetry Group). *Radiother. Oncol.* 32 (2):174–179.

Wang, X., Sahoo, N., Zhu, R. X., Zullo, J. R., and Gillin, M. T. 2010. Measurement of neutron dose equivalent and its dependence on beam configuration for a passive scattering proton delivery system. *Int. J. Radiat. Oncol. Biol. Phys.* 76 (5):1563–1570.

Yan, X., Titt, U., Koehler, A. M., and Newhauser, W. D. 2002. Measurement of neutron dose equivalent to proton therapy patients outside of the proton radiation field. *Nucl. Instrum. Meth. A* 476:429–434.

Yudelev, M., Hunter, S. and J. B. Farr. 2004. Thermoluminescence dosimetry in mixed neutron/gamma radiation beam. *Radiat. Prot. Dosim.* 110 (1–4):613–617.

Yukihara, E. G., Gaza, R., McKeever, S. W. S., and Soares, C. G. 2004. Optically stimulated luminescence and thermoluminescence efficiencies for high-energy heavy charged particle irradiation in Al2O3:C. *Radiat. Meas.* 38 (1):59–70.

Yukihara, E. G., and McKeever, S. W. S. 2006. Spectroscopy and optically stimulated luminescence of Al_2O_3:C using time-resolved measurements. *J. Appl. Phys.* 100 (8): 083512–083512-9.

Zeiden, O. A., Sriprisan, S. I., Lopatiuk-Tirpak, O., Kupelian, P. A., Meeks, S. L., Hsi, W. C., Li, Z., Palta, J. R., and Maryanski, M. J. 2010. Dosimetric evaluation of a novel polymer gel dosimeter for proton therapy. *Med. Phys.* 37 (5):2145–2152.

Zhao, L., and Das, I. J. 2010. Gafchromic EBT film dosimetry in proton beams. *Phys. Med. Biol.* 55 (10):N291–N301.

Zullo, J. R., Kudchadker, R. J., Zhu, X. R., Sahoo, N., and Gillin, M. T. 2010. LiF TLD-100 as a dosimeter in high-energy proton beam therapy—Can it yield accurate results? *Med. Dosim.* 35 (1):63–66.

Chapter 6

7. Image-Guided Proton and Carbon Ion Therapy

Lei Dong, Joey P. Cheung, and X. Ronald Zhu

7.1 Image-Guided Charged Particle Therapy Process

7.1.1 Introduction

The remarkable progress in radiation therapy over the last century has been largely due to our ability to more effectively focus and deliver radiation to the tumor. Both photon and charged particle therapies are targeted therapeutic methods for treating localized cancers. The goal is to deliver a lethal high dose to the clinical target volume (CTV) while sparing nearby normal tissues as much as possible. To achieve this goal, charged particle therapies, including proton and carbon ion, require accurate treatment delivery to the intended target. Therefore, it is not surprising that various imaging techniques have been explored to aid accurate target delineation and treatment delivery.

Imaging is employed at virtually every step of the radiation therapy process. This includes diagnosis, assessment of the extent of the disease, delineation of target regions to be irradiated and normal tissues to be protected, treatment planning, setup of patients and alignment of targets for treatment delivery, monitoring and quality assurance for treatment delivery, follow-up and assessment of response to treatments, and evaluation of the efficacy of treatment strategies and their subsequent refinement. Image guidance during treatment is used to minimize problems that arise from daily setup errors and anatomic changes. These include interfractional and intrafractional anatomic variations that have been and are still being identified during fractionated radiotherapy. The primary goal for image-guided radiotherapy (IGRT) is to ensure target accuracy prior to each treatment. Most of these procedures are similar to photon therapy or charged particle therapy. The main differences are in the specific implementations for image guidance in treatment delivery and interpretation of images.

Discrete online imaging systems have perhaps been more widely used for charged particle therapy in its

Proton and Carbon Ion Therapy Edited by C.-M. Charlie Ma and Tony Lomax © 2013 Taylor & Francis Group, LLC. ISBN: 978-1-4398-1607-3.

infancy than photon therapy. This is likely due to the dosimetric benefit for such implementation in conformal therapy. IGRT reduces setup uncertainties that otherwise could diminish the dosimetric benefit of charged particle therapy. Another important factor is that the portal images using charge particles are not physically possible (Miller 1995). Since therapeutic charge particles stop in the patient, there is nothing to image on the beam exit side of the patient (proton computed tomography will be discussed later in this chapter).

Charged particle therapy has a unique advantage over photon therapy due to its sharp distal fall-off, which can be used to spare normal tissues distal to the target volume. However, this unique feature introduces uncertainty in the location of the delivered high-dose region due to CT number/stopping power uncertainties and the potential mass change along the beam direction in day-to-day treatment. The location of the distal fall-off is strongly affected by radiological path-length variations for particle therapy beams due to patient anatomical and positional changes. Therefore, image guidance, immobilization, and perhaps adaptive radiotherapy are more important for charged particle therapy than for photon therapy. Additionally, hypofractionation may be a good choice for charged particle therapy due to its high relative biological effectiveness (RBE). In these cases, image guidance is even more important when fewer treatment fractions are to be used.

7.1.2   Components of Image Guidance

To understand the process of image guidance for treatment delivery, it is necessary to discuss three essential components in an image guided treatment procedure:

1. *In-room imaging*: In-room imaging is necessary to detect target deviations from the original treatment plan. Imaging is used as a sensor in this process to detect any possible setup errors and anatomical changes. Because the primary goal for in-room imaging is to detect positional changes from the original plan, in-room imaging usually has poor image quality in order to limit unnecessary dose to the patient. However, the image quality must still be good enough to serve its purpose in detecting gross changes in target position. Occasionally, high-quality volumetric images can be acquired using an in-room CT scanner; these CT images

could be used for replanning to correct for more complicated shape changes and tissue deformation in a patient's anatomy.

2. *Correction methods*: Image guidance necessitates a treatment intervention, which is to correct target positional deviations. The most common correction method is a simple couch shift, which corrects for the translational differences between the current target and the intended treatment target. A more advanced correction method is to use a 6-degrees-of-freedom (6D) robotic couch, which corrects for both translational shifts and small rotations of the body. For nonrigid changes, such as organ deformation and tumor shrinkage, online image-guided couch correction may not be sufficient. When a correction strategy involves a change in the original treatment plan, it is usually termed as "adaptive radiotherapy" instead of "image guidance." Adaptive radiotherapy, which will be discussed in more detail later, can correct for complicated geometric and volumetric changes in the target or nearby critical organs.

3. *Image registration and interpretation*: Image registration and interpretation is a computer-assisted decision-making process to determine the necessary corrections to bring the target closer to the original treatment plan. Image registration and alignment evaluation play a critical role in image-guided setup. The ideal image registration algorithm should be robust, accurate, and take full advantage of image information and available correction methods. Image registration also plays an important role in monitoring changes in patient's anatomy. When a perfect alignment becomes impossible, image interpretation becomes important to determine the best available correction.

These three components are synergistic towards the eventual goal for image-guided treatment delivery. Different IGRT approaches represent variations of these three essential components. We will discuss the details of typical clinical implementations later.

7.1.3   A Typical In–Room Imaging Setup for Charged Particle Therapy Systems

The first hospital-based proton therapy unit at the Loma Linda University Medical Center used kilovolt imaging for patient setup (Slater et al. 1998; Slater et al. 1992; Slater, Miller and Archambeau 1988). Similarly, modern charged particle therapy systems, including the IBA and Hitachi proton therapy facilities (Vargas et al. 2008; Smith et al. 2009), all use the kilovolt projection X-ray method for daily image guidance and treatment portal verification.

Figure 7.1 illustrates a typical kilovolt X-ray imaging system configuration, the Hitachi image-guided proton therapy system installed at the MD Anderson Cancer Center. In this rotating gantry room, the treatment nozzle can rotate around the patient and deliver proton beams in any coplanar angle. During imaging verification, the nozzle X-ray tube can slide into the beam line and provide a "beam's-eye-view" X-ray image. Nozzle X-rays can be used to verify beam-shaping devices such as block aperture; therefore, it is necessary for it to be located near the virtual source position as the proton beam. In addition to the nozzle X-ray imaging system, a separate gantry-mounted X-ray imaging system (the cage X-ray system) perpendicular to the nozzle X-ray beam direction is provided for orthogonal/stereoscopic X-ray imaging. Due to the fact that a proton/carbon therapy gantry is much bigger and heavier than a photon therapy gantry, minimizing gantry rotation is necessary to minimize wear and tear on the machine. A dual-source orthogonal X-ray imaging system can localize a 3-D object without the need to rotate the gantry and thus makes the imaging process more efficient.

Unlike traditional photon therapy systems, the nozzle X-ray source position on a charged particle therapy system is typically much farther away from the isocenter compared to the orthogonal cage X-ray source. Figure 7.2 explicitly shows the difference in the source-to-axis distance (SAD) between the in-beam (nozzle) X-ray and the orthogonal (cage) X-ray system. A typical nozzle X-ray is at a SAD of 2.7 mm, whereas the cage X-ray has a typical SAD of 1.0 mm, which is the same as linac-based photon therapy systems. Different X-ray source locations will require the treatment planning system to produce correct digitally reconstructed radiographs (DRRs) at the corresponding SAD or SID (source-to-imager distance) for each X-ray source. Due to this distinction, the traditional method of labeling the DRR by gantry angle is insufficient. Indication of X-ray source may be needed to match with the correct DRR. It is important to have an in-house convention to label the difference between the nozzle X-ray and the cage X-ray to avoid confusion.

Unlike photon therapy, most particle therapy treatment couches can provide six-degrees-of-freedom movement for more flexible setup corrections. These robotic treatment couches can typically correct translational

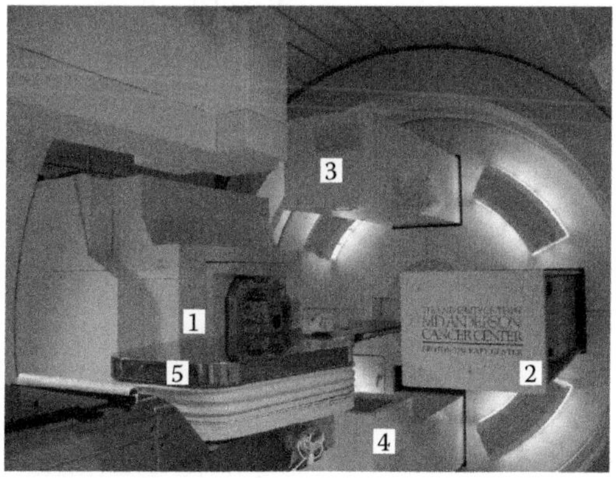

FIGURE 7.1 Typical in-room imaging configuration for image-guided proton therapy. The proton beam will come out from the nozzle (1), which is mounted on a gantry and can rotate 360° around a patient. When performing imaging guidance, the X-ray tube can slide into the proton beam line and provide a "beams-eye-view" for verification of either the beam-shaping device (block apertures) or the patient's anatomy relative to the beam axis. The image detector for the nozzle X-ray (2) is also mounted on the gantry and can extend out for image acquisition. The cage X-ray imaging system [(3) and (4)] is mounted on the gantry orthogonal to the proton beamline. A 6-degrees-of-freedom couch top (5) can shift the patient in three dimensions and can perform a small rotational correction around all three major axes.

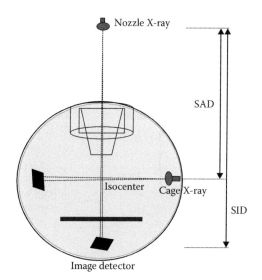

FIGURE 7.2 Diagram illustrating different X-ray imaging locations and distances in a typical charged particle therapy system. Because the virtual particle source position is much further than in a typical photon therapy system, the nozzle X-ray tube is usually much farther away from the isocenter than the orthogonal cage X-ray tube. Using the dual-X-ray imaging system, image guidance can be done using the simultaneous orthogonal imaging method without rotating the gantry.

shifts to within 1 mm accuracy and correct for small rotations up to 3°–6°. Note that it is not recommended to correct for large rotations because patients may feel unbalanced on an inclined surface resulting in unexpected patient immobilization issues. Additionally, when rotation is introduced, the sequential order of correction may be important when both translation and rotation are needed.

7.1.4 General Workflow for Image-Guided Particle Therapy

Daily image guidance can be considered as a closed-loop feedback system, which takes input image information and compares it to the previously defined treatment plan, and then makes the best available correction to the patient position or designs a new (adaptive) plan. This overall process is illustrated in Figure 7.3. The process starts with treatment simulation and initial treatment planning. Once the treatment plan is designed, the first goal of image guidance is to move the patient to correct for any setup errors in the patient position. The green lines in Figure 7.3 indicate the information flow of a rigid alignment process, in which orthogonal X-ray imaging is used to define target position relative to the treatment beam. Because projection X-rays typically cannot show the soft tissue target directly, it is common to use target surrogates such as bony landmarks or implanted fiducial markers to indicate the target location. It is important to note that proper image registration and geometric transformation algorithms are necessary to adequately obtain the correct type of corrections to reposition the patient. For example, 6D image registration algorithms should only be used for with a 6D-capable couch. Corrections will not be optimal if only the 3-D translational shifts are used for the correction but the image registration is performed using all 6 degrees of freedom from the rigid registration algorithm (Murphy 2007).

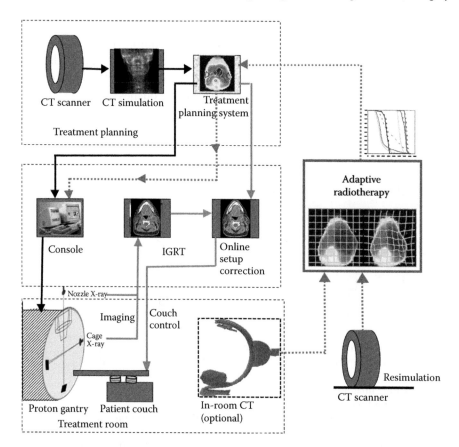

FIGURE 7.3 Workflow process for the image-guided charged particle therapy process. In the proposed workflow, image guidance to reposition the patient is shown in the green lines. In-room projection X-rays are typically equipped in charged particle therapy. Future designs will include volumetric imaging techniques, such as CBCT or CT-on-rails, which would allow for online adaptive radiotherapy to correct for nonrigid changes in patient's anatomy. Charged particle therapy demands both target positioning accuracy and constant radiological pathlength in the beam direction. Adaptive radiotherapy can be performed offline using a conventional CT scanner for resimulation.

One important distinction between photon therapy and charged particle therapy is the range uncertainty in the beam path. In charged particle therapy, anatomical changes outside the target may still be important for accurate treatment delivery. Due to this, a perfect target alignment under image guidance for charged particles may not be sufficient to guarantee target coverage. For this reason, IGRT for charged particle therapy should also include the verification of global anatomical changes of the patient, despite the fact that correction strategies may be limited due to practicality. Ideally, complicated anatomical changes should lead to online adaptive radiotherapy, which is defined as the modification of treatment plan to adapt to the changes in patient's anatomy prior to each treatment. However, time-trend anatomical changes, such as tumor shrinkage or weight loss, could be corrected using an offline adaptive radiotherapy approach with active monitoring of the patient on a daily or weekly basis.

7.1.5 Margins and Treatment Verification in the IGRT Process

Various uncertainties exist in treatment planning and treatment delivery for charged particle therapy. These uncertainties are usually handled by applying a large treatment margin to ensure target coverage during treatment. In photon therapy, the margin is usually handled by the concept of the planning target volume (PTV) because photon dose distributions are less dependent on the change in the patient's anatomy. For photon therapy, it is assumed that spatial dose distributions from the photon plan may not be noticeably affected by the geometric change in the target or patient's anatomy (Stroom et al. 1999; Cho et al. 2002; van Herk et al. 2000). Cho et al. (Cho et al. 2002) conducted a simulated study showing that, for the majority of clinical cases, change in photon dose distribution due to a small misplacement error of the target is negligible and that simple expansions from the CTV are adequate if a pure rigid setup error for the entire body is assumed. However, other studies have found that simple geometric expansions of the CTV are inadequate for proton therapy treatment planning if internal organ motion relative to the rest of anatomy occurs (Moyers et al. 2001; Engelsman and Kooy 2005). The difficulty of applying a geometric concept of the PTV to particle beam therapy is due to the fact that their dose distribution can vary when the patient's anatomy in the beam path is changed. The effect can be quite

significant if tissue heterogeneity is also present near the beam path. Setup error combined with tissue heterogeneity (e.g., air cavity, bony structures, or skin surface irregularities) can affect the position of the distal falloff for charged particle therapy. In addition to anatomical changes, there are other uncertainties, such as CT number/stopping power accuracy, which can affect the range of the planned treatment. Therefore, in addition to correcting for setup errors, IGRT should include the verification of anatomical changes and inherent uncertainties in stopping power calculations (Moyers et al. 2010; Schaffner and Pedroni 1998).

One solution to extend the concept of the PTV to charged particle therapy is the use of a beam-specific PTV (Park et al. 2012). For most current treatment techniques using passively scattered proton therapy, each treatment beam usually delivers a uniform dose to the entire target. In this situation, a beam-specific PTV can be designed to distinguish the impact of setup error and range uncertainties both lateral and parallel to the beam direction. This is illustrated in Figure 7.4 (left) for a treatment field A at an oblique angle. Global setup errors along the beam direction (shown by the red arrow) will have little impact on the dose distribution (assuming negligible range loss due to air scatter and minimal inverse-square beam divergence effect); therefore, the PTV due to setup error will only be affected if the error is in the lateral direction, as shown in the dotted volume in Figure 7.4 (left). However, range uncertainties along the beam direction will affect both the distal and proximal margins. The combined PTV is the oval shaped volume, which includes both setup error and range uncertainty. The range uncertainty is usually estimated in water-equivalent thickness. If there is tissue heterogeneity near the PTV region, the range uncertainty should be converted to an equivalent physical distance based on local tissue density in the region. This will make the proton PTV (including the range uncertainty) similar to the concept of photon PTV. A clinical example is shown on the right side of Figure 7.4.

Beam-specific PTVs in charged particle therapy can be treated as if they were the PTV in photon therapy. Target coverage can be evaluated by comparing the prescribed isodose line with the beam-specific PTV. Including range uncertainty in target verification is an important aspect for charged particle therapy. A perfectly and geometrically aligned target may not receive an adequate dose in charged particle therapy if there are significant density changes outside the target. IGRT in photon therapy is focused on target alignment. With

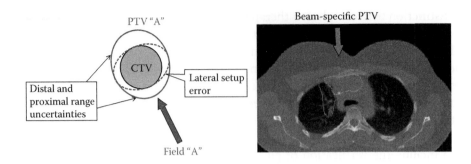

FIGURE 7.4 **(See color insert.)** Beam-specific PTV concept. To accommodate both setup errors and range uncertainties, a beam-specific PTV can be used for setup verification. Because the charged particle beam will not be affected by a setup error parallel to the beam direction, setup error will only contribute to a lateral margin, as indicated by the dotted red line. However, range uncertainties due to organ motion in the beam path as well as CT imaging uncertainties will make up the distal and proximal margins. The combined PTV is shown in the solid red line for Field A. The beam-specific PTV also depends on local tissue heterogeneities. An example of beam-specific PTV for a lung cancer patient is shown to the right. The beam angle is shown by the red arrow. The CTV is shown in green and the beam-specific PTV is shown in red.

IGRT in charged particle therapy, however, one to worry about both target alignment and range uncertainty outside the target (along the beam path).

The use of beam-specific PTV for target alignment is still a relatively new concept. Although beam-specific PTV may be a better approach to handle both setup errors and range uncertainties, true assessment of the particle beam range does require volumetric imaging inside a treatment room. Unfortunately, most existing charged particle therapy facilities do not currently provide in-room volumetric imaging. There is a strong demand for such capability because volumetric imaging is becoming popular in photon therapy. Clinical benefits of image guidance learned from photon therapy will help accelerate IGRT technologies for charged particle therapy.

7.2 Patient Immobilization

7.2.1 General Considerations

One major advantage of charged particle therapy is that there is almost no dose delivered beyond the end of range. However, the position of the distal falloff (at the end of range) depends on the consistency of the patient's anatomy along the beam path. Due to organ motion, weight loss, and normal day-to-day physiological changes in a patient's anatomy (e.g., bladder, rectum, and stomach fillings), the radiological path length may vary along the beam direction, which could strongly impact the accuracy of dose delivery, especially for high-RBE particle therapy beams. Therefore, immobilization for charged particle therapy is traditionally considered much more important for patient setup than in photon therapy. In addition to making a patient's position more consistent, special attention should also be paid when a particle beam passes through a part of the immobilization device.

To maintain a constant beam path length during day-to-day treatment, there are several general tips in selecting immobilization devices and beam angles for treatment:

1. Treatment beams should avoid penetrating through the immobilization device as much as possible. One reason is because the CT number for the immobilization device may not be accurately known and the material composition of the immobilization device may be different from human tissue. In addition, the position of the immobilization device may not be reproducible relative to the human body during day-to-day setup. It is preferred that charged particle beams should reach the patient's skin directly without penetrating a layer of the immobilization device. Sometimes the covering sheet (to help keep the patient warm) or patient's hair can cause a certain degree of range uncertainties in daily treatment. When a beam must decidedly go through an immobilization device, care should be taken to model the thickness, density, and composition of the material correctly to minimize range uncertainties.

2. The immobilization device should be made of smooth surfaces, absent of abrupt corners or high-density compositions if possible. When IGRT is used to set up a patient's internal target,

the immobilization device can deviate from its marked position, causing relative shifts of the immobilization device to the treatment beam. If the beam goes through the immobilization device, these irregular surfaces and density variations may perturb the dose distribution downstream in day-to-day repositioning of the patient.

3. Indexing the immobilization device relative to the treatment couch is very useful when setting up the device on the table. Immobilization devices can be labeled appropriately so that the patient can be set up consistently at the same location of the treatment couch. This will minimize the large variation of patient's initial position before an IGRT correction. Good indexing of the immobilization device can also help set up a patient more quickly as well.

The first two considerations are somewhat unique for charged particle therapy. The selection of immobilization device should also be considered based on potential treatment beam directions.

If the charged particle beam has to penetrate through the treatment couch, the geometry, thickness, and density of the couch should be carefully measured and included in the treatment planning process. Many immobilization devices, such as vacuum bags and face masks, are made of low density materials and are difficult to model accurately because CT numbers are typically not calibrated accurately for low density materials. Folds and wrinkles in the immobilization device, such as the vacuum bag, can also cause density irregularities. Therefore, it is best to minimize these uncertainties and try to avoid treating through immobilization devices. If these beam angles cannot be avoided, additional margins to compensate for these uncertainties should be considered when designing such treatment beams. Examples of patient immobilization are discussed in the following site-specific sections. They serve as examples of application for the principles described above.

7.2.2 Effect of Couch Edge

One good example for considering the dosimetric effect of an external object in the beam path is the edge of the treatment couch. At times, the optimal beam angle based on the tumor location and normal tissue considerations may go through the edge of the couch. Due to the difficulty in positioning the patient on the exact same location on the couch, the couch edge relative to the patient may vary on a daily basis. Image-guided setup (IGRT) will not reduce this problem because IGRT only focuses on aligning the target and usually ignores the relative patient position on the couch. Therefore, the couch edge can have a large day-to-day variation from its planned position after IGRT correction to the target.

Studies at MD Anderson Cancer showed that even with the indexing of immobilization device, the daily lateral couch shifts relative to the patient can be more than 1 cm (1 SD). Figure 7.5 shows an example of the dosimetric effect due to daily variation of the couch edge. The picture on the left shows the couch position during treatment simulation, from which a treatment plan was designed. The picture on the right shows a hypothetical 1-cm patient shift on the couch into this beam. Due to inconsistency in positioning the patient on the part of couch, the portion of the couch in the beam is 1 cm short, resulting in a beam overshoot towards the distal end of the target. The opposite situation occurs if the patient is shifted in the other direction. Although it is good practice to think about the daily variation of external device in the beam path, it is important to remember that the selection of beam

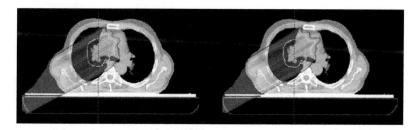

FIGURE 7.5 (See color insert.) Example showing the dosimetric impact of the couch edge if the patient is not exactly positioned on the same location of the couch. One of the treatment beams go through the edge of the couch before reaching the tumor target. Left: Nominal (planned) position with beam central axis going through couch edge. Right: Same beam setup as before but with the patient shifted 1 cm laterally on the couch towards the beam. As a result, the portion of the couch in the beam is 1 cm short, resulting beam overshoot in the distal end. The isocenter is set to the center of the CTV in both situations.

angles should not compromise the quality of a treatment plan. Sometimes it may be necessary to select beam angles that go through external devices in order to spare more normal tissues. If multiple beam angles are used, the range uncertainty caused by the variation of external device may not be as significant as shown in the single beam arrangement. Therefore, the couch edge effect should not be exaggerated. Nevertheless, it is still a good practice to think about the variation of external immobilization device during treatment planning.

It is worth noting that a digital couch model was used in the CT image for the patient in Figure 7.5. An in-house image processing software was used to remove the CT couch and replace it with the treatment couch, which was measured and modeled with the correct CT numbers and geometry at MD Anderson. If automatic image processing is not available, manual editing and density overrides are often used to correct the couch shape and density.

7.2.3 Immobilization for Prostate Patient Treatment

Prostate cancer is the most treated site in the United States for proton therapy. There are many strategies in setting up patients for prostate treatment, but the most common beam arrangements are laterally opposed beams. The use of lateral beams reduces range uncertainties due to irregular rectum or bladder fillings. Additionally, there are no major critical structures immediately lateral to the prostate.

Loma Linda University Medical Center was the first hospital-based proton facility to treat prostate cancer in the United States (Slater et al. 2004; Slater et al. 1998; Rossi et al. 1998). Patients are immobilized in a half cylinder of polyvinyl chloride with custom-shaped foam material. This immobilization device not only minimizes body movement but also maintains a constant distance from the surface to the distal target volume. A balloon is inserted into the rectum and filled with water to distend the posterior rectal wall away from the treatment field. Studies have shown that patients generally tolerated the rectal balloon well (Ronson et al. 2006; Bastasch et al. 2006). At the University of Florida, patients are immobilized with vacuum bags (not extending into the field), and beam angles are mostly anterior oblique fields or posterior oblique fields (Mendenhall et al. 2012; Vargas et al. 2009). A rectal balloon is used if the daily prostate movement

is greater than 5 mm (as measured by implanted fiducial markers) (Mendenhall et al. 2012). At the MD Anderson Cancer Center, patients are immobilized only to the lower extremities by using a knee and foot positioning device (CIVCO Medical Solutions, Orange City, IA). The positions of the legs have an impact on the consistency of the pelvic bone and femoral head positions, which will be in the beam path for lateral beam arrangements. Figure 7.6 shows the setup position and immobilization device for a typical case at the MD Anderson Cancer Center. A rectal balloon is used on all prostate patients to provide internal immobilization of the prostate.

Many studies have looked at the effectiveness of using rectal balloons for prostate radiation therapy. D'Amico et al. and McGary et al. showed that the prostate motion was reduced with an inserted rectal balloon for photon therapy (McGary et al. 2002; D'Amico et al. 2001). However, a conflicting study by Van Lin et al. reported that rectal balloons might not be as completely effective at immobilizing the prostate (Van Lin, Van Der Vight et al. 2005). They performed a comparative study with 22 patients using an endorectal balloon and 30 patients without a balloon. The interfraction prostate gland position was measured using implanted gold markers and stereoscopic X-ray imaging. It was found that the balloon did not reduce the day-to-day variation of the prostate relative to bony structures. It was also speculated that the presence of the rectal balloon can also cause additional rectal wall contraction and possible pelvic muscle tension alterations, causing prostate motion (Van Lin et al. 2005a; Husband et al. 1998). Although it is still controversial whether the rectal balloon can reduce interfractional prostate

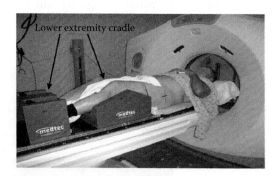

FIGURE 7.6 Prostate immobilization at MD Anderson Cancer Center. Patients are only immobilized to the lower extremities. The knee and foot positioning device help stabilize the pelvic bony structure. A rectal balloon (not shown) helps stabilize the prostate and rectum. Bilateral proton beams are typically used to treat prostate patients.

position, there is strong evidence that the use of a rectal balloon can result in a dosimetric benefit for the rectum by pushing the posterior rectal wall away from the high dose region (Vlachaki, Teslow and Ahmad 2007; Vargas et al. 2007; van Lin et al. 2007; D'Amico et al. 2006; Van Lin et al. 2005a; Hille et al. 2005; Sanghani et al. 2004; Patel et al. 2003; Teh et al. 2002; Teh et al. 2001). A study by van Lin et al. (van Lin et al. 2007) demonstrated reduced late mucosal changes when a rectal balloon was used in prostate radiotherapy. The endorectal balloon significantly reduced the rectal wall volume exposed to doses >40 Gy. Late rectal toxicity (grade ≥1, including excess of bowel movements and slight rectal discharge) was reduced significantly in the balloon group. A total of 146 endoscopies and 2336 mucosal areas were analyzed. Wachter et al. used three CT scans performed at the start, middle, and end of the treatment for a group of 10 consecutive patients with and without a rectal balloon (Wachter et al. 2002) and found that rectum filling variations ($p = 0.04$) and maximum anterior-posterior displacements of the prostate ($p = 0.008$) were reduced significantly with the use of a balloon, leading to a reduction in DVH variations during treatment.

Vargas et al. studied the dosimetric benefits of proton therapy when using a rectal balloon or simply 100 mL of saline water (Vargas et al. 2007). They found lowered doses to the rectum and the rectal wall in both situations. However, they found that only 33% of patients will have a small but significant advantage for using rectal balloon over the water alone technique.

They did not study the reproducibility of the water alone approach.

Figure 7.7 demonstrates the dosimetric advantage of using a rectal balloon in pushing the posterior rectum away from the high dose region. At MD Anderson Cancer Center, a specially designed rectal balloon for prostate immobilization is used. The inflated balloon has an asymmetric shape with a flat surface and groove to "cup" the posterior portion of the prostate (Figure 7.7c). The shaft to hold the balloon is softer than the typical rectal balloon used in MRI diagnostic imaging to decrease discomfort during insertion. The tip of the balloon also has a separate air communication hole to allow gas to pass through the balloon. The amount of water inserted is patient-specific (typically between 60 and 100 cc). Patients usually tolerate the balloon well.

7.2.4 Immobilization for Lung Patient Treatment

For cancers of the thorax, it is often necessary to simulate and treat the patient with their arms raised above their shoulders so that the treatment region can be directly exposed to the particle beams and to eliminate CT artifacts from the humerus bone. This setup is illustrated in Figure 7.8. A wing board (with a T-bar to let the patient hold their hands) is typically used with a small vacuum bag and head holder for shoulder and neck immobilization. The vacuum bag should not extend down into the thoracic region, nor should it extend beyond the position of the T-bar. A knee and

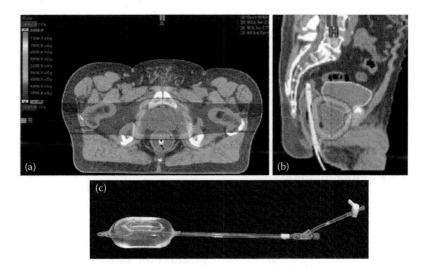

FIGURE 7.7 (See color insert.) Typical treatment plan for prostate treatment using bilateral passively scattered proton therapy. The posterior rectum was displaced away from the high dose region by the rectal balloon, as shown in the axial (a) and sagittal (b) planes. The rectal balloon was specially designed for prostate immobilization with asymmetric flat surface and groove (c).

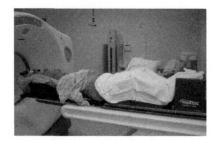

FIGURE 7.8 For lung cancer patients, a knee and foot positioning device are useful to stabilize the patient's lower body. A wing board with T-bar is used to position the patients in an arm-up position. A small vacuum bag and head holder is also used to immobilize the shoulders, head and neck.

foot positioning device is used to immobilize the lower body, although the foot cradle may not be necessary. For lung cancer treatment, it is often necessary to use posterior beams; therefore, the treatment couch should be modeled during treatment planning, as discussed earlier (Figure 7.5).

7.2.5 Immobilization for Craniospinal Irradiation

Craniospinal irradiation for pediatric and adult patients is an example of a special application of charged particle therapy. Many studies have demonstrated the clinical advantage of proton therapy for medulloblastoma (Fossati, Ricardi and Orecchia 2009; Miralbell, Lomax and Russo 1997; Archambeau et al. 1992; Suit et al. 2003). With a single posterior proton beam, the spine is treated to the full dose while essentially no dose is delivered to tissues anterior to the vertebral bodies. Because of convenience and reproducibility considerations, this technique is best applied for young patients in the supine position to allow easier administration of general anesthesia. Figure 7.9 illustrates the immobilization technique used at the MD

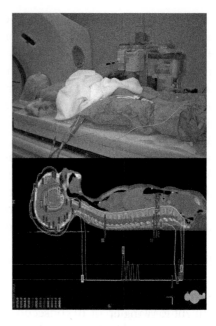

FIGURE 7.9 **(See color insert.)** Setup technique for craniospinal irradiation for pediatric patients. Top: Patient is immobilized with a vacuum bag for the legs and a head holder and thermoplastic face mask for the head and neck. Bottom: Dose distribution is highly conformal to the treatment site while delivering essentially no dose to the tissues anterior to the vertebral bodies.

Anderson Cancer Center. The patient is positioned on two sheets of Styrofoam with a vacuum bag to immobilize the patient's legs. The patient's head rests in a head holder with a thermoplastic face mask. The cranial fields use posterior oblique (typically 15° posterior from the lateral direction) proton beams, which are matched to the single posterior field. The slightly posterior beams provide an effective method delivering dose to the cribriform plate while sparing the global. The bottom of Figure 7.9 shows the dose distribution of the cranial boost field and the spine posterior treatment field. Junction shifts are applied to minimize setup uncertainties in the match lines. Daily image guidance is used to reduce setup error.

7.3 In-Room Image-Guided Technology

7.3.1 Stereoscopic kV X-Ray Imaging

Due to long construction times (5–7 years), most existing charged particle facilities were designed at least 5 years ago when commercial cone-beam computed tomography (CBCT)-based volumetric imaging did not exist. Therefore, the current standard for image guidance in charged particle therapy is stereoscopic kilovolt Xray imaging technique, as illustrated in

Figure 7.1. This technique is based on the stereoscopic principle that a 3-D (rigid) object can be identified from a pair of planar (2-D) images, which can be used to localize the tumor for treatment setup.

This adoption of kilovolt imaging in proton therapy parallels the rapid adoption of kilovolt imaging into routine clinical practice for photon therapy. Due to workflow simplicity, stereoscopic kilovolt X-ray image guidance is still the mainstream IGRT procedure for daily patient setup today and will likely be that way in the near future (Song et al. 2008; Kan et al. 2008; Soete et al. 2006; Linthout et al. 2006; Fuller et al. 2006; Fox et al. 2006; Sorcini and Tilikidis 2006; Huntzinger et al. 2006). Additionally, the radiation dose from kilovolt X-ray imaging is typically small (in the range of 0.01–0.1 cGy per image) (Song et al. 2008; Kan et al. 2008).

Stereoscopic kilovolt X-ray imaging can be used to align bony landmarks or implanted radio-opaque markers (fiducials) as surrogates of soft tissue targets. The limitation of this approach is that it ignores the soft tissue outside the target region, which is an important factor that affects range uncertainty in charged particle therapy. Nevertheless, projection kilovolt imaging has been found to be adequate for proton therapy treatment of the prostate by several studies and may be adequate for other sites as well (Vargas et al. 2008; Vargas et al. 2009; Wang et al. 2005; Arjomandy 2011; Daftari et al. 2010).

The major uncertainties in 2-D–2-D orthogonal X-ray imaging systems include (1) the assumption that the 3-D rigid target can be identified from its projections, and (2) the software capability to resolve out-of-plane rotations. In 2-D projection radiographs, the patient's anatomy along one direction is stacked into a single view. This makes it difficult (without prior knowledge) to distinguish whether an object in the image is proximal or distal to the isocenter. This projection ambiguity can lead to a large target localization error. The development of 2-D–3-D image registration algorithms can mitigate these uncertainties by taking advantage of the reference planning CT, which represents the entire 3-D volume of the patient. Reference DRRs can be created with various possible out-of-plane rotations and compared with individual 2-D projection images acquired during patient setup (Figl et al. 2010; Jans et al. 2006; Wu et al. 2009). Advancement in computer hardware, such as massive parallel graphics processing units, allows DRRs to be generated on-the-fly. Using this 2-D–3-D image registration technique, out-of-plane rotations and translational shifts can be simultaneously identified with the best matching DRRs. Unfortunately; few systems are using this newer computing technique.

7.3.2   Use of Implanted Fiducial Markers

Due to the inability of kilovolt stereoscopic X-ray imaging to detect soft tissue targets directly, target surrogates, in the form of implanted fiducial markers or surgical clips, are often used to detect target position.

These target surrogates need to be visible under X-ray projection imaging (radio-opaque), and should interfere with the therapeutic beam as little as possible.

The site that has been most studied regarding fiducial localization is the prostate. Radio-opaque markers can be directly implanted into the prostate and used to detect and monitor prostate positions during the course of treatment (Balter et al. 1995; Shimizu et al. 2000; Kitamura et al. 2002; Litzenberg et al. 2002; Langen et al. 2003; Van den Heuvel et al. 2003). Typically, three gold markers are implanted into the prostate under transrectal ultrasound guidance prior to CT simulation. Studies have shown that there is negligible seed migration within the prostate over the entire course of definitive radiotherapy (Poggi et al. 2003; Kupelian et al. 2005b; Dehnad et al. 2003; Pouliot et al. 2003). However, there are small, detectable movements in individual seed locations, perhaps resulting from topographic changes in the gland secondary to seed placement, anatomic changes in bladder and rectum, or treatment itself. Prostate edema may also deform the gland and make the seeds appear as if there has been migration. These seed migrations are much smaller than the shifts they provide to guide for aligning the prostate position. Multiple markers also provide redundancy in presenting the prostate position (Kudchadker et al. 2009).

In addition to fiducial seeds, linear fiducial markers have been developed to mark the prostate radiographically. One such example is the Visicoil™, which is made from a small flexible gold wire of 0.75 mm diameter and a fixed length (Visicoil™, Core Oncology, Santa Barbara, CA). The Visicoil™ can be implanted to each lobe of the prostate glands or in the periprostatic tissue (Letourneau et al. 2005; Gates et al. 2007; Teh et al. 2007).

One concern with the use of fiducial markers in charged particle therapy is the high atomic number (Z) in these fiducial markers, which are typically gold. High-density objects may perturb dose distributions along the beam path, potentially resulting in underdosage of the target (Newhauser et al. 2007). These high-Z materials are also known to cause significant CT imaging artifacts that can also cause dose calculation and target delineation errors. Cheung et al. studied the advantages of using a commercially available alternative lower-Z carbon-coated zirconium dioxide fiducial for proton therapy of the prostate (Cheung et al. 2010). They found that the use of gold fiducials for proton therapy resulted in unacceptably large dose perturbations while the carbon-coated zirconium

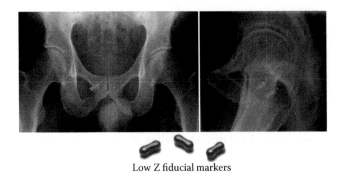

Low Z fiducial markers

FIGURE 7.10 Example of low-Z fiducial markers used in kilovolt X-ray image-guided prostate proton therapy. The low-Z fiducial marker (Carbon Medical Technologies, St. Paul, MN) is made of a zirconium dioxide (ZrO_2) core with a carbon coating outside with dimensions of 1 mm × 3 mm. The markers are clearly visible in X-ray radiographs, although not as visible as gold markers.

dioxide fiducials resulted in much smaller dose perturbations while also greatly reducing the CT artifacts. Studies have been performed to compare different types of fiducial markers, including Visicoil™, for proton therapy and their radiographic visibility (Huang et al. 2011; Giebeler et al. 2009). Figure 7.10 shows example radiographs with two low-Z fiducial markers used in prostate proton therapy. The 1 mm × 3 mm low-Z fiducial marker is made of a zirconium dioxide (ZrO_2) core with a carbon coating outside (Carbon Medical Technologies, St. Paul, MN). Although the visibility of the marker is not as good as gold markers, the low-Z marker can still be easily identified.

7.3.3 In-Room CT and CBCT

Three-dimensional "volumetric" imaging using computed tomography (CT) inside a treatment room represents the latest development in the IGRT armamentariums (Jaffray et al. 2002). With the patient immobilized on the treatment couch, true 3-D (voxel-by-voxel) information can be acquired with a CT scanner in the treatment room or an integrated gantry-mounted CBCT scanner just prior to the start of treatment. Such volumetric imaging information allows for more accurate evaluation of the patient's anatomy, particularly for soft tissue target. In-room CT images can be used to align the soft tissue target directly without using target surrogates (Court and Dong 2003). Perhaps more importantly for radiation therapy, in-room CT images can be used to calculate dose distributions based on anatomy captured in the treatment position, which allows for image-guided online or offline adaptive radiotherapy. Adaptive radiotherapy,

which will be discussed in more detail later, provides more comprehensive corrections than simple image guidance (Li et al. 2011; Court et al. 2005; Yan et al. 2005; Yan et al. 1998).

Unfortunately, the mainstream adoption of in-room CT for particle therapy is not as remarkable as for photon therapy. This is perhaps due to the long design and construction cycle for particle therapy facilities. Nevertheless, indications show that commercial vendors have plans to implement in-room CT/CBCT in their future products. Although in-room CT imaging is not currently available in most particle therapy facilities, several research institutions have implemented their own CT-image guided treatment procedures. At the National Institute of Radiological Sciences (NIRS) in Chiba, Japan, a fixed horizontal beam line is used for carbon-ion therapy (Kamada et al. 1999). In order to plan and verify patient treatment in seated position, a modified CT scanner (Xvision/GX, Toshiba, Tokyo) was installed inside the treatment room, which provides "horizontal" CT cross-section slices for patients in the vertical (seated) position. Figure 7.11a shows a picture of the scanner and the treatment chair. The gantry of this CT scanner was mounted horizontally and can be lifted vertically through a ceiling mounted mechanism during CT scanning. The axis of the CT scanner intersects with the horizontal beam line and

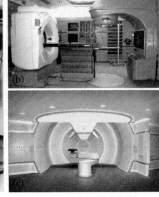

FIGURE 7.11 Examples of in-room CT scanners for CT-guided heavy-ion particle and photon therapies. (a) Vertical scanning CT scanner was installed at the National Institute of Radiological Sciences (NIRS), Chiba, Japan, for patients receiving treatment in seated position. (Kamada, T., H. Tsujii et al. 1999. A horizontal CT system dedicated to heavy-ion beam treatment. *Radiother. Oncol.* 50(2): 235–237.) (b) CT-on-rails system was installed at the MD Anderson Cancer Center with a linear accelerator for photon therapy. The system will share the same patient couch after a 180° couch rotation. (c) Prototype gantry-mounted CBCT imaging system for image-guided proton therapy (Varian Oncology Systems, Palo Alto, CA).

the rotation axis of the treatment chair. Because this is a modified commercial CT scanner, the CT scanning parameters are similar to a conventional CT scanner. The scanner has been used to align the patient with the horizontal beamline as well as to calculate dose distributions for planning and dose verification during the course of treatment.

There are several versions of in-room CT implementations. The first commercial CT-linac system was installed in 2000 at the Morristown Memorial Hospital, New Jersey (Wong et al. 2001). The system consists of a linac and a moveable CT scanner that can slide along a pair of rails (CT-on-rails). A unique feature of the system is that the couch remains stationary during the CT scan while the entire CT gantry moves on rails across the patient. A similar implementation was produced by Varian Oncology Systems (Palo Alto, CA) and installed at the MD Anderson Cancer Center (Figure 7.11b). The mechanical and operational accuracies of this system were evaluated by Court et al. (Court et al. 2003). With proper implementation, the system can achieve <1 mm accuracy in controlled phantom validations.

A kilovolt imaging system capable of radiography, fluoroscopy, and CBCT would be an attractive solution for IGRT applications. Recent effort in integrating such a system with a medical linear accelerator has produced enormous interest in proton therapy. CBCT imaging involves acquiring multiple kilovolt radiographs as the gantry rotates through at least 180° of rotation. Jaffray and his coworkers led the effort in implementing kilovolt cone-beam imaging with a large flat-panel detector (Jaffray et al. 1999; Jaffray and Siewerdsen 2000; Jaffray et al. 2000; Siewerdsen and Jaffray 2000; Jaffray and Siewerdsen 2001; Siewerdsen and Jaffray 2001; Jaffray et al. 2002; Siewerdsen et al. 2004). Kilovoltage X-rays are generated by a conventional (kilovolt) X-ray tube mounted on a retractable arm at 90° to the treatment beam line. A flat-panel X-ray detector is mounted opposite the kilovolt tube. Figure 7.11c shows a prototype implementation on Varian's latest proton therapy system, which will be available soon. An extension of CBCT is the 4-D CBCT, in which projections are sorted by the patient's respiratory signal before reconstruction. 4-D CBCT provides both 3-D and respiratory motion information, which can be used to improve image guidance for treating moving targets (Sonke et al. 2009; Sonke et al. 2005).

Some of the major limitations in CBCT-guided IGRT include (1) slow image acquisition and impact of intrafractional motion (in order to acquire multiple projection images, the gantry has to rotate around the patient), and (2) poor image quality in CBCT at low dose mode. Due to scatter (in CBCT geometry), in-room CBCT images do not have the same image quality as with diagnostic kilovolt CT images. At times, the soft tissue image quality can be so poor that they are insufficient to use for direct target alignment. Moseley et al. found that direct soft tissue targeting using CBCT only corrects 70% of cases within the 3 mm tolerance (Moseley et al. 2007). Inaccurate Hounsfield numbers in CBCT also presents a problem for accurate dose calculation in charged particle therapies. However, several technological developments are in progress to further improve image quality in CBCT. In-room CT-guided radiotherapy is considered as an ideal IGRT approach because images acquired can be directly compared with the planning CT.

Although it is not considered an in-room CT-guided application, a remote CT-based setup procedure was implemented at the Paul Scherrer Institute (PSI) in Villigen, Switzerland (Bolsi et al. 2008). The procedure uses a conventional CT scanner to provide daily setup correction relative to a docking couch system, which can be wheeled to the treatment room and detached and coupled to the treatment table. Optional in-room X-ray imaging can be performed to verify a patient's position. By using posttreatment repeat CT imaging, Bolsi et al. demonstrated that the systematic error was less than 0.6 mm and that random errors were below 1.5 mm for bite-block and 2.5 mm for regular mask immobilization for head and neck treatments. The residual setup error was 2.8 mm (SD) for abdominal/pelvic regions. The outside CT scanner reduced overall positioning time in the treatment room and thereby increased the beam utilization time. Residual setup error using this approach seems to be larger than in-room CT-guided setup; however, the process improved the overall workflow and can still provide patient-specific anatomic information for adaptive radiotherapy.

7.3.4 Proton Radiography, Proton CT, and Ultrasound Imaging Techniques

Unlike therapeutic X-ray beams, which penetrate through patient's body to form a transmission image and can be used to verify treatment field and beam aperture with the patient's anatomical information, proton beams do not usually exit the patient's body unless their energy is intentionally increased for the purpose of proton radiography. Proton radiography is a planar imaging technique, which uses a high-energy proton beam that carries both positional and range information when exiting through the patient body. However, due to multiple Coulomb scattering, proton radiographs typically have lower inherent spatial resolution than X-ray radiographs (Schneider and Pedroni 1994) and therefore have a harder time resolving beam shape and bony anatomy than X-ray radiographs. Nevertheless, the advantage of proton radiographs is that they can give important range information about the radiological path length of the beam. The information could be used to verify proton range in a patient. A recent simulation study showed that proton radiographs can produce range-enhanced anatomical images with much higher contrast than X-ray imaging, which could be used to verify internal target position and range variation. Depauw et al. used Monte Carlo simulation to compare X-ray radiographs with proton radiographs for a thoracic cancer patient (Depauw and Seco 2011). One example is shown in Figure 7.12. Results showed that the proton radiography could produce images with good spatial resolution and excellent soft tissue contrast, resulting in better tumor localization within a soft tissue background region such as the lung. However, proton radiography is not routinely used for target verification because practical issues in generating high energy beams, detecting such beams, and inherent scatter are still major obstacles to use this method in routine clinical practice.

Similar to proton radiography, proton CT (pCT) uses high-energy proton projection images (proton radiographs) for reconstructing computed tomography imaging. One unique advantage of pCT is its ability to measure proton stopping power ratios *in vivo* with much higher accuracy. It is well known that one of the major uncertainties in particle therapy is the uncertainty of the stopping power ratio measurement in patient using the X-ray CT imaging method (Schneider, Pedroni and Lomax 1996; Schneider et al. 2005; Moyers et al. 2001; Moyers et al. 2009). However,

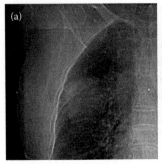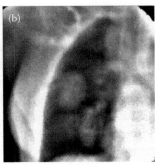

FIGURE 7.12 A comparison of an actual X-ray radiograph of a lung cancer patient (a) and simulated proton beam radiograph using a 490-MeV pencil scanning proton beam (b). (Depauw et al. 2011. *Phys. Med. Biol.* 56:2407–2421.)

it is still challenging to perform quantitative proton CT imaging and many groups are currently investigating its use, implementation, and application (Zygmanski et al. 2000; Johnson et al. 2003; De Assis et al. 2005; Schulte et al. 2005; Li et al. 2006; Petterson et al. 2007; Penfold et al. 2009).

Ultrasound is a noninvasive, nonradiographic, relatively easy, rapid, and real-time imaging technique for soft tissue targeting in radiotherapy. Ultrasound imaging was first used in radiation therapy for prostate localization in conjunction with the initial introduction of intensity modulated radiation therapy (IMRT) (Kuban et al. 2005; Molloy et al. 2011). The basic principles of this imaging technique limit its use to soft tissue targeting for tumors in the pelvic, abdominal, and breast locations. Unfortunately, ultrasound-based IGRT was not generally considered an accurate image guidance approach for many reasons, including inherently poor image quality, large interobserver variations (Langen et al. 2003), uncertainties with the probe pressure (McGahan, Ryu and Fogata 2004), different speeds of sound propagation in media, and so forth. A review of various uncertainties and recommended QA procedures can be found in the recent publication by the Task Group 154 Report of American Association of Physicists in Medicine (AAPM) (Molloy et al. 2011).

7.4 Image-Guided Adaptive Radiotherapy

7.4.1 Impact of Anatomical Changes and Adaptive Radiotherapy

There is a growing body of literature reporting data on interfractional and intrafractional variations and tumor shrinkage (Liu et al. 2007; Barker et al. 2004; Zhang et al. 2007; Wang, Willett and Yin 2007; Dieleman et al. 2007; de Crevoisier et al. 2007; Chen et al. 2007; Kavanagh, McGarry and Timmerman 2006; Jiang 2006; Hysing et al. 2006; Cai et al. 2007; Britton et al. 2007; Hui et al. 2008; Loo et al. 2011; Hansen et al. 2006; Kupelian et al. 2005a). With the growing popularity of volumetric image-guided radiation therapy, anatomical changes of patients can be reviewed on a daily basis. Some of these large anatomical changes discovered during routine treatment will prompt dosimetric evaluation, which will most likely lead to replanning (i.e., adaptive radiotherapy). As discussed earlier, particle therapy beams are much more sensitive to anatomy changes along the beam direction compared to photon therapy. Therefore, monitoring these changes using serial volumetric imaging is critical for particle therapy.

Figure 7.13 shows an example of such a case for a lung cancer proton therapy patient. After approximately 4 weeks of treatment, the patient's tumor shrank significantly with large cavities developed in the tumor volume. This large change of density caused the original proton beams to penetrate deeper into the normal tissue distal to the target, resulting in suboptimal dose distributions. The patient required a revised proton plan (i.e., using adaptive radiotherapy). However, it is important to consider the type of imaging to use for adaptive radiotherapy for charged particles since charged particle beams are more sensitive to CT number uncertainties and inaccuracies. CT numbers are usually calibrated in a conventional CT scanner with reasonable accuracy. However, CBCT images can have highly variable CT numbers due to patient size and difference in beam hardening effects of the kilovolt X-ray source. In addition, motion artifacts in CBCT can cause additional errors in CT number or stopping

power conversion. As a result, images acquired from CBCT images may not be directly usable for dose calculation in charged particle therapy. Figure 7.14 illustrates a lung cancer patient who received two CT scans on the same treatment day. The top panel shows the conventional CT scan and the bottom panel shows the CBCT scan. An IMRT plan (left panels) and a proton plan (right panels) were calculated on this set of CT images. It is apparent that the IMRT dose distribution calculated on the CBCT is similar to the IMRT plan calculated on the conventional CT. However, the proton dose distribution calculated on the CBCT shows notable differences from the same proton plan calculated on the conventional CT. This example illustrates that protons or charged particle therapies are more sensitive to CT number inaccuracies and that CBCT images may not be directly used for proton dose calculation. Additionally, care should be taken to ensure that the correct CT number to stopping power ratio calibration curves are used when calculating the dose on a CT image.

Nevertheless, in-room CT or CBCT is generally not available in most current charged particle therapy facilities. Experience from photon therapy will help in identifying patients who might experience large anatomical changes during their course of treatment. The use of the simulation CT scanner for midcourse or weekly anatomy assessment is perhaps a practical

Original proton plan | Dose recalculated on fourth week's anatomy

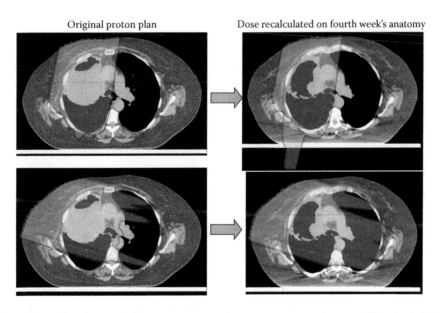

FIGURE 7.13 Significant tumor shrinkage and density variations in lung cancer proton therapy will lead to inferior dose distributions. In this example, the original two-beam passively scatter proton plan is shown on the left. After approximately 4 weeks of treatment, the same beams are used to recalculate the dose on the anatomy shown to the right. Tumor shrinkage and cavitation has led to an overshoot of the proton beams, resulting in a significant increase in dose to the normal tissues distal to the target.

Chapter 7

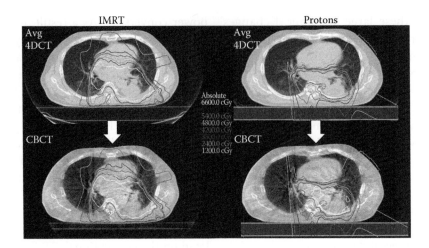

FIGURE 7.14 Example of dose calculation on a conventional CT image (top panel) and a CBCT image (bottom panel) of the same patient on the same day using intensity modulated photon therapy (IMRT; left column) and proton therapy (right column). Inaccurate CT numbers and motion artifacts in CBCT caused more dose deviations in the proton plan (bottom right panel) than in the photon plan (bottom left panel).

approach because tumor shrinkage is typically a slow and trending event. At the MD Anderson Cancer Center, all lung cancer patients receive a midcourse repeat CT scan to evaluate internal anatomy changes. Selected lung cancer patients participating in clinical trials also received weekly 4-D computed tomography (4DCT) scans. Large changes in anatomy will trigger a dosimetric evaluation, which may or may not lead to a new adaptive plan. In clinical practice, the use of deformable image registration to map the original target volume and critical structures onto the new anatomy has proven to be effective (Gao et al. 2006; Lu et al. 2006; Wang et al. 2008; Brock 2010; Wang et al. 2005). Although deformable image registration may not be perfectly accurate, they generally reduce the overall contouring time for replanning. Contouring from scratch on repeat CT images may not be practical.

7.4.2 Four-Dimensional Dose Calculation and Motion Management

In addition to day-to-day changes in patient anatomy, intrafractional internal organ motion will also have a large impact in charged particle therapy. To aid treatment planning and dosimetric evaluation of mobile tumor, 4DCT can be used to explicitly measure the temporal movement of internal organs (Vedam et al. 2003; Keall 2004; Pan et al. 2004; Rietzel et al. 2005; Rietzel, Pan and Chen 2005). Conventional CT scans performed using a single free-breathing CT image set usually only captures a random snapshot of a breathing

phase. Therefore, conventional free-breathing CT scans can be very unpredictable for tumors or normal tissues that have large motion.

To quantify the actual dose a patient receives, a "4-D" dose calculation is needed. A 4-D dose distribution represents the composite of dose distributions delivered in the sequence of respiratory phases. In order to calculate a 4-D dose, it is necessary to map dose distributions computed for each phase of the 4DCT onto a reference 3-D computed tomography (3DCT) study. This is accomplished by using a deformable mapping algorithm to map the dose distributions from each individual phase to a reference phase, which is typically the end-expiration phase. The details of the 4-D dose calculation method are described in a paper by Kang et al. (Kang et al. 2007).

An example of the 4-D dose distribution computed in this manner is illustrated in Figure 7.15. The top left figure shows a treatment plan that was designed on a free-breathing CT scan. Each of the breathing phases are indicated by the T-phase, from T0 (inspiration) to T50 (expiration) and going back to T90 (close to the inspiration phase again). In order to simulate dose delivery on each of the breathing phases, the same treatment plan is calculated on each T0-T90 phases to generate dose distribution at each individual phase. The deformable mapping technique is then applied to generate the actual cumulative dose distribution when the same plan (developed using a free-breathing 3DCT) is used for all breathing phases of the 4DCT. The final dose distributions, as represented by the yellow prescription dose line, are different in each phase

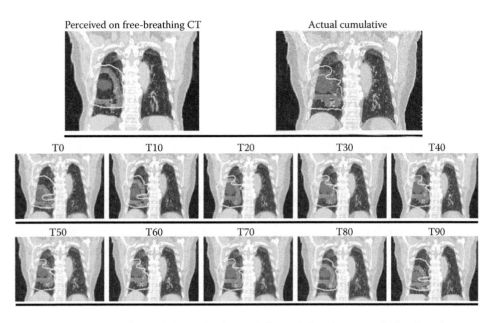

FIGURE 7.15 Dose distribution perceived on a proton plan (upper left panel) based on a single free-breathing CT, may be different from the actual dose distribution after accounting for respiratory motion. Dose distributions may also be different for each phase of the breathing cycle (bottom two rows). In this illustration, the treatment plan designed for the free-breathing CT was applied to each phase of a 4-D CT image set. T0 represents the end-inhale phase and T50 represents the end-exhale phase. The actual cumulative dose distribution (upper right panel) combines all 10 phases using deformable mapping. The prescription isodose lines are shown in light color. The GTV and PTV are shown in small (darker) and large (lighter) shaded areas.

of respiration, indicating the impact of organ motion to the proton dose distribution. The large panel in the upper right of Figure 7.15 shows the cumulative dose distribution. For comparison, the upper left panel shows the dose distribution that would be computed if only the free-breathing CT were available. Note that failure to account for respiratory motion can result in underdosage to the CTV, overdose to adjacent normal tissues, or both. Therefore, accounting for organ motion is an important part of treatment planning for treating a moving target.

In practice, 4-D dose calculation can be cumbersome and is generally not available on existing commercial treatment planning systems. To design compensators in passive scatter proton therapy to account for organ motion, 4DCT scans need to be acquired to quantify the 3-D motion. One strategy is to design separate compensators for the inspiration and expiration phases, and then combine the compensator matrix by accommodating the deepest distal depth of the target (Engelsman, Rietzel and Kooy 2006). Alternatively, Kang et al. proposed an empirical method that has been validated with 4-D dose calculation (Kang et al. 2007). This method uses the time-averaged 4DCT (calculated by averaging 10 breathing phase CTs voxel-by-voxel) as the base CT image for computing the proton dose distribution. This time-averaged CT images

accounts for the average motion effect if the patient is to be treated in a natural/free-breathing condition. In addition, Kang et al. found that overriding the density of the internal gross tumor volume (iGTV) will provide a further margin to cover the movement of the high-density GTV without overdosing the normal tissue (Kang et al. 2007).

Typical motion management methods used in photon therapy, such as margin-based motion accommodation, gating and breath-hold, can also be used for charged particle therapies. However, due to irregular breathing and patient compliance, gating and breath-hold usually carry large uncertainties and extra margins may be needed.

Although conventional motion management may be useful for passive scattered particle therapy, it is important to note that motion management for scanning particle therapy presents a significant challenge. Actively scanned beams could be out of sync with the internal motion of the patient and therefore the interplay between the scanning motion and the target motion may significantly impact the dose delivered. The precise synchronization of spot position, beam energy, and patient's motion can present a significant technological challenge for the manufacturer and the patient compliance for breathing regularity. Newer technologies, such as fast raster scanning systems (Furukawa et al. 2007;

Chapter 7

Furukawa et al. 2010), and repainting may reduce the impact of organ motion. However, accurate motion sensing and irregular breathing motion can still present a practical challenge. Issues related to scanning particle therapy were discussed in a special workshop and published by Knopf et al. (Knopf et al. 2010).

7.5 Summary and Future Development

To take full advantage of charged particle therapy in sparing normal tissues, tight margins are necessary. Therefore, it is not surprising that IGRT was used in charged particle therapies routinely from the beginning in the clinical practice. However, advanced techniques in IGRT, such as CBCT and various real-time motion management strategies, have not been as well developed or implemented for particle beam therapy as in photon therapy. Nevertheless, protons or other heavy ion therapies demand high-precision treatment delivery and may benefit from these advanced IGRT techniques. The potential benefits of IGRT far outweigh the technical challenges and workflow limitations.

It is important to note that immobilization plays a bigger role in charged particle therapies. Consistency in body shape and anatomy are more important than in photon therapy. The impact of the beam passing through immobilization devices will also have a larger impact to the distal dose conformality.

Range uncertainties in patients are a major concern for charged particle therapies. This is why volumetric imaging and repeat imaging are vital to assess the patient's anatomical changes on a routine basis. We expect that adaptive radiotherapy using repeat CT scans will be quite routine in future clinical practice.

Since dose distributions are highly affected by organ motion and anatomical changes, tools such as deformable image registration will be developed and used for computing cumulative dose distributions. In return, more accurate estimation of delivered dose will result in more accurate dose response curves, which will further enhance our knowledge about charged particle therapy. With further technology development, we expect that online replanning may be possible, which would further improve our ability to manage interfractional motion on a fraction-by-fraction basis.

Motion management for scanning particle therapy perhaps presents the biggest challenge. Intensity and energy modulated scanning particle therapy provides superior dose conformality than passive scatter technology. However, this new technology is more sensitive to anatomical changes and organ motion. Various approaches are proposed, which include margin-based approaches, rescanning, gating, beam tracking, and so forth. However, each approach has its own limitations and uncertainties associated with the technology. Better modeling of motion and faster hardware in delivering the scanning beam are needed.

In reality, the selection of an appropriate image-guidance solution is a complex process that is a compromise of clinical objective, product availability, existing infrastructure, workflow, and manpower. The deployment of a new technology requires a thorough understanding of the complete clinical process and the necessary infrastructure to support data collection, analysis, and intervention. IGRT in charged particle therapy provides unique challenges due to the inherent characteristics of charged particle beams. Understanding various sources of uncertainties is important; and user training is critical for the proper use of IGRT.

References

Archambeau, J. O., Slater, J. D., Slater, J. M., and Tangeman, R. 1992. Role for proton beam irradiation in treatment of pediatric CNS malignancies. *Int. J. Radiat. Oncol. Biol. Phys.* 22 (2):287–294.

Barker, J. L. Jr., Garden, A. S., Ang, K. K., O'Daniel, J. C., Wang, H., Court, L. E., Morrison, W. H., Rosenthal, D. I., Chao, K. S. C., Tucker, S. L., Mohan, R., and Dong, L. 2004. Quantification of volumetric and geometric changes occurring during fractionated radiotherapy for head-and-neck cancer using an integrated CT/linear accelerator system. *Int. J. Radiat. Oncol. Biol. Phys.* 59 (4):960–970.

Bastasch, M. D., Teh, B. S., Mai, W. Y., McGary, J. E., and Grant, W. H., 3rd, and Butler, E. B. 2006. Tolerance of endorectal balloon in 396 patients treated with intensity-modulated radiation therapy (IMRT) for prostate cancer. *Am. J. Clin. Oncol.* 29 (1):8–11.

Bolsi, A., Lomax, A. J., Pedroni, E., Goitein, G., and Hug, E. 2008. Experiences at the Paul Scherrer Institute with a remote patient positioning procedure for high-throughput proton radiation therapy. *Int. J. Radiat. Oncol. Biol. Phys.* 71 (5):1581–1590.

Britton, K. R., Starkschall, G., Tucker, S. L., Pan, T., Nelson, C., Chang, J. Y., Cox, J. D., Mohan, R., and Komaki, R. 2007. Assessment of gross tumor volume regression and motion changes during radiotherapy for non-small-cell lung cancer as measured by four-dimensional computed tomography. *Int. J. Radiat. Oncol. Biol. Phys.* 68 (4):1036–1046.

Brock, K. K. 2010. Results of a multi-institution deformable registration accuracy study (MIDRAS). *Int. J. Radiat. Oncol. Biol. Phys.* 76 (2):583–596.

Cai, J., Read, P. W., Altes, T. A., Molloy, J. A., Brookeman, J. R., and Sheng, K. 2007. Evaluation of the reproducibility of lung motion probability distribution function (PDF) using dynamic MRI. *Phys. Med. Biol.* 52 (2):365–373.

Chen, J., Lee, R. J., Handrahan, D. and Sause, W. T. 2007. Intensity-modulated radiotherapy using implanted fiducial markers with daily portal imaging: Assessment of prostate organ motion. *Int. J. Radiat. Oncol. Biol. Phys.* 68 (3):912–919.

Cheung, J., Kudchadker, R. J., Zhu, X. R., Lee, A. K., and Newhauser, W. D. 2010. Dose perturbations and image artifacts caused by carbon-coated ceramic and stainless steel fiducials used in proton therapy for prostate cancer. *Phys. Med. Biol.* 55 (23):7135–7147.

Cho, B. C., van Herk, M., Mijnheer, B. J., and Bartelink, H. 2002. The effect of set-up uncertainties, contour changes, and tissue inhomogeneities on target dose-volume histograms. *Med. Phys.* 29 (10):2305–18.

Court, L. E., and Dong, L. 2003. Automatic registration of the prostate for computed-tomography-guided radiotherapy. *Med. Phys.* 30 (10):2750–2757.

Court, L., Rosen, I., Mohan, R., and Dong, L. 2003. Evaluation of mechanical precision and alignment uncertainties for an integrated CT/LINAC system. *Med. Phys.* 30 (6):1198–1210.

Court, L. E., Dong, L., O'Daniel, J., Wang, H., Mohan, R., Lee, A. K., Cheung, R., Bonnen, M. D., and Kuban, D. 2005. An automatic CT-guided adaptive radiation therapy technique by online modification of multileaf collimator leaf positions for prostate cancer. *Int. J. Radiat. Oncol. Biol. Phys.* 62 (1):154–163.

D'Amico, A. V., Manola, J., Loffredo, M., Lopes, L., Nissen, K., O'Farrell, D. A., Gordon, L., Tempany, C. M., and Cormack, R. A. 2001. A practical method to achieve prostate gland immobilization and target verification for daily treatment. *Int. J. Radiat. Oncol. Biol. Phys.* 51 (5):1431–1436.

D'Amico, A. V., Manola, J., McMahon, E., Loffredo, M., Lopes, L., Ching, J., Albert, M. et al. 2006. A prospective evaluation of rectal bleeding after dose-escalated three-dimensional conformal radiation therapy using an intrarectal balloon for prostate gland localization and immobilization. *Urology* 67 (4):780–784.

Daftari, I. K., Mishra, K. K., O'Brien, J. M., Tsai, T., Park, S. S., Sheen, M., and Phillips, T. L. 2010. Fundus image fusion in EYEPLAN software: An evaluation of a novel technique for ocular melanoma radiation treatment planning. *Med. Phys.* 37 (10):5199–207.

De Assis, J. T., Yevseyeva, O., Evseev, I., Lopes, R. T., Schelin, H. R., Klock, M. C. L., Paschuk, S. A., Schulte, R. W., and Williams, D. C. 2005. Proton computed tomography as a tool for proton therapy planning: Preliminary computer simulations and comparisons with x-ray CT basics. *X-Ray Spectrom.* 34 (6):481–492.

de Crevoisier, R., Melancon, A. D., Kuban, D. A., Lee, A. K., Cheung, R. M., Tucker, S. L., Kudchadker, R. J., Newhauser, W. D., Zhang, L., Mohan, R., and Dong, L. 2007. Changes in the pelvic anatomy after an IMRT treatment fraction of prostate cancer. *Int. J. Radiat. Oncol. Biol. Phys.* 68 (5):1529–1536.

Dehnad, H., Nederveen, A. J., van der Heide, U. A., van Moorselaar, R. J., Hofman, P., and Lagendijk, J. J. 2003. Clinical feasibility study for the use of implanted gold seeds in the prostate as reliable positioning markers during megavoltage irradiation. *Radiother. Oncol.* 67 (3):295–302.

Depauw, N., and Seco, J. 2011. Sensitivity study of proton radiography and comparison with kV and MV x-ray imaging using GEANT4 Monte Carlo simulations. *Phys. Med. Biol.* 56 (8):2407–2421.

Dieleman, E. M., Senan, S., Vincent, A., Lagerwaard, F. J., Slotman, B. J., and van Sornsen de Koste, J. R. 2007. Four-dimensional computed tomographic analysis of esophageal mobility during normal respiration. *Int. J. Radiat. Oncol. Biol. Phys.* 67 (3):775–780.

Engelsman, M., and Kooy, H. M. 2005. Target volume dose considerations in proton beam treatment planning for lung tumors. *Med. Phys.* 32 (12):3549–3557.

Engelsman, M., Rietzel, E. and Kooy, H. M. 2006. Four-dimensional proton treatment planning for lung tumors. *Int. J. Radiat. Oncol. Biol. Phys.* 64 (5):1589–1595.

Figl, M., Bloch, C., Gendrin, C., Weber, C., Pawiro, S. A., Hummel, J., Markelj, P., Pernus, F., Bergmann, H., and Birkfellner, W. 2010. Efficient implementation of the rank correlation merit function for 2D/3D registration. *Phys. Med. Biol.* 55 (19):N465–N471.

Fossati, P., Ricardi, U., and Orecchia, R. 2009. Pediatric medulloblastoma: Toxicity of current treatment and potential role of protontherapy. *Cancer Treat. Rev.* 35 (1):79–96.

Fox, T. H., Elder, E. S., Crocker, I. R., Davis, L. W., Landry, J. C., and Johnstone, P. A. 2006. Clinical implementation and efficiency of kilovoltage image-guided radiation therapy. *J. Am. Coll. Radiol.* 3 (1):38–44.

Fuller, C. D., Thomas, C. R., Schwartz, S., Golden, N., Ting, J., Wong, A., Erdogmus, D., and Scarbrough, T. J. 2006. Method comparison of ultrasound and kilovoltage x-ray fiducial marker imaging for prostate radiotherapy targeting. *Phys. Med. Biol.* 51 (19):4981–4993.

Furukawa, T., Inaniwa, T., Sato, S., Shirai, T., Mori, S., Takeshita, E., Mizushima, K., Himukai, T., and Noda, K. 2010. Moving target irradiation with fast rescanning and gating in particle therapy. *Med. Phys.* 37 (9):4874–4879.

Furukawa, T., Inaniwa, T., Sato, S., Tomitani, T., Minohara, S., Noda, K., and Kanai, T. 2007. Design study of a raster scanning system for moving target irradiation in heavy-ion radiotherapy. *Med. Phys.* 34 (3):1085–1097.

Gao, S., Zhang, L., Wang, H., De Crevoisier, R., Kuban, D. D., Mohan, R., and Dong, L. 2006. A deformable image registration method to handle distended rectums in prostate cancer radiotherapy. *Med. Phys.* 33 (9):3304–3312.

Gates, L. L., Gladstone, D. J., Kasibhatla, M. S., Marshall, J. F., Seigne, J. D., Hug, E., and Hartford, A. C. 2007. Prostate Localization Using Serrated Gold Coil Markers. *Int. J. Radiat. Oncol. Biol. Phys.* 69 (3, Suppl. 1):S382–S382.

Giebeler, A., Fontenot, J., Balter, P., Ciangaru, G., Zhu, R., and Newhauser, W. 2009. Dose perturbations from implanted helical gold markers in proton therapy of prostate cancer. *J. Appl. Clin. Med. Phys.* 10 (1):2875.

Hansen, E. K., Bucci, M. K., Quivey, J. M., Weinberg, V., and Xia, P. 2006. Repeat CT imaging and replanning during the course of IMRT for head-and-neck cancer. *Int. J. Radiat. Oncol. Biol. Phys.* 64 (2):355–362.

Hille, A., Schmidberger, H., Tows, N., Weiss, E., Vorwerk, H., and Hess, C. F. 2005. The impact of varying volumes in rectal balloons on rectal dose sparing in conformal radiation therapy of prostate cancer. A prospective three-dimensional analysis. *Strahlenther Onkol.* 181 (11):709–716.

Huang, J. Y., Newhauser, W. D., Zhu, X. R., Lee, A. K., and Kudchadker, R. J. 2011. Investigation of dose perturbations and the radiographic visibility of potential fiducials for proton radiation therapy of the prostate. *Phys. Med. Biol.* 56 (16):5287–5302.

Chapter 7

Hui, Z., Zhang, X., Starkschall, G., Li, Y., Mohan, R., Komaki, R., Cox, J. D., and Chang, J. Y. 2008. Effects of interfractional motion and anatomic changes on proton therapy dose distribution in lung cancer. *Int. J. Radiat. Oncol. Biol. Phys.* 72 (5):1385–1395.

Huntzinger, C., Munro, P., Johnson, S., Miettinen, M., Zankowski, C., Ahlstrom, G., Glettig, R. et al. 2006. Dynamic targeting image-guided radiotherapy. *Med. Dosim.* 31 (2):113–125.

Husband, J. E., Padhani, A. R., MacVicar, A. D, and Revell, P. 1998. Magnetic resonance imaging of prostate cancer: comparison of image quality using endorectal and pelvic phased array coils. *Clin. Radiol.* 53 (9):673–681.

Hysing, L. B., Kvinnsland, Y., Lord, H., and Muren, L. P. 2006. Planning organ at risk volume margins for organ motion of the intestine. *Radiother. Oncol.* 80 (3):349–354.

Jaffray, D. A., Drake, D. G., Moreau, M., Martinez, A. A., and Wong, J. W. 1999. A radiographic and tomographic imaging system integrated into a medical linear accelerator for localization of bone and soft-tissue targets. *Int. J. Radiat. Oncol. Biol. Phys.* 45 (3):773–789.

Jaffray, D. A., and Siewerdsen, J. H. 2000. Cone-beam computed tomography with a flat-panel imager: Initial performance characterization. *Med. Phys.* 27 (6):1311–1323.

Jaffray, D. A., Siewerdsen, J. H., Wong, J. W., and Martinez, A. A. 2002. Flat-panel cone-beam computed tomography for image-guided radiation therapy. *Int. J. Radiat. Oncol. Biol. Phys.* 53 (5):1337–1349.

Jaffray, D. A., and Siewerdsen, J. H. 2001. A volumetric cone-beam CT system based on a 41 × 41 cm2 flat-panel imager. Paper read at Proceedings of SPIE—The International Society for Optical Engineering, 2001.

Jaffray, D. A., Siewerdsen, J. H., Edmundson, G. E., Wong, J. W., and Martinez, A. 2000. Cone-beam CT: Applications in image-guided external beam radiotherapy and brachytherapy. *Annual International Conference of the IEEE Engineering in Medicine and Biology—Proceedings* 3:2044.

Jans, H. S., Syme, A. M., Rathee, S., and Fallone, B. G. 2006. 3D interfractional patient position verification using 2D-3D registration of orthogonal images. *Med. Phys.* 33 (5):1420–1439.

Jiang, S. B. 2006. Radiotherapy of mobile tumors. *Semin Radiat Oncol* 16 (4):239–248.

Johnson, L., Keeney, B., Ross, G., Sadrozinski, H. F. W., Seiden, A., Williams, D. C., Zhang, L., Bashkirov, V., Schulte, R. W., and Shahnazi, K. 2003. Initial studies on proton computed tomography using a silicon strip detector telescope. *Nucl. Instrum. Meth. A* 514 (1–3):215–223.

Kamada, T., Tsujii, H., Mizoe, J. E., Matsuoka, Y., Tsuji, H., Osaka, Y., Minohara, S., Miyahara, N., Endo, M., and Kanai, T. 1999. A horizontal CT system dedicated to heavy-ion beam treatment. *Radiother. Oncol.* 50 (2):235–237.

Kan, M. W., Leung, L. H., Wong, W., and Lam, N. 2008. Radiation dose from cone beam computed tomography for image-guided radiation therapy. *Int. J. Radiat. Oncol. Biol. Phys.* 70 (1):272–279.

Kang, Y., Zhang, X., Chang, J. Y., Wang, H., Wei, X., Liao, Z., Komaki, R. et al. 2007. 4D Proton treatment planning strategy for mobile lung tumors. *Int. J. Radiat. Oncol. Biol. Phys.* 67 (3):906–914.

Kavanagh, B. D., McGarry, R. C., and Timmerman, R. D. 2006. Extracranial radiosurgery (stereotactic body radiation therapy) for oligometastases. *Semin. Radiat. Oncol.* 16 (2):77–84.

Keall, P. 2004. 4-dimensional computed tomography imaging and treatment planning. *Semin. Radiat. Oncol.* 14 (1):81–90.

Kitamura, K., Shirato, H., Seppenwoolde, Y., Onimaru, R., Oda, M., Fujita, K., Shimizu, S., Shinohara, N., Harabayashi, T., and Miyasaka, K. 2002. Three-dimensional intrafractional movement of prostate measured during real-time tumor-tracking radiotherapy in supine and prone treatment positions. *Int. J. Radiat. Oncol. Biol. Phys.* 53 (5):1117–1123.

Knopf, A., Bert, C., Heath, E., Nill, S., Kraus, K., Richter, D., Hug, E., et al. 2010. Special report: Workshop on 4D-treatment planning in actively scanned particle therapy—recommendations, technical challenges, and future research directions. *Med. Phys.* 37 (9):4608–4614.

Kuban, D. A., Dong, L., Cheung, R., Strom, E., and De Crevoisier, R. 2005. Ultrasound-based localization. *Semin. Radiat. Oncol.* 15 (3):180–191.

Kudchadker, R. J., Lee, A. K., Yu, Z. H., Johnson, J. L., Zhang, L., Zhang, Y., Amos, R. A., Nakanishi, H., Ochiai, A., and Dong, L. 2009. Effectiveness of using fewer implanted fiducial markers for prostate target alignment. *Int. J. Radiat. Oncol. Biol. Phys.* 74 (4):1283–1289.

Kupelian, P. A., Ramsey, C., Meeks, S. L., Willoughby, T. R., Forbes, A., Wagner, T. H., and Langen, K. M. 2005a. Serial megavoltage CT imaging during external beam radiotherapy for non-small-cell lung cancer: observations on tumor regression during treatment. *Int. J. Radiat. Oncol. Biol. Phys.* 63 (4):1024–1028.

Kupelian, P. A., Willoughby, T. R., Meeks, S. L., Forbes, A., Wagner, T., Maach, M. and Langen, K. M. 2005b. Intraprostatic fiducials for localization of the prostate gland: Monitoring inter-marker distances during radiation therapy to test for marker stability. *Int. J. Radiat. Oncol. Biol. Phys* 62 (5):1291–1296.

Langen, K. M., Pouliot, J., Anezinos, C., Aubin, M., Gottschalk, A. R., Hsu, I. C., Lowther, D. et al. 2003. Evaluation of ultrasound-based prostate localization for image-guided radiotherapy. *Int. J. Radiat. Oncol. Biol. Phys.* 57 (3):635–644.

Letourneau, D., Martinez, A. A., Lockman, D., Yan, D., Vargas, C., Ivaldi, G., and Wong, J. 2005. Assessment of residual error for online cone-beam CT-guided treatment of prostate cancer patients. *Int. J. Radiat. Oncol. Biol. Phys.* 62 (4):1239–1246.

Li, T., Liang, Z., Singanallur, J. V., Satogata, T. J., Williams, D. C., and Schulte, R. W. 2006. Reconstruction for proton computed tomography by tracing proton trajectories: A Monte Carlo study. *Med. Phys.* 33 (3):699–706.

Li, T., Thongphiew, D., Zhu, X., Lee, W. R., Vujaskovic, Z., Yin, F. F., and Wu, Q. J. 2011. Adaptive prostate IGRT combining online re-optimization and re-positioning: A feasibility study. *Phys. Med. Biol.* 56 (5):1243–1258.

Linthout, N., Verellen, D., Tournel, K., and Storme, G. 2006. Six dimensional analysis with daily stereoscopic x-ray imaging of intrafraction patient motion in head and neck treatments using five points fixation masks. *Med. Phys.* 33 (2):504–513.

Litzenberg, D., Dawson, L. A., Sandler, H., Sanda, M. G., McShan, D. L., Ten Haken, R. K., Lam, K. L., Brock, K. K., and Balter, J. M. 2002. Daily prostate targeting using implanted radiopaque markers. *Int. J. Radiat. Oncol. Biol. Phys.* 52 (3):699–703.

Liu, H. H., Balter, P., Tutt, T., Choi, B., Zhang, J., Wang, C., Chi, M. et al. 2007. Assessing respiration-induced tumor motion and internal target volume using four-dimensional computed tomography for radiotherapy of lung cancer. *Int. J. Radiat. Oncol. Biol. Phys.* 68 (2):531–540.

Loo, H., Fairfoul, J., Chakrabarti, A., Dean, J. C., Benson, R. J., Jefferies, S. J., and Burnet, N. G. 2011. Tumour shrinkage and contour change during radiotherapy increase the dose to organs at risk but not the target volumes for head and neck cancer patients treated on the TomoTherapy HiArt system. *Clin. Oncol.* 23 (1):40–47.

Lu, W., Olivera, G. H., Chen, Q., Ruchala, K. J., Haimerl, J., Meeks, S. L., Langen, K. M., and Kupelian, P. A. 2006. Deformable registration of the planning image (kVCT) and the daily images (MVCT) for adaptive radiation therapy. *Phys. Med. Biol.* 51 (17):4357–4374.

McGahan, J. P., Ryu, J., and Fogata, M. 2004. Ultrasound probe pressure as a source of error in prostate localization for external beam radiotherapy. *Int. J. Radiat. Oncol. Biol. Phys.* 60 (3):788–793.

McGary, J. E., Teh, B. S., Butler, E. B., and Grant, W. 3rd. 2002. Prostate immobilization using a rectal balloon. *J. Appl. Clin. Med. Phys.* 3 (1):6–11.

Mendenhall, N. P., Li, Z., Hoppe, B. S., Marcus, R. B. Jr., Mendenhall, W. M., Nichols, R. C., Morris, C. G., Williams, C. R., Costa, J., and Henderson, R. 2012. Early outcomes from three prospective trials of image-guided proton therapy for prostate cancer. *Int. J. Radiat. Oncol. Biol. Phys.* 82 (1):213–221.

Miller, D. W. 1995. A review of proton beam radiation therapy. *Med. Phys.* 22 (11 Pt 2):1943–1954.

Miralbell, R., Lomax, A. and Russo, M. 1997. Potential role of proton therapy in the treatment of pediatric medulloblastoma/primitive neuro-ectodermal tumors: Spinal theca irradiation. *Int. J. Radiat. Oncol. Biol. Phys.* 38 (4):805–11.

Molloy, J. A., Chan, G., Markovic, A., McNeeley, S., Pfeiffer, D., Salter, B., and Tome, W. A. 2011. Quality assurance of U.S.-guided external beam radiotherapy for prostate cancer: Report of AAPM Task Group 154. *Med. Phys.* 38 (2):857–871.

Moseley, D. J., White, E. A., Wiltshire, K. L., Rosewall, T., Sharpe, M. B., Siewerdsen, J. H., Bissonnette, J. P. et al. 2007. Comparison of localization performance with implanted fiducial markers and cone-beam computed tomography for on-line image-guided radiotherapy of the prostate. *Int. J. Radiat. Oncol. Biol. Phys.* 67 (3):942–953.

Moyers, M. F., Miller, D. W., Bush, D. A., and Slater, J. D. 2001. Methodologies and tools for proton beam design for lung tumors. *Int. J. Radiat. Oncol. Biol. Phys.* 49 (5):1429–1438.

Moyers, M. F., Sardesai, M., Sun, S., and Miller, D. W. 2009. Ion stopping powers and CT numbers. *Med. Dosim.* 35 (3):179–194.

Repeated Author. 2010. Ion stopping powers and CT numbers. *Med. Dosim.* 35 (3):179–194.

Murphy, M. J. 2007. Image-guided patient positioning: If one cannot correct for rotational offsets in external-beam radiotherapy setup, how should rotational offsets be managed? *Med. Phys.* 34 (6):1880–1883.

Newhauser, W., Fontenot, J., Koch, N., Dong, L., Lee, A., Zheng, Y., Waters, L., and Mohan, R. 2007. Monte Carlo simulations of the dosimetric impact of radiopaque fiducial markers for proton radiotherapy of the prostate. *Phys. Med. Biol.* 52 (11):2937–2952.

Pan, T., Lee, T. Y., Rietzel, E., and Chen, G. T. Y. 2004. 4D-CT imaging of a volume influenced by respiratory motion on multislice CT. *Med. Phys.* 31 (2):333–340.

Park, P. C., Zhu, X. R., Lee, A. K., Sahoo, N., Melancon, A. D., Zhang, L., and Dong, L. 2012. A beam-specific planning target volume (PTV) design for proton therapy to account for setup and range uncertainties. *Int. J. Radiat. Oncol. Biol. Phys.* 82 (2):e329–e336.

Patel, R. R., Orton, N., Tome, W. A., Chappell, R., and Ritter, M. A. 2003. Rectal dose sparing with a balloon catheter and ultrasound localization in conformal radiation therapy for prostate cancer. *Radiother. Oncol.* 67 (3):285–294.

Penfold, S. N., Rosenfeld, A. B., Schulte, R. W., and Schubert, K. E. 2009. A more accurate reconstruction system matrix for quantitative proton computed tomography. *Med. Phys.* 36 (10):4511–4518.

Petterson, M., Blumenkrantz, N., Feldt, J., Heimann, J., Lucia, D., Seiden, A., Williams, D. C. et al. 2007. Proton radiography studies for proton CT. Paper read at IEEE Nuclear Science Symposium Conference Record.

Poggi, M. M., Gant, D. A., Sewchand, W., and Warlick, W. B. 2003. Marker seed migration in prostate localization. *Int. J. Radiat. Oncol. Biol. Phys.* 56 (5):1248–1251.

Pouliot, J., Aubin, M., Langen, K. M., Liu, Y. M., Pickett, B., Shinohara, K., and Roach, M. 3rd. 2003. (Non)-migration of radiopaque markers used for on-line localization of the prostate with an electronic portal imaging device. *Int. J. Radiat. Oncol. Biol. Phys.* 56 (3):862–866.

Rietzel, E., Chen, G. T. Y. Choi, N. C., and Willet, C. G. 2005. Four-dimensional image-based treatment planning: Target volume segmentation and dose calculation in the presence of respiratory motion. *Int. J. Radiat. Oncol. Biol. Phys.* 61 (5):1535–1550.

Rietzel, E., Pan, T. S., and Chen, G. T. Y. 2005. Four-dimensional computed tomography: Image formation and clinical protocol. *Med. Phys.* 32 (4):874–889.

Ronson, B. B., Yonemoto, L. T., Rossi, C. J., Slater, J. M., and Slater, J. D. 2006. Patient tolerance of rectal balloons in conformal radiation treatment of prostate cancer. *Int. J. Radiat. Oncol. Biol. Phys.* 64 (5):1367–1370.

Rossi, C. J. Jr., Slater, J. D., Reyes-Molyneux, N., Yonemoto, L. T., Archambeau, J. O., Coutrakon, G., and Slater, J. M. 1998. Particle beam radiation therapy in prostate cancer: Is there an advantage? *Sem. Radiat. Oncol.* 8 (2):115–123.

Sanghani, M. V., Ching, J., Schultz, D., Cormack, R., Loffredo, M., McMahon, E., Beard, C., and D'Amico, A. V. 2004. Impact on rectal dose from the use of a prostate immobilization and rectal localization device for patients receiving dose escalated 3D conformal radiation therapy. *Urol. Oncol.* 22 (3):165–168.

Schaffner, B., and Pedroni, E. 1998. The precision of proton range calculations in proton radiotherapy treatment planning: experimental verification of the relation between CT-HU and proton stopping power. *Phys. Med. Biol.* 43 (6):1579–1592.

Schneider, U., and E. Pedroni. 1994. Multiple coulomb scattering and spatial resolution in proton radiography. *Med. Phys.* 21 (11):1657–1663.

Schneider, U., Pedroni, E., and Lomax, A. 1996. The calibration of CT Hounsfield units for radiotherapy treatment planning. *Phys. Med. Biol.* 41 (1):111–124.

Schneider, U., Pemler, P., Besserer, J., Pedroni, E., Lomax, A., and Kaser-Hotz, B. 2005. Patient specific optimization of the relation between CT-hounsfield units and proton stopping power with proton radiography. *Med. Phys.* 32 (1):195–199.

Schulte, R. W., Bashkirov, V., Klock, M. C. L., Li, T., Wroe, A. J., Evseev, I., Williams, D. C. and Satogata, T. 2005. Density resolution of proton computed tomography. *Med. Phys.* 32 (4):1035–1046.

Shimizu, S., Shirato, H., Kitamura, K., Shinohara, N., Harabayashi, T., Tsukamoto, T., Koyanagi, T., and Miyasaka, K. 2000. Use of an implanted marker and real-time tracking of the marker for the positioning of prostate and bladder cancers. *Int. J. Radiat. Oncol. Biol. Phys.* 48 (5):1591–1597.

Siewerdsen, J. H., and Jaffray, D. A. 2000. Optimal x-ray imaging geometry for flat-panel cone-beam computed tomography. *Annual International Conference of the IEEE Engineering in Medicine and Biology—Proceedings* 1:106–107.

Siewerdsen, J. H., and Jaffray, D. A. 2001. Cone-beam computed tomography with a flat-panel imager: Magnitude and effects of x-ray scatter. *Med. Phys.* 28 (2):220–231.

Chapter 7

Siewerdsen, J. H., Moseley, D. J., Jaffray, D. A., Richard, S., and Bakhtiar, B. 2004. The influence of antiscatter grids on soft-tissue detectability in cone-beam computed tomography with flat-panel detectors. *Med. Phys.* 31 (12):3506–3520.

Slater, J. D., Rossi, C. J. Jr., Yonemoto, L. T., Bush, D. A., Jabola, B. R., Levy, R. P., Grove, R. I., Preston, W., and Slater, J. M. 2004. Proton therapy for prostate cancer: the initial Loma Linda University experience. *Int. J. Radiat. Oncol. Biol. Phys.* 59 (2):348–352.

Slater, J. D., Yonemoto, L. T., Rossi, C. J. Jr., Reyes-Molyneux, N. J., Bush, D. A., Antoine, J. E., Loredo, L. N., Schulte, R. W. Teichman, S. L., and Slater, J. M. 1998. Conformal proton therapy for prostate carcinoma. *Int. J. Radiat. Oncol. Biol. Phys.* 42 (2):299–304.

Slater, J. M., Archambeau, J. O., Miller, D. W., Notarus, M. I., Preston, W., and Slater, J. D. 1992. The proton treatment center at Loma Linda University Medical Center: rationale for and description of its development. *Int. J. Radiat. Oncol. Biol. Phys.* 22 (2):383–389.

Slater, J. M., Miller, D. W., and Archambeau, J. O. 1988. Development of a hospital-based proton beam treatment center. *Int. J. Radiat. Oncol. Biol. Phys.* 14 (4):761–775.

Smith, A., Gillin, M., Bues, M., Zhu, X. R., Suzuki, K., Mohan, R., Woo, S. et al. 2009. The M. D. Anderson proton therapy system. *Med. Phys.* 36 (9):4068–4083.

Soete, G., Verellen, D., Tournel, K., and Storme, G. 2006. Setup accuracy of stereoscopic X-ray positioning with automated correction for rotational errors in patients treated with conformal arc radiotherapy for prostate cancer. *Radiother. Oncol.* 80 (3):371–373.

Song, W. Y., Kamath, S., Ozawa, S., Ani, S. A., Chvetsov, A., Bhandare, N., Palta, J. R., Liu, C., and Li, J. G. 2008. A dose comparison study between XVI and OBI CBCT systems. *Med. Phys.* 35 (2):480–486.

Sonke, J.-J., Zijp, L., Remeijer, P., and Van Herk, M. 2005. Respiratory correlated cone beam CT. *Med. Phys.* 32 (4):1176–1186.

Sonke, J. J., Rossi, M., Wolthaus, J., van Herk, M., Damen, E., and Belderbos, J. 2009. Frameless Stereotactic Body Radiotherapy for Lung Cancer Using Four-Dimensional Cone Beam CT Guidance. *Int. J. Radiat. Oncol. Biol. Phys.* 74 (2):567–574.

Sorcini, B., and Tilikidis, A. 2006. Clinical application of image-guided radiotherapy, IGRT (on the Varian OBI platform). *Cancer Radiother.* 10 (5):252–257.

Stroom, J. C., de Boer, H. C., Huizenga, H., and Visser, A. G. 1999. Inclusion of geometrical uncertainties in radiotherapy treatment planning by means of coverage probability. *Int. J. Radiat. Oncol. Biol. Phys.* 43 (4):905–919.

Suit, H., Goldberg, S., Niemierko, A., Trofimov, A., Adams, J., Paganetti, H., Chen, G. T. et al. 2003. Proton beams to replace photon beams in radical dose treatments. *Acta Oncol.* 42 (8):800–808.

Teh, B. S., Mai, W. Y., Uhl, B. M., Augspurger, M. E., Grant, W. H. 3rd, Lu, H. H., Woo, S. Y., Carpenter, L. S., Chiu, J. K. and Butler, E. B. 2001. Intensity-modulated radiation therapy (IMRT) for prostate cancer with the use of a rectal balloon for prostate immobilization: acute toxicity and dose-volume analysis. *Int. J. Radiat. Oncol. Biol. Phys.* 49 (3):705–712.

Teh, B. S., Paulino, A. C., Lu, H. H., Chiu, J. K., Richardson, S., Chiang, S., Amato, R., Butler, E. B. and Bloch, C. 2007. Versatility of the Novalis system to deliver image-guided stereotactic body radiation therapy (SBRT) for various anatomical sites. *Technol. Cancer Res. Treat.* 6 (4):347–354.

Teh, B. S., Woo, S. Y., Mai, W. Y., McGary, J. E., Carpenter, L. S., Lu, H. H., Chiu, J. K., Vlachaki, M. T., Grant, W. H. III, and Butler, E. B. 2002. Clinical experience with intensity-modulated radiation therapy (IMRT) for prostate cancer with the use of rectal balloon for prostate immobilization. *Med. Dosim.* 27 (2):105–113.

Van den Heuvel, F., Powell, T., Seppi, E., Littrupp, P., Khan, M., Wang, Y., and Forman, J. D. 2003. Independent verification of ultrasound based image-guided radiation treatment, using electronic portal imaging and implanted gold markers. *Med/Phys/* 30 (11):2878–2887.

van Herk, M., Remeijer, P., Rasch, C., and Lebesque, J. V. 2000. The probability of correct target dosage: dose-population histograms for deriving treatment margins in radiotherapy. *Int. J. Radiat. Oncol. Biol. Phys.* 47 (4):1121–1135.

Van Lin, E. N. J. T., Hoffmann, A. L., Van Kollenburg, P., Leer, J. W., and Visser, A. G. 2005a. Rectal wall sparing effect of three different endorectal balloons in 3D conformal and IMRT prostate radiotherapy. *Int. J. Radiat. Oncol. Biol. Phys.* 63 (2):565–576.

Van Lin, E. N. J. T., Van Der Vight, L. P., Witjes, J. A., Huisman, H. J., Leer, J. W., and Visser, A. G. 2005b. The effect of an endorectal balloon and offline correction on the interfraction systematic and random prostate position variations: A comparative study. *Int. J. Radiat. Oncol. Biol. Phys.* 61 (1):278–288.

van Lin, E. N., Kristinsson, J., Philippens, M. E., de Jong, D. J., van der Vight, L. P., Kaanders, J. H., Leer, J. W., and Visser, A. G. 2007. Reduced late rectal mucosal changes after prostate three-dimensional conformal radiotherapy with endorectal balloon as observed in repeated endoscopy. *Int. J. Radiat. Oncol. Biol. Phys.* 67 (3):799–811.

Vargas, C., Falchook, A., Indelicato, D., Yeung, A., Henderson, R., Olivier, K., Keole, S., Williams, C., Li, Z., and Palta, J. 2009. Proton therapy for prostate cancer treatment employing online image guidance and an action level threshold. *Am. J. Clin. Oncol.* 32 (2):180–186.

Vargas, C., Mahajan, C., Fryer, A., Indelicato, D., Henderson, R. H., McKenzie, C., Horne, D. et al. 2007. Rectal dose-volume differences using proton radiotherapy and a rectal balloon or water alone for the treatment of prostate cancer. *Int. J. Radiat. Oncol. Biol. Phys.* 69 (4):1110–1116.

Vargas, C., Wagner, M., Indelicato, D., Fryer, A., Horne, D., Chellini, A., McKenzie, C. et al. 2008. Image guidance based on prostate position for prostate cancer proton therapy. *Int. J. Radiat. Oncol. Biol. Phys.* 71 (5):1322–13228.

Vedam, S. S., Keall, P. J., Kini, V. R., Mostafavi, H., Shukla, H. P., and Mohan, R. 2003. Acquiring a four-dimensional computed tomography dataset using an external respiratory signal. *Phys. Med. Biol.* 48 (1):45–62.

Vlachaki, M. T., Teslow, T. N., and Ahmad, S. 2007. Impact of endorectal balloon in the dosimetry of prostate and surrounding tissues in prostate cancer patients treated with IMRT. *Med. Dosim.* 32 (4):281–286.

Wachter, S., Gerstner, N., Dorner, D., Goldner, G., Colotto, A., Wambersie, A., and Potter, R. 2002. The influence of a rectal balloon tube as internal immobilization device on variations of volumes and dose-volume histograms during treatment course of conformal radiotherapy for prostate cancer. *Int. J. Radiat. Oncol. Biol. Phys.* 52 (1):91–100.

Wang, H., Dong, L., O'Daniel, J., Mohan, R., Garden, A. S., Ang, K. K., Kuban, D. A., Bonnen, M., Chang, J. Y., and Cheung, R. 2005. Validation of an accelerated 'demons' algorithm for deformable image registration in radiation therapy. *Phys. Med. Biol.* 50 (12):2887–2905.

Wang, H., Garden, A. S., Zhang, L., Wei, X., Ahamad, A., Kuban, D. A., Komaki, R. et al. 2008. Performance evaluation of automatic anatomy segmentation algorithm on repeat or four-dimensional computed tomography images using deformable image registration method. *Int. J. Radiat. Oncol. Biol. Phys.* 72 (1):210–219.

Wang, Z., Willett, C. G., and Yin, F. F. 2007. Reduction of organ motion by combined cardiac gating and respiratory gating. *Int. J. Radiat. Oncol. Biol. Phys.* 68 (1):259–266.

Wong, J. R., Cheng, C. W., Grimm, L., and Uematsu, M. 2001. Clinical implementation of the world's first Primatom, a combination of CT scanner and linear accelerator, for precise tumor targeting and treatment. *Phys. Medica* 17 (4):271–276.

Wu, J., Kim, M., Peters, J., Chung, H., and Samant, S. S. 2009. Evaluation of similarity measures for use in the intensity-based rigid 2D-3D registration for patient positioning in radiotherapy. *Med. Phys.* 36 (12):5391–5403.

Yan, D., Lockman, D., Martinez, A., Wong, J., Brabbins, D., Vicini, F., Liang, J., and Kestin, L. 2005. Computed tomography guided management of interfractional patient variation. *Semin. Radiat. Oncol.* 15 (3):168–179.

Yan, D., Ziaja, E., Jaffray, D., Wong, J., Brabbins, D., Vicini, F., and Martinez, A. 1998. The use of adaptive radiation therapy to reduce setup error: a prospective clinical study. *Int. J. Radiat. Oncol. Biol. Phys.* 41 (3):715–720.

Zhang, X., Dong, L., Lee, A. K., Cox, J. D., Kuban, D. A., Zhu, R. X., Wang, X., Li, Y., Newhauser, W. D., Gillin, M., and Mohan, R. 2007. Effect of anatomic motion on proton therapy dose distributions in prostate cancer treatment. *Int. J. Radiat. Oncol. Biol. Phys.* 67 (2):620–629.

Zygmanski, P., Gall, K. P., Rabin, M. S., and Rosenthal, S. J. 2000. The measurement of proton stopping power using proton-cone-beam computed tomography. *Phys. Med. Biol.* 45 (2):511–528.

Chapter 7

8. Treatment Planning for Scanned Particle Beams

Tony Lomax

8.1 Introduction

The more sophisticated a radiotherapy delivery technology is, the more dependent it is on the quality and sophistication of the treatment planning software associated with it. This can be seen in the way that treatment planning developments have mirrored technological developments. It was not too many years ago that the majority of radiotherapy treatments were planned using hand calculations. Although this seems extremely primitive by today's standards, when the majority of treatments were delivered using rectangular open fields, this approach was sufficient to correctly prescribe a dose, and even make a reasonable effort at estimating the dose distribution in the patient, at least in two dimensions. With the introduction of wedges and apertures, things became somewhat more complicated, and although hand plans could still be calculated (e.g., using the concept of equivalent field sizes to calculate monitor units), it became increasingly necessary to provide computer-based programs

for better estimating the delivered dose to the patient. This development was made even more necessary with the routine introduction of computed tomography (CT) scanners into radiotherapy at the beginning of the 1980s. Now, not only were data for the patient outline available to the treatment planner but also an accurate and realistic approximation of the internal anatomy of the patient, together with a measure of the likely photon attenuation for different organs and internal structures. The era of 3-D treatment planning was born.

Since those early times in computer-aided treatment planning, the technology has advanced rapidly, in some cases now driving the technological development as computer algorithms become more sophisticated and accurate. Certainly, the advent of hardware for the delivery of intensity-modulated radiotherapy (IMRT) through the use of motorized and computer controlled multileaf collimators was at least partially accelerated through the prior development of optimization approaches for the calculation of arbitrarily complex photon fluence patterns. It is not an overstatement to talk of a revolution in radiotherapy with the introduction of IMRT [and its natural

Proton and Carbon Ion Therapy Edited by C.-M. Charlie Ma and Tony Lomax © 2013 Taylor & Francis Group, LLC. ISBN: 978-1-4398-1607-3.

offspring, *volumetric modulated arc therapy* (VMAT)], and a revolution predominantly driven by treatment planning developments in the late 1980s and 1990s.

For particle therapy, it is perhaps the other way around. Due to the physical characteristics of particles [good dose localization in depth around the Bragg peak (BP) and a sharp dose falloff distal to the BP], it was evident to the early pioneers of particle therapy that the full potential of this modality could only be achieved through the use of computer-assisted planning processes. Indeed, it has been argued that many of the tools used in treatment planning (beam's-eye view (BEV) displays, dose volume histograms, etc.) were originally developed for particle therapy planning. Although it is perhaps difficult to verify this, it is certainly true that such tools were in routine use in treatment planning for particle therapy already in the early 1980s at the Harvard Cyclotron unit. Indeed, without CT data, BEV displays, accurate 3-D dose calculations, and modern plan evaluation tools, it is unlikely that particle therapy, and in particular proton therapy, would have been so successful from such an

early date, as sophisticated planning methods such as beam patching (which, although somewhat arduous by modern planning standards, nevertheless provide the tools to construct highly conformal dose distributions around complex tumors) could never have been possible without them.

Interestingly, this history of the very early deployment of advanced treatment planning methods in particle therapy is also true of optimization. At the Paul Scherrer Institute (PSI), radiotherapy treatments using pions were introduced routinely in the early 1980s, and a core component of the planning system for this approach (Pedroni 1981) was a gradient-based optimization algorithm very similar in concept to that later published for photons (Bortfeld et al. 1990).

In this chapter, we will review the current state of the art in treatment planning for particle therapy, concentrating on protons, but mentioning difference for other particles such as carbon-ions as appropriate. We will begin by describing the overall planning process for particle therapy, and follow this with more detailed discussions of each component of this process.

8.2   Treatment Planning Process for Particle Therapy

An overview of the treatment planning process for particle therapy is shown in Figure 8.1. Given the highly conformal nature of particle therapy, the first step in the planning process is the creation and production of patient immobilization devices. Ideally, such devices are produced before the initial imaging, such that the primary imaging study used for planning (X-ray CT) is performed with all immobilization devices in place (see Section 8.3). In addition, complimentary imaging modalities such as magnetic resonance imaging (MRI) or positron emission tomography (PET) will be acquired and need to be fused with the primary imaging study. After fusion, these studies are used by the responsible medical doctor to define the target volumes (clinical target volumes, critical structures, other anatomical structures of interest, etc.) on which the subsequent treatment plan or plans will be based. These data are then passed to the treatment planning team, who will then select field directions that best allows the resultant delivered dose to "focus" into the defined tumor volume or volumes (Section 8.5). Perhaps of the whole planning process, the selection of "good" field directions for particle therapy is the aspect of the planning process that most differs from that for photon therapy. An integral part of this process is the ability to estimate, view, and evaluate the resultant

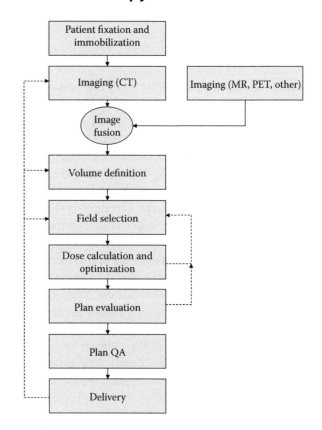

FIGURE 8.1   Schematic overview of the treatment planning process.

dose distribution, through the use of a dose calculation engine, a process that will also include an optimization process if particle beam scanning is being used (Section 4.6). Once a plan (or plans) have been calculated, they are evaluated (Section 4.7) before being delivered to the patient. During therapy, the patient is monitored and in some cases the plan adapted to deal with anatomical changes. This will be discussed in Section 4.8.

As indicated in Figure 8.1, there are a number of loops in this process. As in conventional planning, the process of field selection and dose calculation is almost always an iterative process. This means that the planner will typically try different field directions and combinations of fields in order to find a satisfactory plan. For each arrangement, the dose distributions must be calculated (and optimized in the case of scanned beam particle therapy) and then evaluated, initially by the planner and then by both the responsible medical doctor and medical physicist before being released for delivery.

However, in many cases, this isn't necessarily the end of the planning processes. Perhaps the volumes need to be modified, or even the primary imaging study (CT), to reflect changes to the target (i.e., tumor shrinkage) or to important anatomical compartments (i.e., bladder filling, nasal cavity changes, weight changes, etc.). Such adaptive techniques are particularly important for particle therapy, predominantly due to the sensitivity of the dose distribution on the precise range of the BP in the patient. Thus, plan modifications during treatment are likely to be a more frequent occurrence than for conventional therapy. The planning procedures (and perhaps man/women power for calculating and validating plans) should reflect this more iterative approach.

In the next sections, we will look at these individual steps in more detail.

8.3   Patient Immobilization and Imaging

Particle therapy using protons or heavier particles is a highly conformal and precise form of therapy. It is also a form of therapy whose precision is dependent on an exact knowledge of the range of the BP in the patient. Thus, an accurate (and physically and clinically relevant) model of the patient is essential, as is the ability to be able to accurately position the patient in relation to the beam on a fraction-by-fraction basis. Obviously, if either of these requirements is compromised, then the conformation of the resultant delivered dose to the target will be subsequently compromised. In this section we will look at the requirements for patient immobilization and treatment planning related to imaging for particle therapy.

8.3.1   Patient Immobilization

In order to achieve a precise delivery, there are two main requirements for patient immobilization: the prevention of long time-scale interfraction changes (i.e., changes in patient positioning in relation to the beam on a fraction-by-fraction basis) and the prevention of short time-scale motions of the patient during the delivery of a fraction. Generally, the first of these requirements is the most demanding, particularly as a patient's fractionated treatment will take place every working day over many weeks. During this time, significant changes can take place, from weight loss or gain (the former being a typical problem for patients being treated with concomitant chemotherapy)

through to more mundane problems such as changes in haircut or clothes. An interesting example of this is the treatment of young, pediatric patients. At our institute, we use (where possible) a vacuum bite-block system for immobilization of the patients in the head region. However, during the course of treatment of younger children, problems can sometimes occur due to growth of the teeth, which can be significant over the 4- to 5 weeks of patient treatment. This can lead to noticeable changes in patient positioning, as well as problems in fitting the bite block to the patient.

Typical immobilization techniques include vacuum cushions that take the shape of the supine (or prone) patient, together with customized head pieces and thermoplastic masks or bite-block systems. Generally, the type of device used are functionally similar to those used for photon therapy, but there are a number of additional issues that have to be considered for particle therapy.

As particle range is such an important issue in particle therapy, the material and form of immobilization devices should be considered. As the conversion from CT Hounsfield units to proton stopping power is not always well defined for all materials, the materials used for immobilization should be thin or of low density, such that the effect on the dose calculation be minimized should the stopping power not be well known. In addition, large folds or fixation devices that form sharp steps should also be avoided. In the worst case, such steplike forms in the immobilization devices can

FIGURE 8.2 PSI vacuum bite-block system. This is fixed by a single arm (seen on the left-hand side of the patient), which allows free access for the beam from the other (ipsilateral) side and also allows for the air gap between gantry nozzle (not shown) and patient to be minimized from the ipsilateral side.

act like major density inhomogeneities in the dose calculation, the consequences of which will be dealt with in Section 4.5. In addition, any high-density components (such as metal) can potentially cause reconstruction artifacts in the planning CT, and therefore it should be avoided that such parts overlap with the scan range of the CT. Finally, if nonuniform or metal parts of the immobilization devices can't be avoided, then these should be positioned where possible such that the likely field directions won't pass through. For instance, the bite-block system used at PSI uses a single, patient-specific positioning arm, which although metal, can be positioned out of the scan range of the CT. In addition, we use only one such arm from one side of the patient (usually the contralateral side) such that there is a completely free access to the tumor region from the ipsilateral side (see Figure 8.2).

8.3.2   Imaging for Particle Therapy

The most important imaging modality for particle therapy is undoubtedly X-ray CT. Although CT studies are acquired using X-rays (typically with energies of 120 KeV or so) and not protons, X-ray CT nevertheless provides the most relevant information for particle therapy, at least from the point of view of the dose calculation. This is because CT data are related to tissue density, and density is one of the main determinants of particle stopping power. Nevertheless, the conversion from CT Hounsfield units (HUs) (essentially a measure of the X-ray attenuation coefficient for

the tissue) to relative particle stopping power is nontrivial, as there is unfortunately no one-to-one relationship, as materials with different HU values in CT may have the same relative stopping power, and vice versa (Schneider, Pedroni and Lomax 1996; Schneider, Bortfeld and Schlegel 2000).

In practice, the calibration curve converting CT-HU values to proton stopping powers is generated for biologically relevant materials, for example using the stoichiometric approach proposed by Schneider and colleagues (Schneider, Pedroni and Lomax 1996), which has subsequently been shown to be accurate for tissue samples at the 1% level for soft-tissues and 2% for bone (Schaffner and Pedroni 1998). However, such calibration curves are only strictly valid for CT studies acquired using exactly the same acquisition parameters as that used for the determination of the calibration curve (e.g., X-ray energy, reconstruction filter, and even field of view). Thus, if a CT study is acquired with different parameters, the HU units measured for different tissues could also be different, and on importing the CT data into the treatment planning system, acquisition energy, the reconstruction filter and field of view of the studies should be checked for consistency.

As discussed above, using biologically optimized calibration, nonbiological materials will not necessarily transform to the correct stopping power. This can have some consequences when irradiating through materials used for the immobilization, or perhaps more important, when irradiating through the patient couch. This can be up to a few centimeters thick, and if the stopping power transformation is not correct, this can lead to potentially significant range errors in the dose calculation. For instance, if we assume the treatment couch is 2 cm thick and the error in the calculation of the stopping power based on a biologically optimized calibration curve is 10%, this can already lead to a 2-mm systematic range error in the position of the BPs for beams passing through the couch. Consequently, if nonbiological structures of substantial thickness are present in the CT, and are likely to be in the path of incident beams, then consideration should be given to outlining these structures and assigning a HU value that, based on the calibration curve being used, will give the correct stopping power. This can be determined either by measurement, or if the chemical composition and density of the material is known, by calculation (Schneider, Pedroni and Lomax 1996).

One last point to be considered with the CT imaging for particle therapy is the influence of metal implants on the image quality. A common implant material is titanium, which has a relative stopping power of a little over 3 but is dense enough to saturate in the normal CT HU range. Thus, any conventional CT-stopping power (SP) conversion curve will break down for more dense metals. This is an even larger problem for gold, commonly found in teeth, which has a relative SP of over 20. In both cases, the HU values in CT will saturate at the same value (often 3095), although clearly the stopping powers are very different. Thus it is imperative to know the type of any metal present in a CT, especially if particle fields will be passing through these objects. If in doubt, it is simply best to avoid irradiating through these structures.

A related problem of metal implants is the reconstruction artifacts often present around these structures. These will appear as streak artifacts resulting from problems in the reconstruction algorithm due to these highly dense materials. In severe cases, such artifacts can have very low or high local HU values, which are completely false and can thus detrimentally affect the accuracy of subsequent dose calculations (Jäkel and Reiss 2007). Although there is no correct way to compensate for these, a general approach is to outline these artifacts as well as possible, and assign an average HU value for the assumed underlying tissue (Newhauser et al. 2008).

Before moving away from the subject of imaging for particle therapy, some mention must also be made of other imaging modalities that are important for treatment planning. As in conventional therapy, both MRI and PET imaging are important imaging modalities for defining the target volume, and at least for MRI, definition of soft tissue anatomy. Thus, the ability to fuse such data sets with the primary planning CT data set is an important prerequisite in order to realize the full potential of particle therapy. Indeed, the ability of intensity-modulated particle therapy (IMPT) (see Section 8.6.7) to "dose paint" to selected areas of high tumor cell activity implies that advanced imaging techniques based on MRI and PET, such as functional and biological imaging, will inevitably become increasingly important factors in particle therapy in the future.

8.4 Volume Definition

When defining volumes for particle therapy, there is little reason to digress away from standard procedures and recommendations for volume definition as defined in ICRU reports 50 and 62 (ICRU 1999; ICRU 2007). Certainly, this is true for the definition of gross tumor volumes (GTVs) and clinical target volumes (CTVs), as well as for the definition of organs at risk (OARs). Nevertheless, there are a few areas where volume definition may be somewhat different for particle therapy, and these will be outlined in the following sections.

8.4.1 Target Volumes

The ICRU recommendations for the definition of GTVs and CTVs are quite clear, and will be simply paraphrased here. In principle, the GTV is the extent of tumor that is visible in the available imaging studies, whereas the CTV is a generally larger volume that should additionally encompass all expected regions of microscopic tumor spread that will not usually be visible from imaging studies. In neither of these definitions is any particular type of radiotherapy treatment assumed or implied. Thus, strictly from the ICRU definitions, there is no reason for the GTV or CTV to be different for particle therapy than for any other form of radiotherapy. The same however, is not necessarily true for the definition of the internal target volume (ITV) and planning target volume (PTV).

Let's first look at the PTV. The PTV is a further expansion of the CTV with the intention of allowing for interfraction variability of the target position and shape during therapy. In conventional (photon-based) therapy, it is often sufficient to make an additional, automatic expansion around the CTV taking into account assumed, or measured, changes along each of the major patient axes (AP, L-R, S-I). However, with particles such an approach is, unfortunately, not always sufficient. Particularly when the tumor is in a region of low-density tissue (e.g., in the lung), then changes in tumor position can also cause substantial changes in the *range* of particles, complicating the issue of PTV definition.

Consider Figure 8.3. This shows a schematic, water-equivalent tumor in low-density surroundings (top row) with a conventional PTV defined to allow for a lateral displacement of the tumor. Also shown is an estimate of the 95% isodose line for a single scanned proton field. Note how the doses outside of the lateral edge of the tumor would shoot through into the lung due to the lower densities here. In the bottom, we have assumed that the patient is now displaced to the

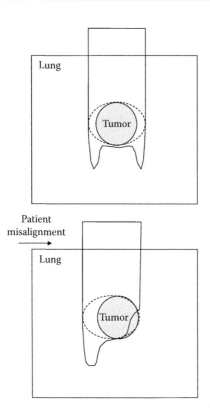

FIGURE 8.3   Schematic representation of the effect of a conventional PTV (dotted contour) on a water equivalent tumor in the lung region. The top picture shows the nominal condition (patient correctly aligned with beam), the bottom with the patient misaligned to the right in comparison with the field.

right in relation to the beam (i.e., the tumor is abutted against the right hand side of the PTV) and again an estimate of the 95% isodose contour for the same field is shown. Note that now, on the right-hand side of the tumor, there is a clear underdosage due to the change in range of the protons as the tumor moves into part of the field that was originally calculated through lung. I hope this simple example shows the problem of PTV definition in particle therapy. If the field width is simply expanded laterally such as to cover the PTV, then this will clearly be insufficient to ensure coverage of the target (at the 95% level) on its own due to the resultant changes in range experienced by the different particles delivering dose to the different parts of the tumor. It should also be clear that the conventional PTV expansion is perhaps not the optimal method from the point of view of normal tissue sparing and that it is quite likely that field specific solutions to the PTV problem may be required for particle therapy. Indeed, in many proton centers using passive scattering, the concept of PTV is not used, with positional uncertainties being dealt with through direct manipulation of the field specific beam-shaping devices (Moyers et al. 2001).

A similar problem occurs for the definition of the ITV. If defined at all, in conventional therapy, this will often be an ellipsoid with the major axis lengths being determined from assumed (or measured) motions along the major axis of the patient. Alternatively, and an increasingly more common technique, 4-D-CT data sets can be used to more accurately determine the ITV through methods such as maximum intensity projections (MIPs), in which, for each voxel of the 4-D-CT data set, the maximum value from each phase of the study is determined and stored in the composite MIP volume. The ITV can then be directly defined using contouring or segmentation techniques directly applied on this MIP image. However, again Figure 8.3 shows the problem of range changes, with this figure simply being interpreted now as *intra*fraction motion, and the PTV as ITV. Again, it is also hopefully clear that, as with the PTV problem described above, it is very likely that field-specific solutions need to be developed for this problem (Cabal and Jäkel 2010; Boye, Knopf and Lomax 2010).

8.4.2   Organs at Risk

For the definition of OARs, there are essentially no major differences between particle and conventional therapy. On the other hand, it can be the case that somewhat more OARs will be defined in particle therapy, simply because the normal tissue sparing characteristics of particle therapy can sometimes mean that it is possible to spare additional OARs than can be spared in conventional therapy. (This is *not* ICRU convention, however, where the same rules apply to OARs as to the PTV and CTV; i.e., that structures should be defined without prior assumptions as to the modality of treatment to be used).

However, is should be noted that in ICRU 62, there is also a recommendation to apply the safety margin concept to OARs, in order to create planning organs at risk volumes (PRVs). If these are to be used, then it is likely that the margins to be used *will* differ between modalities, and for particle therapy may need to be field-specific for the same reasons as for PTVs/ITVs. Although the concept of PRV is very rarely used, at our institute we do expand OAR volumes by a few millimeters when performing IMPT (see Section 8.6.7) in order to help the optimization algorithm achieve the stated prescription dose, but also to pull the sometimes steep dose gradients away from the surface of the organ in question. Thus, in some ways, these expansions can be considered to be partial PRVs.

8.4.3   Technical Volumes

Although there are no equivalent volumes defined in ICRU Reports 50 and 62, in practice there is sometimes also the need to define what can be called technical volumes (TVs). TVs commonly used at our facility are regions of CT reconstruction artifacts, volumes encompassing hardware devices for which CT HUs need to be changed (see Section 8.3.2) and modified PTV volumes for assisting the planning process.

As discussed above, in the presence of metal implants or gold teeth, considerable reconstruction artifacts can be present in the planning CT. As these can affect the accuracy of the dose calculation (ICRU 1999; ICRU 2007), it is good practice to "correct" these regions. Although this is by no means an exact science, the regions of the CT affected by such artifacts, and which, to the judgment of the responsible medical doctor also correspond to areas of soft-tissue, should be outlined such that the HU values within this region can be subsequently set to an average soft-tissue HU value. This should however be performed with care, because if these defined regions mistakenly also include bone (or even worse, the metal itself), then the correction process itself can introduce other errors. It is therefore recommended that the definition of these regions be performed by a medical doctor who has the necessary anatomical knowledge.

A similar approach sometimes needs to be taken for nonbiological structures that may be in the CT, and through which treatment fields may pass (see Section 8.3.2). An example is the treatment couch. To achieve the best range precision from the planning CT study, the HU values of the couch should be altered to give the correct proton stopping power by carefully outlining the couch throughout the CT and modifying its HU values appropriately. Similar volumes can be defined for structures that one may wish to remove from the CT as they will not be there during treatment. Examples include earrings (although ideally these should be removed before the planning CT is acquired) or marking devices that may be included in the planning CT to aid identification of important regions. For instance, at our institute, in cases where the patient was previously operated on, the surgical scar is sometimes marked to be included in the CTV region in order to reduce the risk of tumor recurrence due to seeding of tumor cells in the surgical path. So that it is visible in the planning CT, this is typically a thin piece of metal wire. Ideally, before the planning process begins (but after the volumes have been defined!), this should be removed from the planning CT by outlining it and setting all HU values inside to that of air.

Finally, it is sometimes useful to be able to define a technical PTV (techPTV) to help the planner pull doses off critical structures (much like the use of blocks in conventional therapy) or to avoid dose hot spots at the skin surface for superficial target volumes.

8.5   Field Selection

At the highest level, field selection for particle therapy is similar to that for conventional therapy. Beam incidences will be selected primarily to avoid critical structures and to best focus the dose into the target volume while attempting to minimize dose to surrounding normal tissues. However, the fact that the delivered dose can effectively be stopped at a given range through the careful positioning of distal BPs brings a number of additional considerations for beam incidence selection.

8.5.1   Clinical Considerations

First, and perhaps most important, with particle therapy it is possible to deliver a more or less homogenous dose distribution to the selected target volume from a single direction. This has two main consequences: (1) in some cases, it is possible to irradiate the target satisfactorily from a single field direction (even though this may not be optimum from the point of view of plan robustness), and (2) even when multiple field incidences are required, these do not necessarily have to be distributed around the patient. For instance, all incident fields could be applied from the same side or even same quadrant. A beautiful example of this is shown in Figure 8.4. This shows a rather superficial and posterior meningioma, irradiated using three fields—a lateral and two lateral-superior oblique fields separated by only 30°.

Second, with particle therapy the steep distal dose falloff can be used to additionally spare normal tissues in a way that is not possible with photons. Although many people working in particle therapy often state that the BP should not be used to spare critical structures due to possible uncertainties in its position, in fact every particle treatment to a greater or lesser

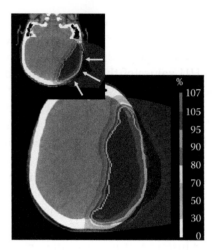

FIGURE 8.4   **(See color insert.)** Example dose distribution from the PSI spot scanning system for a meningioma. This is a three-field plan, with the field arrangement as indicated in the insert. All fields are incident from the same quadrant such as to utilize the BP fall-off in order to spare the uninvolved (contralateral) part of the brain as much as possible.

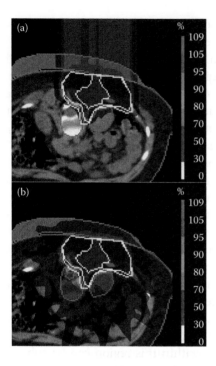

FIGURE 8.5   **(See color insert.)** (a) Single-field proton plan used to irradiate this large superficial desmoids tumor in a 12-year-old boy. Note the use of a techPTV (interior yellow contour) to spare the spinal cord and partially the kidney. For comparison, a nine-field photon IMRT is shown at the bottom (b). Proton plan delivers six times lower dose to the normal tissues than the proton plan.

extent uses the BP to spare normal tissues and critical structures. Although it is certainly expedient to avoid having sharp distal dose gradients stopping directly at the surface of a critical structure, it is perfectly reasonable to use the distal falloff to shield distally placed critical structures when there is enough distance between them to allow for possible range uncertainties. An example of this is shown in Figure 8.5a, which shows the proton dose distribution (top) calculated and delivered to a superficial desmoid tumor in a 12-year-old boy. This single field proton plan, incident from the posterior aspect, clearly spares the spinal cord and the majority of the kidney even though this organ is directly within the beam's cross section. Indeed, Figure 8.5a nicely shows the use of a techPTV for shaping the dose around the spinal cord and applying a distal safety margin in order to pull the distal gradient of the field away from the cord.

On a related subject, there is a significant advantage with particles to take the shortest path to the tumor, in order to reduce the integral dose to all normal tissues. Again, this is nicely shown in Figures 8.4 and 8.5a. For the second case, there was minimal irradiation of any normal tissues through the use of a single proton field incident from the posterior aspect of the patient, which resulted in a *six* times reduction in the nontarget tissue integral dose delivered to the patient in comparison with a 9 field photon IMRT plan, shown for comparison on the right-hand side in Figure 8.5b.

8.5.2   Physical Considerations

As with anything in life, the advantages of particle therapy also bring some additional potential disadvantages that also must be considered when selecting beam incidences. These can be divided into two issues; the effect of density heterogeneities and range uncertainties. We will describe each of these in the following paragraphs, before discussing how beam incidence selection can be used to mitigate the compromising consequences of these issues.

Particles have well-defined ranges, which are predominantly determined by their energy and by the density of the tissue or medium through which they travel. In a homogenous medium, published range tables can be used to very accurately predict the position of a pristine (monoenergetic) BP in water. Even when the initial particle beam has a spectrum, as long as this is known, it is a rather simple matter to reconstruct the composite BP position rather accurately. However, this is not the case in inhomogenous regions. If a particle beam is passing parallel to an interface between two mediums (or tissues) of different densities (and therefore relative stopping powers), then inevitably some protons will pass predominantly through the higher-density material, while some will pass through the lower-density material. Indeed, some will pass through both, being scattered in or out of these different materials as they undergo Coulomb scattering events. One can well imagine the end effect. Beyond this interface, different protons have lost substantially different amounts of energy, and the BP is severely degraded, or broadened, due to the wide spectrum of residual energies of the remaining protons. Thus, the wonderful shape of the pristine BP shown in most text books about particle therapy will be lost, or at least severely distorted. An example of this is effect is shown in Figure 8.6. Here a Monte Carlo calculation has been used to calculate a single, finite width pencil beam passing completely through a CT data set of a patient's head, at the level of the skull base. This is one of the more (density) heterogeneous areas of the head, and the distortion of the pencil beam at the BP end is clearly visible in the resulting dose distribution, calculated in a simulated block behind the patient. Also shown in the figure is a plot of the *integral* dose delivered in infinite planes perpendicular to the pencil beam direction as a function of depth. In a homogenous medium, such a plot would give us the classic BP shape. The plot in Figure 8.6 is a long way from such a beautiful form, and, given the

generally inhomogenous nature of humans, is perhaps closer to what is actually being delivered in a patient than the pristine BPs normally shown.

So how do density heterogeneities affect beam incidence in particle therapy? Well, the answer of course is to, wherever possible, select incident fields that as much as possible avoid passing through areas of complex density heterogeneities, and in particular avoid passing along high-gradient density heterogeneities that are perpendicular to the beam direction. Although this is not always possible, it is certainly good practice to avoid such directions, and for this purpose we have added a heterogeneity index tool to the field selection tool in our treatment planning system, which provides the planner with an indication of the heterogeneity of any selected or defined beam path. In the absence of other constraints, this tool can be used to select optimum angles that can help to minimize the effects of density heterogeneities.

Let's now move on to the problem of range uncertainty. Above I said that the range of particles in a homogenous medium can be easily derived. However, this is not strictly correct. It can be easily derived *if* the medium is homogenous *and* the stopping power of the medium is well known. In the patient, neither is necessarily the case.

As discussed above, patients are generally not homogenous, and as discussed in Section 8.3.2, the Hounsfield values of the planning CT have to be converted to stopping powers using a previously defined calibration curve. The derivation of this curve, although rather accurate, can inevitably never be perfect, and thus the stopping powers of the different tissues can only be estimated. In the work of Schaffner and Pedroni (1998), it was found that the use of the stoichiometric approach of Schneider, Pedroni and Lomax (1996) resulted in an accuracy of about 1% in soft tissues and about 2% in bone. Indeed, given the other errors associated with CT imaging (image noise, beam hardening, etc.), it is reasonable to assume that the stopping powers for biological tissues *in vivo* can currently only be determined to about the 3% level. Although this is quite accurate, one should remember that for a BP placed 10 cm under the surface of the patient, this already corresponds to an uncertainty in its actual position of about 3 mm. Given the steepness of the distal falloff of the pristine BP, this can translate into a quite significant uncertainty in the delivered dose in this region. Of course, in the clinical case, the actual uncertainty can be quite considerably larger, as the actual range can also be affected

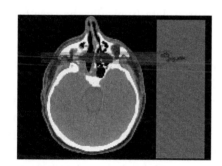

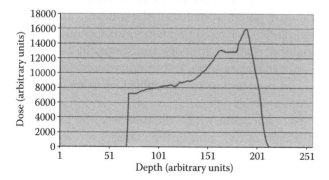

FIGURE 8.6   (See color insert.) Effect of density heterogeneities on the shape of proton pencil beam (top) and integral BP (bottom). In many cases within the patient, the actual delivered depth dose curve will be quite different from the normal textbook BP.

by patient positioning, changes in patient weight (Albertini et al. 2008) and changes of internal anatomy (e.g., changes in the filling of cavities or shrinkage of tumor volume). All these issues indicate that, during planning, the planner should always be aware of the likely uncertainties in range, and it is for this reason that many institutes are extremely reluctant to use single, highly weighted beams that "stop" with their distal end directly against one or more critical structures. However, as discussed above, this dose *not* mean that the distal falloff can never be used, and the example of Figure 8.5a shows how the sharp falloff has been utilized, but only together with the use of effectively a PRV around the spinal cord which was used to construct a techPTV that was pulled a few millimeters away from the cord to allow for such uncertainties.

So, in summary, when constructing particle therapy treatments, the advantageous characteristics of particles means that homogenous doses can be delivered from single-incident beam directions, making beam selection on the one hand rather simple. However, the issues of density hetereogeneities and range uncertainties in practice can restrict the choice of angles. In all cases, it is good practice to treat using multiple beam angles in order to spread the effects of density heterogeneities over different regions and to mitigate the effect of range uncertainties should the beam under- or overshoot in relation to the nominal plan. For instance, with enough angular separation between beams, any potential errors in the delivered dose can be limited to different regions of the target or normal tissues, thus blurring out such effects somewhat.

8.6 Dose Calculation and Optimization

8.6.1 Dose Calculation Algorithms

The dose calculation is the central, physics-driven element of the treatment planning process. Without the ability to calculate and predict dose in three dimensions within the patient, highly precise and conformal radiotherapy would simply not be possible. Indeed, we are fortunate in radiotherapy that the interaction of the medical agent (in our case radiation) with the patient can be quite accurately modeled based on mature and validated physics models, providing us with a tool that is, I believe, unprecedented in any other area of medicine. For instance, state-of-the-art dose calculations can predict the delivered dose at millimeter resolutions and with absolute accuracies of a few percentages. This provides radiotherapy with the ability to predict and record a patient's treatment that is unrivaled in other areas of therapeutic medicine.

Dose calculation algorithms for proton therapy are broadly similar in concept to those for photon therapy, while obviously having to exploit different physics models for the interaction of particles with tissue. The algorithms can be divided into the following categories: ray casting, pencil beam, and Monte Carlo techniques. Each of these will be briefly described in the following sections.

8.6.2 Ray-Casting Algorithms

Ray-casting algorithms are conceptually rather simple and have the advantage of being extremely fast. They consist of a description of the depth dose curve for the element being calculated (this can be for a single, quasi-monoenergetic proton beam of finite width in scanning, or a broad, SOBP beam for passive scattering) and some description of the lateral falloff of the beam as a function of the distance from the field edge (for broad-beam algorithms) or the central axis (for finite pencil beams). The input data to these can be measured data (typically the case for broad-beam algorithms) or analytically derived and validated by measurement (for scanning beams). For instance, for the finite pencil beam mode used for the treatment planning of scanned proton therapy at our institute, the depth dose curve of a quasi-monoenergetic beam is modeled using an analytical dose model based on the Bethe–Bloch formulism (Scheib and Pedroni 1992; Schaffner et al. 1997; Pedroni et al. 2005), which is fitted to measured depth dose data in water. For the lateral dose falloff, a model for the transport of protons through water is used that takes into account multiple Coulomb scattering theory and results in a Gaussian beam profile whose width (sigma) increases as a function of penetration in the patient or other materials through which it passes.

For each point where the dose is to be calculated, the depth of the point in the material or tissue needs to be known, as well as its lateral position in relation to the field edge or central axis of the finite pencil beam. Again, taking the PSI model as an example, the dose delivered at a point from a single, finite pencil beam would then simply be given by calculating the dose along the central axis of the beam at the given depth, and then, if the point is laterally displaced a certain distance from the central axis, calculating the drop-off in dose due to this displacement, using the Gaussian approximation of the lateral spread of the beam model whose sigma is also derived from the estimated sigma at the depth of the dose calculation point. Thus, proton dose calculations (in this form at least) are really quite simple, and can be performed with standard mathematical functions such as the Gaussian, making them easy to implement and fast to execute. Indeed, at least in a homogenous medium like water, such an algorithm is quite accurate against measurements or even more sophisticated dose calculations such as Monte Carlo.

However, the real test of a dose calculation engine is in the calculation of dose distributions in inhomogenous mediums such as a patient. This is where the name of this type of algorithm comes from. In the ray-casting approach, density heterogeneities are corrected for by simply scaling the depth of the dose calculation point in the patient to its water equivalent depth, taking into account the integral of all the relative stopping power values (derived from the planning CT data) along the beam's path from the exit of the nozzle to the dose calculation point itself. The effect of this ray-casting approach on a single, finite pencil beam can be seen in Figure 8.7. As the water equivalent depth of each dose calculation point is taken into account

in this approach (rather than just the central axis of the pencil beam itself), the effect of this algorithm is to "sheer" the Bragg peak behind the density heterogeneity, a result that tends to overestimate the effect of the heterogeneity. In reality, there will be additional scattering of protons behind the density heterogeneity, as well as in- and out-scattering from the high density material itself, which will tend to smear out the distortions, as shown by the Monte Carlo based calculation of the same situation shown in Figure 8.7b. However, it should be noted that the nearer the density heterogeneity is to the end of range of the pencil beam, the better is the ray-casting approximation, simply because there is less in- and out-scatter after the heterogeneity to smooth the dose across the sheer interface. On the other hand, the sharpness of the dose gradient across the interface after the density heterogeneity is also independent of where the density heterogeneity is along the beam. That is, the dose distribution profile at the depth indicated by the broken line will be calculated to be more or less the same regardless of whether the density heterogeneity is right at the surface of the patient (as shown by the dotted line) or deeper within the patient (the solid line), a result which is clearly unphysical.

Algorithms such as these have been described in (Scheib and Pedroni 1992; Schaffner et al. 1997), and as has been shown by Tourovsky et al. (2005), can be remarkably good in comparison to Monte Carlo calculations in clinical cases and even to measurements in complex anthropomorphic phantoms (Albertini et al. 2011a).

8.6.3 Pencil Beam Algorithms

Pencil beam algorithms can be considered to be the next step up in complexity, although the basic physics is the same as that for the ray-casting approach. As with that method, the input will be depth dose curves (although these will generally be for quasi monoenergetic beams rather than SOBPs) and some model of the lateral spread of the beam due to scattering. However, as its name implies, the dose calculation is in all cases based on the summation of dose from individual pencil beams, by the summation of dose from many initially infinitely small pencil beams. Thus, in order to model a broad beam with this approach, it must first be decomposed into a large number of small pencil beams, weighted such as to sum up to a single broad beam (see Figure 8.8a). Similarly, to model a finite-sized pencil beam for scanning, the shape of the

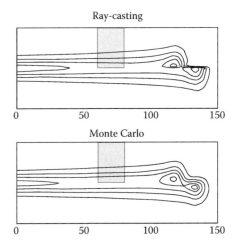

FIGURE 8.7 Analytical (ray-casting) and Monte Carlo dose calculations of a single proton pencil beam whose central axis is passing directly along a bone-water interface. Note the effective shearing of the BP behind the density heterogeneity due to the ray-casting approach and the smoother splitting of the BP predicted by the Monte Carlo code. (Adapted from Schaffner, B. and Pedroni, E. 1998. *Phys. Med. Biol.* 43: 1579–1592.)

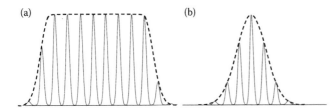

FIGURE 8.8 Decomposition of a (a) flat field or (b) physical pencil beam into a set of smaller proton beamlets for performing high-resolution analytical dose calculations.

physical pencil beam leaving the treatment nozzle is also decomposed into an appropriately weighted set of smaller beams, often called beamlets (see Figure 8.8b). The dose calculation is then identical to that described for ray-casting above, except for the fact that density heterogeneities are dealt with by scaling only the central axis depth of each individual beamlet (Schaffner et al. 1997; Hong et al. 1996). This can be contrasted to the ray-casting approach, where the depth of the dose calculation point itself is taken into account.

Figure 8.9 shows the same situation as Figure 8.7, but with the dose now calculated using such a pencil beam algorithm. The dose now more closely matches the dose calculated by the Monte Carlo approach, indicating that this type of algorithm is, at least in some circumstances, more applicable than the simpler ray-casting approach. However, as with the ray-casting approach, this model has no way of differentiating between density heterogeneities close to the surface or at depth, with the shape of the resulting dose

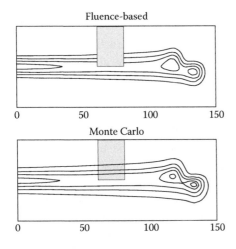

FIGURE 8.9 Analytical (beamlet-based) and Monte Carlo dose calculations of a single-proton pencil beam whose central axis is passing directly along a bone-water interface. (Adapted from Schaffner, B. and Pedroni, E. 1998. *Phys. Med. Biol.* 43: 1579–1592.)

distribution being independent of its position. Indeed, although pencil beam algorithms are certainly better than ray-casting algorithms when heterogeneities are close to the surface, this is not always the case when they are close to the end of range of the beam, where the ray-casting approach can still be more accurate (Schaffner et al. 1997). Nevertheless, this algorithm is probably the most commonly applied in current commercial particle treatment planning systems.

8.6.4 Secondary Particle Distributions

Both the ray-casting and pencil beam algorithms outlined above are algorithms that analytically model the interactions of *primary* protons as they pass through tissue or other mediums. However, there is another effect that needs to be taken into account to truly model the physics of particle interactions correctly. In addition to depositing energy and being scattered though multiple Coulomb scattering processes, primary protons also have a small, but finite, probability of interacting directly with atomic nuclei. It is not the purpose of this chapter to go into these effects in detail, but suffice to say that when this occurs, the proton can be completely lost from the primary beam, and a number of secondary particles produced. For example, as a rule of thumb, 1% of primary protons are lost due to such interactions every 1 cm of tissue traversed. Most of the energy deposited by these interactions (at least for protons) is deposited locally to the interaction in the form of heavy secondary particles. However, there are also secondary protons produced, which can have energies almost up to the energy of the initial primary proton. In addition, the secondary protons can have a much larger angular spread than the primary proton beam, leading to a low, but significant 'halo' of secondary protons around the primary beam. Although this halo doesn't always have a significant effect on the final dose distribution, it can have an effect on the overall dose, particularly for small fields (Pedroni et al. 2005), and can sometimes lead to an additional accumulation of dose of a few percent in narrow dose "valleys" for IMPT fields. Consequently, this halo should be accounted for, and is usually modeled as a second Gaussian distribution that can be inferred from experiment (Pedroni et al. 2005) or Monte Carlo calculations (Soukup, Fippel and Alber 2005).

The effect of nuclear interactions is a more significant effect for heavier ions like carbon. First, the probability for a nuclear interaction is higher (the cross sections are larger) and the ranges of the heavier secondary particles resulting from the interactions are also considerably longer than for protons. Whereas for protons, at the clinical scale, the heavier secondary particles can be considered to deposit their energy more or less at the point of the interaction (Paganetti 2002), for carbon, these can have a considerable range. Indeed, these ranges are so large that they have a distinct effect on the BP curve. Figure 8.10 shows the depth dose curve for a 330 MeV/u carbon ion beam in water. It shows the well-known BP shape up to the peak itself, together with a considerably higher peak-to-plateau ratio compared to protons, but also a distinct tail of dose extending a few centimeters beyond the distal falloff region. This is known as the fragmentation tail, and is the result of secondary particles resulting from nuclear reactions. So although carbon ions exhibit considerably sharper BPs than protons, as well as significantly reduced lateral scatter, the ability to completely spare normal tissues beyond the distal edge is somewhat compromised by this effect, particularly as the tail is generally made up of high RBE particles.

The most accurate form of dose calculation for particle therapy is to use Monte Carlo techniques. Monte Carlo methods are used in many areas of science to model effects that can be well described using statistical approaches and that are characterized by very many individual events. This is certainly the case for particle therapy, where many billions of particles will be applied per field, and each particle interacts with matter under well-described statistical conditions. For instance, both energy loss and scatter can be represented as probability functions, which can be subsequently sampled using random number generators. Thus, using Mone Carlo approaches, individual particles can be modeled and tracked through any medium, if the characteristics of the material are known, and hence the relevant probability functions

derived. In addition, as long as the physics of the interactions are known, then also the secondary particle distributions described in the previous section can also be much more accurately calculated using Monte Carlo codes simply by following the main secondary particles resulting from nuclear interactions as well. In essence, particles can be tracked using this approach at any resolution, although the finer the resolution of the tracking, the more calculation time is required per particle. As many millions of particles are generally calculated to achieve the required statistical accuracy in the final dose distribution, Mone Carlo based calculations are generally extremely time consuming, especially if secondary particles are tracked as well. Nevertheless, there is no denying that, given their particle-by-particle modeling approach, Monte Carlo calculations have to be considered the gold standard dose calculation approach.

Many authors have described Monte Carol approaches to in-patient dose calculations for particle therapy (Tourovsky et al. 2005; Jiang and Paganetti 2004; Paganetti et al. 2008; Yepes et al. 2009) and many of these papers show the flexibility and completeness of the Monte Carlo method for modeling particle interactions and dose distributions at many different levels and due to many different interactions and physics principles.

8.6.5 Relative Biological Effectiveness

Due to the very different interaction mechanisms of particles to photons, there is also a different biological effect resulting from the deposited energy. Put another way, for particles such as protons and heavier ions, the same deposited physical dose (in J/Kg or Gray) can lead to an enhanced biological effect on tissue than the same dose of photons. This biological enhancement is the relative biological effectiveness (RBE) and is an important factor in all particle therapy, but particularly for therapy with heavier ions such as carbon. In principle, RBE is related to the linear energy transfer (LET) metric, which can be considered to be a measure of the density of ionization caused by a particle along a short track (typically μm). In the BP region of a particle pencil beam, the rate of change of energy loss gets quite high (this after all gives us the BP) and thus the LET increases, which in turn can lead to increased RBE values. For example, the RBE for a proton beam in the plateau region is close to 1 (so more or less equivalent to photon irradiation) where the energy is high and energy deposition rate relatively low. As the protons

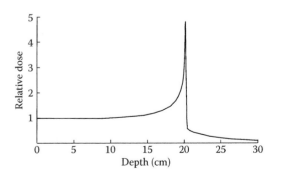

FIGURE 8.10 Depth dose curve for 330-MeV/u carbon ions. Note the higher peak to plateau ratio than for protons, the sharper BP, and the fragmentation tail beyond the distal fall-off of the BP.

slow down however, the LET increases such that in the BP, the RBE can be 1.1–1.2. However, the highest values are in the distal falloff regions, where the average proton energy is at its lowest, and here RBE can reach values of 1.5–1.6. This sounds quite dramatic, and indeed in some circumstances it can be. However, one should remember that the dose where the RBE is at its highest (in the distal falloff region) is also quite low, mitigating somewhat the increased RBE values in this region. In practice, for protons, most centers use a generic RBE value of 1.1 over the whole dose distribution, indicating that the convention is that proton therapy is globally 10% more biologically effective than photon therapy.

However, the effect of RBE becomes much more important in heavy ion therapy, for instance using carbon ions. Here values of 3–4 can be reached in the BP, and have to be taken into account at the dose calculation level. The two main centers worldwide delivering carbon-ion therapy (NIRS in Chiba, Japan and GSI/HIT facility in Heidelberg, Germany) have different ways of dealing with this. The Japanese facility applies a relatively simple depth dependent RBE correction, with the RBE increasing as a function of depth (in relation to the position of the most distal BP), while at the German facility, a comprehensive model for RBE is applied at the dose optimization stage (see Section 8.6.7) such that the resulting pencil beam fluences are already optimized taking into account the varying RBE. This later approach appears to be the most accurate method; however, it is complicated by the fact that RBE is dependent not only on the LET but also on the dose, dose rate, tissue type, and end point for the biological effect in question.

8.6.6 Role of Optimization in Particle Therapy

Arguably, one of the biggest leaps in the quality of radiotherapy came with the introduction of optimization techniques for the calculation of arbitrarily complex intensity/fluence maps into the planning process. The idea of optimization regimes can be traced a long way back, but certainly one of the first clinical uses of the mathematical-based optimization of delivered fluences was for the pion therapy project at PSI. This was based on the scanning of a set pion spots formed from the application of many radially distributed pencil beams—a sort of gammaknife approach with pions. This spot was scanned through the patient by a 3-D motion of the patient couch, with the pion fluence at each spot being determined in the treatment

planning system through the use of a gradient-based, least squares optimization algorithm (Pedroni 1981). Indeed, the mathematics used for this operation, although described in a slightly different form, is essentially the same as that subsequently published by Bortfeld et al. (1990). The pion therapy project at PSI ran for about 10 years (from 1981–1992), with more than 500 patients being treated, each of whose plans were calculated using this optimization regime.

So, like many things, the optimization of delivered fluences has a long (and pioneering) history in particle therapy. It also has a great future. Particularly, optimization methods are essential for scanned proton beam delivery (at least if one ignores the special case of uniform scanning), in that, to utilize the full flexibility provided by pencil beam scanning, the definition of the individual fluences of all physical pencil beams in a field can only be feasibly performed using optimization.

For scanned particle therapies, there are essentially two approaches to optimizing fluencies: single-field uniform dose (SFUD) and IMPT. Although these are the terms commonly in use, it is perhaps more meaningful to think of these two approaches from the point of view of the mechanics of the optimization process. SFUD could also be categorized as *single-field optimization*, whereas IMPT is a *multiple-field optimization* technique. This really is the main difference between the two.

In SFUD optimization, the optimization algorithm is applied to a single field only, essentially in complete isolation from any other fields that may be calculated for the same plan (Lomax et al. 1996). As such, the optimization's main goal is to find the set of individual pencil beam weights that best covers the selected target volume with a *homogenous* dose. Hence its name, in that the result of SFUD optimization should be a single-field dose distribution that delivers a (more or less) homogenous dose across the target. An SFUD plan is then the simple combination of a set of such fields. In essence, SFUD planning can be thought of as being the scanned particle therapy equivalent of open-field planning in photons, or SOBP planning in particle therapy when using passive scattering techniques.

In contrast, for IMPT plans, the optimization algorithm is applied to all pencil beams from all selected fields *simultaneously* (Lomax 1999). In principle, this means that individual IMPT field dose distributions can be of arbitrary complexity, even though the resultant total dose across the target volume is homogenous. In practice, in the absence of any normal tissue constraints, the individual fields of an IMPT optimization can be similar to those of a SFUD optimization, even though small differences can still be found (Lomax 2008a). The major difference comes when applying IMPT optimization in complex cases where dose constraints are also defined on one or more neighboring critical structures. In this case, IMPT will work like IMRT with photons, selectively reducing the fluence of pencil beams passing through (or stopping in) these critical structures, with the missing dose in other areas of the target being compensated for by the other fields in the IMPT plan.

IMPT optimization has one major difference to IMRT techniques however, and that is the number of degrees of freedom available to the optimization process. For IMRT with photons, it is only possible to modulate the applied fluence of a 2-D field—the depth dose curve of the field being determined completely by the characteristics of the linear accelerator energy. In contrast, with particles, the BPs can be placed throughout the target volume in three dimensions and also from every incident field direction. Thus, in its most general implementation at least, the IMPT optimizer can modulate BP fluences in three dimensions from each incident field, bringing much more flexibility in the resultant dose distribution than is possible with photon-based IMRT.

However, one does not *have* to perform IMPT this way, and an alternative technique is to just place single BPs on the distal edge of the target volume from each field, and then compensate for the necessarily inhomogenous dose within the target volume using the optimization algorithm. This approach has been named distal edge tracking (DET), and has been the subject of a number of publications (Lomax 1999; Deasy, Shephard and Mackie 1997; Oelfke and Bortfeld 2000; Albertini, Hug and Lomax 2010). It has been also

shown that it could have a number of advantages over normal (3-D) IMPT, although in order to achieve dose homogeneity in the target, with DET one has to resort to a larger number of angularly spaced angles similar to the beam arrangements used in photon-based IMRT (Lomax 1999, 2008a).

As an example, Figure 8.11 shows two plans for the same case. On the left is a 3-D-IMPT plan (utilizing all possible BPs that stop in or close to the target volume) and on the right is a DET plan. The plans are so similar that it is difficult to tell them apart, and nicely illustrates that IMPT is a very degenerate problem, meaning that there are very many possible solutions to the fluence maps applied to the pencil beams that lead to very similar dose distributions.

In the discussion above, for simplicity we have talked about the optimizer trying to obtain a homogenous dose across the target. However, just as is the case for IMRT, this doesn't have to be the case for IMPT. If the optimizer is given the correct "goal" function, then of course, IMPT can be used for simultaneous integrated boost techniques and dose-painting. Indeed, it could be potentially an even better tool for these approaches than conventional IMRT due to its large degeneracy.

8.6.7 Field Patching

IMPT techniques as described above are currently confined to particle centers who have access to scanning technology, which at the time of writing is rather few (PSI, Heidelberg Ion Therapy Centre, Rineker Proton Therapy Centre in Munich, and MD Anderson in Houston). However, also with passive scattering, a type of IMPT is possible, even if it is rather more time consuming both at the treatment planning and delivery stage. This technique is called field patching.

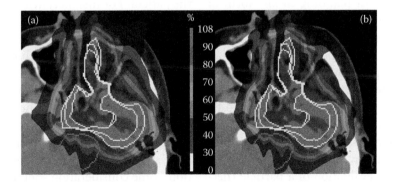

FIGURE 8.11 (See color insert.) (a) Three-dimensional IMPT and (b) DET plans to a large chondrosarcoma. Both plans consist of coplanar five fields, equally distributed around the patient. Both plans are practically identical, although the plan on the right delivers BPs only at the DET, whereas the 3-D-IMPT plan delivers BPs in three dimensions throughout the volume.

With field patching, the sharp lateral edge of a collimated particle beam can be matched to the distal falloff of a compensated orthogonal or near orthogonal field. This allows the planner to construct concave dose distributions around critical structures by covering the full target volume with a number of smaller fields, each of which in themselves only cover a subportion of the target. With an experienced treatment planner, such plans can reach an impressive level of complexity, but often only through the use of many different field directions. In addition, as the technique is dependent on the patching of potentially sharp distal dose edges to the lateral edges of other beams, the homogeneity of dose across the patch can sometimes be poor (particularly when the distally patched field is passing through complex density heterogeneities) and is also extremely sensitive to positional and range errors, which can affect the relative position of the two dose gradients leading to potentially large uncertainties in the delivered dose across the patch. For this reason, such plans are often feathered, in which the range of the distally patched field is deliberately changed on a day-by-day basis in order to smooth out the gradients along the patch line. Alternatively, different combinations of patches are defined in the same plan, with the different combinations being delivered on different days. This can lead to treatment plans consisting of many different fields, for each of which both compensators and collimators have to be produced and then mounted on the gantry for the delivery of each field. Nevertheless, field patching with passive scattering has proven to be extremely successful for the treatment of complex lesions with particles, and for many years provided the only truly highly conformal form of radiotherapy anywhere.

8.6.8 Dose Evaluation

The main role of treatment planning in any form of radiotherapy is to provide an estimate of the delivered dose or dose distribution to the patient. As such, it is both a combination of CAD and a decision-making tool. For the treatment planner, it is a design tool; for the medical doctor it is primarily a decision-making tool. In both cases, the quality of a plan needs to be evaluated, either as part of the design process (Is plan A better than plan B?) or as part of the final acceptance process (Plan A fulfills all my clinical requirements).

To a large extent, the evaluation of a particle therapy plan is very similar to the corresponding process in conventional therapy. The primary tool is the display of the 3-D, calculated dose distribution in relation to the patient's anatomy and to the previously defined target and critical structure volumes. To provide a more quantitative assessment, the use of dose volume histograms (DVHs) are mandatory. Indeed, for a large part of the plan evaluation process, only the dose distribution is required, and for the planning system it is not absolutely necessary to even know if the plan has been calculated for particles or photons. A dose distribution is a dose distribution after all, and this allows for the direct comparison of treatment plans from both modalities, as witnessed by the large number of treatment planning comparison studies published and still being published. However, from the practical point of view, there are a few additional points that need to be considered when evaluating particle therapy plans.

8.6.9 Geometrical Evaluations

Given the dependency of particle therapy on a good knowledge of the exact position and shape of the BP in the patient, a detailed geometric evaluation of every plan is a must. Whereas field selection in conventional therapy is mainly dose distribution driven (equally spaced fields for IMRT, parallel opposed wedged fields for breast, etc.), particle therapy plans are selected at least as much on geometrical considerations as well. As discussed in Section 8.5.2, density hetereogeneities can degrade the quality of the BP (and therefore sharpness of the distal falloff) and can also lead to range errors. Thus, it is prudent to also consider in detail beam incidences when evaluating particle therapy treatment plans. For this, it is extremely useful to be able to display the individual field dose distributions for the plan. At PSI, the plan QA process is so defined that this is mandatory for each plan, and a number of checks are made of each field. One of the main ones is a detailed review of the entrance path of the field *before* it enters the patient. For instance, fields should be checked that they are not passing through external structures that will vary in position day by day. Examples are earrings, belt buckles, and so forth, but also tubes and wires used for monitoring the patient during treatment. This can be a particular issue for the treatment of pediatric patients treated under anesthesia. In addition, the fixation devices themselves, or the treatment couch, can possibly intersect with fields, which again, if possible should be avoided in particle therapy. In addition, some consideration should be made to structures *internal* to the patient. We have already discussed the problems of density heterogeneities, and where possible these should be avoided and the plan evaluated from this point of view. Although there is no absolute right and wrong for this (from the clinical point of view, there may be no alternative to passing through complex density heterogeneities), it is certainly worthwhile to look at this for each field to be aware of the situation. Other issues are to ensure that the fields are not passing through gold teeth (it is extremely difficult to model the correct range through dense metals), and wherever possible, to avoid regions of the planning CT affected by reconstruction artifacts (again which can be caused by gold teeth or other metal implants). If the artifact regions cannot be avoided, then the possible effect on the calculated range should be estimated, and if necessary, action taken to "correct" the artifact as discussed above. Finally, due to the problems of range uncertainty and RBE, the position of the distal edges of fields in relation to critical structures should also be evaluated.

An invaluable tool that we have in our treatment planning system at PSI (for scanning and IMPT) is the ability to display, in three dimensions, the positions of all delivered BPs for an individual field (see Figure 8.12). With this, both the position and relative weights of the BPs of a field can be shown, and gives the planner a good feel for how the field is constructed. This can be also used to identify where the most highly weighted BPs are delivered and thus potentially indicated where the most sensitive areas will be to range and perhaps RBE uncertainties. Finally, such a distribution can be used as a QA tool, whereby the water-equivalent depth of a well-defined BP can be calculated from the treatment planning system. From this, it is possible to calculate the required energy for this peak and check this against the control system files for the

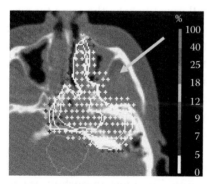

FIGURE 8.12 (**See color insert.**) Example slice through the same case as shown in Figure 8.11, but showing the positions and relative weights of the delivered BPs for one field of the plan (direction shown). Each cross marks the position of a delivered BP and the relative BP peak weight (fluence) is color-coded using the scale on the right.

field at this point. This, for instance, is a QA test that at PSI we make for every field we deliver.

8.6.10 Plan Robustness

Much of what has been discussed in this chapter is related to the possible uncertainties in particle therapy. Indeed, the very advantage of particles (the sharply defined BP) is potentially also its Achilles' heel. As mentioned a number of times already, if the range of the BP in the patient in not accurately known, then the delivered dose distribution could be quite different from that calculated. In practice of course, it is difficult to know the range *in vivo* exactly, and thus, we have to accept a certain amount of uncertainty in this parameter.

Given the importance of range, and its related uncertainty, it seems extremely prudent to be able to perform some robustness checks on the quality of the plan. For instance, how does the dose distribution change when the particle range changes by 3% or when the patient is shifted 2 mm to the left? This is what is called robustness analysis, and is something which, in our opinion, should be available in every treatment planning system (particle or otherwise). Such ideas were pioneered by Goitein in the 1980s (Urie et al. 1991; Goitein 1985) but unfortunately were not taken up by any of the commercial treatment planning firms. Some research groups are now returning to this (Albertini, Hug and Lomax 2010; Meyer et al. 2010; Soukup et al. 2009; Morávek et al. 2009; Lomax 2008b, c) and at PSI we are beginning to build such tools into our routine planning system. As an example, we are looking into the concept of dose-variance distributions as an additional tool for plan evaluation. In this approach, a number of dose distributions are calculated from the nominal distribution taking into account a number of different error conditions (spatial translations of the patient, range errors, etc.). These calculations then provide a number of calculated dose values per point, providing essentially a spectrum of dose values. From this, we calculate the spread of doses at very point, which then becomes the dose-variance distribution (Albertini et al. 2011b). An example of such a distribution is shown in Figure 8.13. As would be expected, for this rather simple geometry, the dose variance is largest at the field edges and low in the middle of the target, indicating that this plan is (in the target at least) extremely robust. Such distributions can also be used to construct dose-variance-volume-histograms, which can then be analyzed in much the same way as conventional DVHs (i.e., on an OAR-by-OAR basis, with

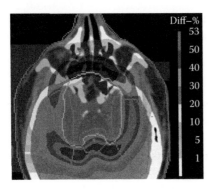

FIGURE 8.13 **(See color insert.)** Dose variance distribution calculated based on the technique described in Morávek et al. (2009) for a SFUD proton plan. Here the colors indicate the possible spread of doses at each point under a given set of error conditions, in this case setup errors along each of the cardinal axes. This figure shows that within the target volume, the dose is very robust (small variance), but at the lateral edges of the field, dose can vary considerably due to the sharp dose gradients in these regions.

the resulting curve now relating to the robustness of the dose within that structure). Such tools we believe can be extremely useful at the planning stage, particularly for IMPT plans, where the large degeneracy can lead to many plans giving very similar dose distributions, but where the robustness of the plan could be very different. It is therefore a shame that at the time of writing, and despite that the first papers in this area were published more than 20 years ago, no commercial particle therapy planning system provides such tools. Hopefully this will change in the near future.

On a final note, there is now a lot of interesting research taking place on robust optimization for particle therapy. As stated above, IMPT is inherently highly degenerate from the point of view of the clinical constraints alone. However, it is not necessarily degenerate when both clinical and robustness constraints are taken into account. Robust optimization therefore attempts to incorporate robustness issues already at the optimization level. Initial results look very promising (Albertini, Hug and Lomax 2010; Unkelbach, Chan and Bortfeld 2007; Unkelbach et al. 2009), and this is likely to be the future of particle therapy treatment planning. It can also be considered the past of treatment planning as well. For many years, planning for passive scanned proton therapy has manually taken robustness into account through the use of feathering of patched fields and multiple, overlapping fields (see Section 8.6.8). Plan robustness has always been a consideration for particle therapy, and will remain so.

8.7 Treatment Delivery and the Importance of Patient Monitoring

This chapter is about treatment planning for particle therapy and so not much will be said about the delivery itself. However, there are some issues to do with the delivery that can affect both the design of a plan and the workload. These will be briefly discussed in this final section.

At the most obvious level, it is impossible to completely decouple treatment planning from the actual delivery as, at the very least, a detailed description of the treatment machine itself (beam characteristics, etc.) have to be known by the TPS. However, at a more subtle level, there should also be *a priori* knowledge of the accuracy with which the patient (or better put, a population of patients) can be set up on the treatment machine. As is described in ICRU 50, such knowledge is essential for a reasonable estimate of the safety margins to be used [see, e.g., van Herk (2004)]. Generally this experience must be built up by each individual clinic depending on their methods for patient fixation and imaging and intervention strategy. For a detailed analysis of this for the PSI facility see the paper by Bolsi et al. (2008). In addition, for particle therapy, one should also have a good estimate for the likely range uncertainties associated with your system. As discussed above, 3% seems a reasonable estimate for the uncertainties from CT data alone (Schaffner and Pedroni 1998; Moyers et al. 2001), but other factors in the CT data may affect this (e.g., artifacts). One should also remember that range uncertainty is almost certainly systematic in nature, as opposed to random. That means that any error in the calculation of range at the time of the planning, and which is based on the CT data set, will very likely be the same for every fraction of delivery. Such an uncertainty is clearly much more problematic than random errors, which will generally smear out over the course of fractionated treatment. This difference is clearly reflected in the famous margin recipe published by van Herk (2004), whereby the effect of systematic errors on the final safety margin is a factor 3.5 times more than those of random errors.

In addition to positional and range errors, there are other patient- and delivery-specific factors that could affect planning. The most important are motion and anatomical changes during treatment.

Motion is a problem for all forms of radiotherapy, but is a much bigger problem for scanned particle beam therapy. It is out of the scope of this chapter to go into all the problems of motion and their possible management. However, suffice to say that when delivering dose using a narrow scanned beam in the presence of motion, so-called interplay effects can result, which can be thought of as being similar to constructive and destructive interference of waves. In the worst case, motion can lead to a considerable degradation of the applied dose homogeneity (Phillips et al. 1992) and is *a priori* difficult to predict due to the sensitivity of the dose distribution on the details of the scanning timing and motion phase, period, and amplitude. However, if motion is thought to be a problem at the planning stage, then some precautions can be taken to try to mitigate these effects, in the absence of more sophisticated methods such as gating, tracking, and so forth. One simple exponent is to try to select field directions that as much as possible are parallel to the main axis of motion. The interplay effect for motions parallel to the beam direction is considerably less than that for motions orthogonal to the beam. An example of this is our policy at PSI of treating spinal axis patients. Due to limitations in our gantry, and to avoid having to treat through the table, such patients are treated in the prone position. As such, due to chest wall motion from breathing, the whole spinal axis moves up and down a few millimeters in the AP direction. As has been shown by Phillips et al. (1992) even motions of a few millimeters can demonstrate substantial interplay effects, and to mitigate this, we generally treat using two to three field directions, all incident from the posterior direction. Such a case is shown in Figure 8.14. The fields are 0° (pure posterior) and ±30° from the posterior. All fields are generally incident then along the major axis of motion, and this field arrangement also has the added advantage of delivering almost no

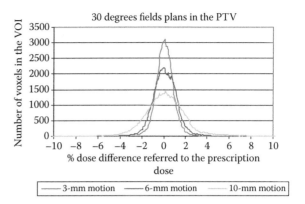

FIGURE 8.15 Motion analysis for the plan shown in Figure 8.14. Using a 4-D dose calculation, sin⁴ patient motions of 3, 6, and 10 mm along the A-P motion have been simulated and the differences between the static and motion cases calculated. The figure shows the difference histograms (the dose differences calculated over all calculation points with the PTV) for the different amplitudes. Even for a peak-to-peak amplitude of 1 cm, the dose differences are no more than ±6% in the worse case, and much smaller for the smaller amplitudes.

dose to the whole abdominal cavity of the patient. This is indeed a great advantage, but it must be stressed that these field directions have been selected predominantly to reduce the effect of motion of the spine. Figure 8.15 shows an analysis of such a field arrangement assuming different amplitudes of motion of the spine, indicating that, even when quite large motions are present, this approach is rather robust to AP motions. Indeed, this example also shows the second, simple expedient for motion management at the planning stage—the use of multiple SFUD fields. As has recently been shown by Knopf, Hug and Lomax (2011), the use of multiple SFUD beams can also be quite effective against the interplay effect for even quite large motions. Thus, knowledge of some of the limitations of the delivery (and patient during delivery) is invaluable knowledge to have already at the planning process.

Anatomical changes during treatment have another effect on the planning process. It is well documented, and well known by anybody who has treated patients with radiotherapy using normal fractionation regimes, that patients can easily change over the many weeks required for fractionated therapy. They can change weight, bodily cavities can fill in or empty out, and tumors can shrink (and unfortunately grow). All these effects are known from conventional therapy, and much work is currently being invested in ways to best deal with these effects (so-called adaptive therapy). However, for conventional therapy, this is mainly focused on preserving as much dose conformity as

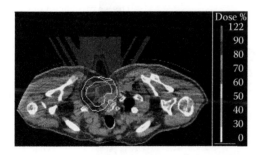

FIGURE 8.14 (See color insert.) Typical narrow-angle IMPT plan for a spinal axis chordoma. This is a three-field plan consisting of a posterior field and two posterior-lateral obliques, angles 30° away from posterior. Although this choice of angles nicely spares the abdominal organs, the choice is primarily driven by the requirement to bring the incident fields predominantly along the principle axes of patient motion, which for a patient in a prone position such as this is along the anterior-posterior direction.

possible by monitoring tumor changes and modifying the plan accordingly to track these changes. However, for particle therapy, due to the sensitivity of the delivered dose on the accurate range of the particles, anatomical changes can quickly have a significant effect on the delivered dose, and an effect that could be large enough that the plan needs to be modified, maybe based on a repeat CT.

The effect of weight loss and gain on IMPT treatments along the spinal axis have already been documented by Albertini et al. (2008). In Figure 8.16 an example of an extreme anatomical change is shown. This shows a single slice through a 3-D spot scanning proton plan applied to an extremely large skull-base chordoma. The CTV and PTV extend well into the nasal areas, as shown on the left. On the right is a slice through the same anatomical region on the first fraction of the treatment (about 3 weeks after the original planning CT). It is very clear that the anatomy now is completely different, and after all the discussions in this chapter about the problem of accurate range calculations, I also hope that the potential effect on the proton plan is also clear. In fact, the difference in dose

distribution is not as large as one may expect. Bottom left is the nominal dose distribution (planned on the original planning CT) and the dose distribution recalculated on the new CT. Although clear differences can be seen, particularly in the normal tissues lateral to the target volume, surprisingly the PTV remains well covered, and the brain stem well spared despite these huge anatomical changes. Nevertheless, a new plan was calculated as soon as possible for this case and the filling of the nasal cavities monitored for the rest of the treatment (they remained like those of the repeat CT for the rest of the treatment). This little scenario points out one of the realities of particle therapy. The highly conformal dose distributions produced by the treatment planning system are very sensitive to anatomical changes, with the consequence that patients may need to be more closely monitored during treatment in order to identify such problems as shown in Figure 8.16 and such that the clinical team can react quickly and replan (maybe after a repeat CT) as soon as possible. In summary, the planning process for proton therapy may not truly be over until the last fraction has been applied.

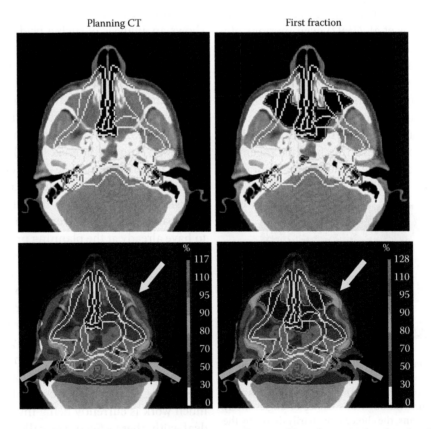

Planning CT   First fraction

FIGURE 8.16 **(See color insert.)** An example of extreme changes in nasal cavity fillings between the planning CT (right column) and the first treatment fraction (left). Nevertheless, despite these massive changes, because of the field arrangement chosen (indicated by the arrows), the differences in calculated dose are still rather small (see bottom two images).

8.8   Summary

In this chapter, we have tried to outline the main aspects of treatment planning for scanned particle beams. As such, one can never hope to go into the details of each of the processes involved in the treatment planning procedure. However, we hope that the issues and features of treatment planning for scanned beams have been sufficiently well reviewed that the reader can delve further into the selected literature of the reference list to find out more details from each of these aspects.

Treatment planning remains a critical foundation stone of high-tech radiotherapy, and the more sophisticated the delivery modality, the more important it is. Having said this, it is also a multivariate and not always easy problem. There are no necessarily right answers to what is the best plan, and the finally selected plan for delivery may well be planner- and doctor-specific. However, we hope that this chapter has provided a few pointers as to some of the thought processes and issues that occur during the planning process, and provides a foundation for further research and development in this exciting and important area.

References

Albertini, F., Bolsi, A., Lomax, A. J., Rutz, H. P., Timmerman, B. and Goitein, G. 2008. Sensitivity of intensity-modulated proton therapy plans to changes in patient weight. *Radiother. Oncol.* 86: 187–194.

Albertini, F., Hug, E. B. and Lomax, A. J. 2010. The influence of the optimization starting conditions on the robustness of intensity-modulated proton therapy plans. *Phys. Med. Biol.* 55: 2863–2878.

Albertini, F., Casiraghi, M., Lorentini, S., Rombi, B. and Lomax, A. J. 2011a. Experimental verification of IMPT treatment plans in an anthropomorphic phantom in the presence of delivery uncertainties *Phys. Med. Biol.* 56: 4415–4431.

Albertini, F., Negreanu, C., Hug, E. B. and Lomax, A. J. 2011b. Is it necessary to plan with safety margins for proton scanning plans? *Phys. Med. Biol.* 56: 4399–4413.

Bolsi, A., Lomax, A. J., Pedroni, E., Goitein, G. and Hug, E. 2008. Experiences at the Paul Scherrer Institute with a remote patient positioning procedure for high-throughput proton radiation therapy. *Int. J. Radiat. Oncol. Biol. Phys.* 71: 1581–1590.

Bortfeld, T., Buerkelbach, J., Boesecke, R. and Schlegel, W. 1990. Methods of image reconstruction from projections applied to conformation radiotherapy *Phys. Med. Biol.* 35: 1423–1434.

Boye, D., Knopf, A. and Lomax, A. J. 2010. 4D dose calculation on a deforming dose grid for scanned proton beams and ITV margin adaption due to proton range variations. (Manuscript in preparation.)

Cabal, G. and Jäkel, O. 2010. Towards a Probabilistic Approach for PTV Definition in Hadron Therapy. Proceedings of the XIVth ICCR, Amsterdam.

Deasy, J. O., Shephard, D. M. and Mackie, T. R. 1997. Distal edge tracking: A proposed delivery method for conformal proton therapy using intensity modulation. In *Proc XIIth ICCR (Salt Lake City)*, D. D. Leavitt and G. S. Starkschall (eds.). Madison, WI: Medical Physics Publishing, pp. 406–409.

Goitein, M. 1985. Calculation of the uncertainty in the dose delivered during radiation therapy. *Med. Phys.* 2: 608–612.

Hong, L., Goitein, M., Bucciolini, M., Comiskey, R., Gottschalk, B., Rosenthal, S., Serago, C. and Urie, M. 1996. A pencil beam algorithm for proton dose calculations. *Phys. Med. Biol.* 41: 1305–1330.

ICRU. 1999. Prescribing, Recording and Reporting Photon Beam Therapy (ICRU Report 50).

ICRU. 2007. Prescribing, Recording and Reporting Photon Beam Therapy. (ICRU Report 62, supplement to ICRU Report 50). J ICRU 7(2).

Jäkel, O. and Reiss, P. 2007. The influence of metal artifacts on the range of ion beams. *Phys. Med. Biol.* 52: 635–644.

Jiang, H. and Paganetti, H. 2004. Adaptation of GEANT4 to Monte Carlo dose calculations based on CT data. *Med. Phys.* 24: 2811–2818.

Karger, C. P., Schulz-Ertner, D., Didinger, B. H., Debus, J. and Jäkel, O. 2003. Influence of setup errors on spinal cord dose and treatment plan quality for cervical spine tumours: A phantom study for photon IMRT and heavy charged particle radiotherapy. *Phys. Med. Biol.* 48: 3171–3189.

Knopf, A., Hug, E. and Lomax, A. J. 2011. Scanned proton radiotherapy for mobile targets—Systematic study on the need and effectiveness of re-scanning in the context of different treatment planning approaches and motion characteristics. *Phys. Med. Biol.* 56: 7257–7271.

Lomax, A. J. 1999. Intensity modulated methods for proton therapy. *Phys. Med. Biol.* 44: 185–205.

Lomax, A.J. 2008a. Intensity modulated proton therapy. In *Proton and Charged Particle Radiotherapy*, T. Delaney and H. Kooy (eds.). Boston: Lippincott, Williams and Wilkins.

Lomax, A. J. 2008b. Intensity modulated proton therapy and its sensitivity to treatment uncertainties—Part 2: The potential effects of inter-fraction and inter-field motions. *Phys. Med. Biol.* 53: 1043–1056.

Lomax, A. J. 2008c. Intensity modulated proton therapy and its sensitivity to treatment uncertainties—Part 1: The potential effects of calculational uncertainties. *Phys. Med. Biol.* 53: 1027–1042.

Lomax, A. J., Pedroni, E., Schaffner, B., Scheib, S., Schneider, U. and Tourovsky, A. 1996. 3D treatment planning for conformal proton therapy by spot scanning. In *Proc. 19th L H Gray Conference*. K. Faulkner (ed.). London: BIR Publishing, pp. 67–71.

Meyer, J., Bluett, J., Amos, R., Levy, L., Choi, S., Nguyen, Q. N., Zhu, X. R., Gillin, M. and Lee, A. 2010. Spot scanning proton beam therapy for prostate cancer: Treatment planning technique and analysis of consequences of rotational and translational alignment errors. *Int. J. Radiat. Oncol. Biol. Phys.* 78(2): 428–434.

Morávek, Z., Rickhey, M., Hartmann, M. and Bogner, L. 2009. Uncertainty reduction in intensity modulated proton therapy by inverse Monte Carlo treatment planning. *Phys. Med. Biol.* 54: 4803–4819.

Moyers, M., Miller, D. W., Bush, D. A. and Slater, J. D. 2001. Methodologies and tools for proton beam design for lung tumors. *Int. J. Radiat. Oncol. Biol. Phys.* 49: 1429–1438.

Chapter 8

Newhauser, W., Giebeler, A., Langen, K. M., Mirkovic, D. and Mohan, R. 2008. Can megavoltage computed tomography reduce proton range uncertainties in treatment plans for patients with large metal implants? *Phys. Med. Biol.* 53: 2327–2344.

Oelfke, U. and Bortfeld, T. 2000. Intensity modulated radiotherapy with charged particle beams: Studies of inverse treatment planning for rotation therapy. *Med. Phys.* 27: 1246–1257.

Paganetti, H. 2002. Nuclear interactions in proton therapy: Dose and relative biological effect distributions originating from primary and secondary particles. *Phys. Med. Biol.* 47: 747–764.

Paganetti, H., Jiang, H., Parodi, K., Slopsema, R. and Engelsmann, M. 2008. Clinical implementation of full Monte Carlo dose calculation in proton beam therapy. *Phys. Med. Biol.* 53: 4825–4853.

Pedroni, E. 1981. Therapy planning system for the SIN-pion therapy facility. *Strahlentherapie* 77: 60–69.

Pedroni, E., Scheib, S., Boehringer, T., Coray, A., Lin, S. and Lomax, A. J. 2005. Experimental characterisation and theoretical modelling of the dose distribution of scanned proton beams: The need to include a nuclear interaction beam halo model to control absolute dose directly from treatment planning. *Phys. Med. Biol.* 50: 541–561.

Phillips, M., Pedroni, E., Blattmann, H., Boehringer, T., Coray, A. and Scheib, S. 1992. Effects of respiratory motion on dose uniformity with a charged particle scanning method. *Phys. Med. Biol.* 37: 223–234.

Schaffner, B. and Pedroni, E. 1998. The precision of proton range calculations in proton radiotherapy treatment planning: Experimental verification of the relation between CT-HU and proton stopping power. *Phys. Med. Biol.* 43: 1579–1592.

Schaffner, B., Lomax, A. J., Pedroni, E. and Tourovsky, A. 1997. Three analytical dose calculation algorithms for the PSI spot scanning system compared to a Monte Carlo simulation. *Proc. World Congress on Medical Physics and Biomedical Engineering*, Nice, September 1997, p. 973.

Scheib, S. and Pedroni, E. 1992. Dose calculations and optimisation for 3D conformal voxel scanning. *Radiat. Environ. Biophys.* 31: 251–256.

Schneider, U., Pedroni, E. and Lomax, A. J. 1996. On the calibration of CT-Hounsfield units for radiotherapy treatment planning. *Phys. Med. Biol.* 41: 111–124.

Schneider, W., Bortfeld, T. and Schlegel, W. 2000. Correlation between CT numbers and tissue parameters needed for Monte Carlo simulations of clinical dose distributions. *Phys. Med. Biol.* 45: 459–478.

Soukup, M., Fippel, M. and Alber, M. 2005. A pencil beam algorithm for intensity modulated proton therapy derived from Monte Carlo simulations. *Phys. Med. Biol.* 50: 5089–5104.

Soukup, M., Söhn, M., Yan, D., Liang, J. and Alber, M. 2009. Study of robustness of IMPT and IMRT for prostate cancer against organ movement. *Int. J. Radiat. Oncol. Biol. Phys.* 75: 941–949.

Tourovsky, A., Lomax, A. J., Schneider, U. and Pedroni, E. 2005. Monte Carlo dose calculations for spot scanned proton therapy. *Phys. Med. Biol.* 50: 971–981.

Unkelbach, J., Chan, T. C. Y. and Bortfeld, T. 2007. Accounting for range uncertainties in the optimization of intensity modulated proton therapy. *Phys. Med. Biol.* 52: 2755–2770.

Unkelbach, J., Bortfeld, T., Martin, B. C. and Soukup, M. 2009. Reducing the sensitivity of IMPT treatment plans to setup errors and range uncertainties via probabilistic treatment planning. *Med Phys* 36: 149–163.

Urie, M. M., Goitein, M., Doppke, K., Kutcher, J. G., LoSasso, T., Mohan, R., Munzenrider, J. E., Sontag, M. and Wong, J. W. 1991. The role of uncertainty analysis in treatment planning. *Int. J. Radiat. Oncol. Biol. Phys.* 21: 91–107.

van Herk, M. 2004. Errors and margins in radiotherapy. *Semin. Radiat. Oncol.* 14: 52–64.

Yepes, P., Randeniya, S., Taddel, P. J. and Newhauser, W. D. 2009. Monte Carlo fast dose calculator for proton radiotherapy: Application to a voxelized geometry representing a patient with prostate cancer. *Phys. Med. Biol.* 54: N21–N28.

Section III

9. Conformal Proton Therapy

Jerry D. Slater and David A. Bush

Proton radiation therapy began nearly six decades ago as an investigative treatment in high-energy physics laboratories. The first hospital-based center is now more than two decades old. Even so, proton therapy is a young modality, still in the process of evolving and maturing. Some of that process is suggested by the various terms used to describe it: "conformal," "image guided," and "intensity modulated" among them, as are detailed in this chapter. This chapter discusses conformal proton therapy and will show that aspects of image guidance and beam modulation are used in conformal delivery as well and have been used virtually since protons began to be utilized clinically, especially in hospital settings.

9.1   History

The hallmark of proton radiation therapy is precision—the ability to deposit the needed dose of ionizing radiation in a discrete volume and spare a great deal of the normal tissue intimate to that volume. The ability to do this has always depended on the prior ability to define the intended volume as precisely as possible. Since the inception of proton therapy at the University of California, Berkeley, in 1954 (Tobias et al. 1956), the clinical sites selected for treatment have depended on some form of imaged guidance; accordingly, targets treated in the early days of proton therapy were relatively small ones that could be localized with relatively routine imaging. Proton beams proved to be successful in managing such conditions as tumors of the pituitary gland (Kjellberg et al. 1968) and arteriovenous malformations (Kjellberg 1986), for example. Ocular melanomas, which also can be localized with imaging and fiducials, is another application for which proton therapy has a long and successful history (Gragoudas et al. 1984;

Munzenrider et al. 1988); even in uncommon cases of recurrence, the highly conformal dose distribution possible with proton beams permits successful retreatment with the continued potential for eye retention (Marucci et al. 2011).

Successful applications of proton therapy to a broader range of anatomic sites depended on precise localization of target volumes in more complex anatomic environments. At our institution, work on developing a heavy-charged-particle treatment facility began in 1970. A feasibility study conducted at that time determined that a facility could not then be built for several reasons, salient among them the need to locate target volumes with better imaging and hence, greater precision. Dr. James M. Slater and colleagues began work on a treatment planning system to guide radiation treatments; the initial system, employed clinically in 1971, used ultrasound images (Figure 9.1; Slater et al. 1974), followed in the mid-1970s by a system using CT scans (Figure 9.2; Neilsen, Slater and Shreyer 1980). These initial attempts at image-guided therapy planning appear crude now, but were early responses to the need to better realize the target volume and thus enhance the precision of

Proton and Carbon Ion Therapy Edited by C.-M. Charlie Ma and Tony Lomax © 2013 Taylor & Francis Group, LLC. ISBN: 978-1-4398-1607-3.

Chapter 9

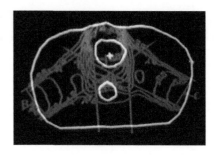

FIGURE 9.1 **(See color insert.)** Therapy planning image from 1973, Loma Linda University Medical Center. The dose distribution and lines of attenuation in tissue were developed from ultrasound scans.

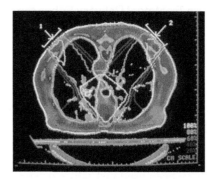

FIGURE 9.2 **(See color insert.)** Therapy planning image from 1978, Loma Linda University Medical Center. This planning image was developed from CT scans. In addition to displaying the anatomy, CT scan-based planning images allowed assessment of density variations as X-ray beams passed through tissue.

treatment. Other planning systems subsequently were developed, notably at Massachusetts General Hospital (MGH; Goitein and Abrams 1983; Goitein et al. 1983). As that image-based therapy planning system developed, the applications of proton therapy expanded at MGH in the 1970s to include clinical problems such as chordomas of the base of the skull (Austin-Seymour et al. 1989) and larger-field applications such as cancer of the prostate (Suit et al. 1982; Shipley et al. 1988). The MGH planning system was used at our proton facility when it began clinical operation in 1990; it was replaced with a system of our own devising in the mid-1990s. The availability of such systems was essential for using proton beams clinically for any but the simplest of setups.

As time passed, clinical capabilities grew significantly. Sites treated with proton beams rose from less than five anatomic sites in the early 1960s to over 50 sites in the twenty-first century. A large part of this growth occurred owing to better imaging capabilities from a variety of sources, and from the development of therapy planning systems that can integrate the imaging data from these sources. Proton therapy followed quickly to take advantage of these developments, as well as advances in other forms of technology. The technological advances that have helped spur the use of proton therapy have included:

- The development of accelerators designed and dedicated for clinical applications, rather than adapting machines designed for physics research
- Positioning and immobilization systems that allow the physician to minimize target motion and maximize patient comfort
- Treatment planning systems that enable the physician to design and review plans in real time
- Alignment and verification technologies to improve image-guided delivery, which has been an integral part of proton therapy since its inception
- Beam delivery systems that allow the physician to manipulate the small Bragg peak (BP), in turn providing the capability for expanding, shaping, modulating, and scanning the proton beam to the precise shape of the target

9.2 An Evolving Modality

As technology has advanced, so too has the practice of proton therapy. The process has been a series of steps building on prior experience, as an overview of proton treatment of prostate cancer at our institution might help to illustrate.

Our experience with protons for prostate cancer began in 1991, a year after the facility opened and after the first of the facility's three gantries was commissioned (Slater, Miller and Archambeau 1988; Slater et al. 1992). Treatment programs were developed within a conservative approach aimed at optimizing proton therapy. Because the Loma Linda University Medical Center (LLUMC) proton treatment facility was designed to treat up to 200 patients per day, radiation oncologists at our institution could accumulate a large, intrainstitutional clinical experience. The initial treatment scheme was not very different from standard photon protocols of the time (we increased the total delivered dose by only 10%). Our first analyses showed that conformal proton radiation therapy at that dose level yielded disease-free survival rates comparable with other forms of local therapy, with only minimal morbidity (Slater et al. 2004). Results were even more favorable when we evaluated patients with early-stage disease (Slater et al. 1999). Further, in reference to the common perception and presumption that younger men (less than 60 years of age) should be treated by surgery rather than radiation for prostate cancer, our studies demonstrated no significant difference when groups were divided by age (Rossi et al. 2004). All of these outcomes were obtained with initial treatment protocols that delivered a modest increase in total dose, compared to photon schemes then in use.

As this total experience accumulated, it gradually became apparent that the precise dose distribution and normal-tissue sparing made possible by the proton beam would enable us to deliver higher total doses to increase the probability of disease control while retaining low rates of side effects. Accordingly, we undertook a phase III randomized trial in collaboration with investigators from MGH. Outcomes of this study demonstrated that men with clinically localized, early-stage prostate cancer had a significantly increased likelihood of biochemical control if they received high-dose conformal radiation, without increasing grade 3 acute or late urinary or rectal morbidity (Zietman et al. 2008). Longer follow-up of this randomized controlled trial continues to show superior long-term cancer control for men with localized prostate cancer receiving high-dose versus conventional-dose radiation, and continues to do so without an increase in grade 3 or higher late urinary or rectal morbidity (Zietman et al. 2010).

As of July 2010, approximately 68,000 patients had been treated with protons worldwide (PTCOG). Over 99% of them were treated with some version of passive scattering or limited scanning that includes daily image target alignment verification. These modes of delivery also have used what might be considered forms of intensity-modulated protons, by breaking the target into subcomponents, each of which is treated with different beams and doses. Indeed, many of the techniques presently used in state-of-the-art photon therapy facilities had their beginnings and development in the charged-particle world.

9.3 Some Clinical Applications of Conformal Proton Therapy

Protons now are employed for treating a wide variety of cancers and other diseases. At our institution, for example, we use protons to treat clinical problems in more than 50 anatomic sites. This chapter will describe our work in four of them: cancers of the lung, breast, liver, and prostate.

9.3.1 Medically Inoperable Non–Small-Cell Lung Cancer

Although surgical resection with lobectomy or pneumonectomy is standard treatment for stage I disease, not all patients are suitable for surgery owing to comorbid conditions such as cardiopulmonary disease and heart disease, conditions frequently seen in patients with lung cancer because of the strong correlation with cigarette smoking. Some otherwise eligible patients will refuse surgical intervention. Most of these patients will be referred for radiotherapy with or without systemic chemotherapy. However, local failure rates often exceed 50%, in many cases owing to delivery of inadequate doses because of the risk of permanent injury to the large volume of normal tissue exposed (Rowell and Williams 2001; Zimmermann et al. 2003). Protons are used to reduce the volume of normal tissue exposed, reduce the risk of side effects owing to normal-tissue exposure, and thus deliver a higher effective dose to the tumor volume.

At our institution, eligible patients are required to have a histologic diagnosis of non-small-cell lung cancer with clinical stage I disease. Patients must be either medically inoperable or have refused surgical resection; all are required to review and sign a study-specific, Institutional Review Board-approved informed consent document. Required pretreatment evaluations include a CT scan of the chest and upper abdomen, pulmonary function testing, and a staging positron emission tomography scan.

Therapy planning is done via a CT scan of the chest while the patient lies in a custom-made, full-body immobilization device. Physiologic respiratory tumor motion is determined with fluoroscopy (motion target). The gross tumor volume is identified on the planning CT scan, with a margin added to account for respiratory motion. If the internal tumor motion is significant and leads to unacceptable exposure of normal lung tissue, a breath-holding technique may be employed to reduce the lung dose. A 3-D treatment plan is then developed to deliver proton beam radiotherapy to the motion target while minimizing the dose delivered to the surrounding lung tissue. No treatment is planned for the mediastinum.

Chapter 9

Initial treatment consisted of a total dose of 60 cobalt Gray equivalent (CGE) in 10 equally divided fractions over 2 weeks (CGE = dose in Gy × 1.1 relative biologic effectiveness). A typical beam arrangement employed three to four beam angles centered on the motion target, with lateral and posterior beams used most frequently. A minimum of two fields were treated each day. All patients were monitored weekly for acute toxicity during treatment.

Clinical results at obtained Loma Linda University were published initially in 2004 (Bush et al. 2004b). Using the Kaplan–Meier method, 3-year analysis revealed a local control rate of 74%, an overall survival rate of 44%, and a disease-specific survival rate of 72%. The metastatic relapse rate was 31% at 3 years. These results compared favorably with summarized published results of photon treatment of similar-stage disease: in a compilation of several such studies, the 3-year overall survival rate was approximately 32%, while the disease-specific survival rate was approximately 43%. Lack of local tumor control at the primary site was a major mode of failure with photon-based therapies, at approximately 41% (8% to 68%). In general, comparative photon regimens delivered the biological equivalent of 60 to 70 Gy to the primary tumor (i.e., substantially less than the biologic proton dose delivered in our series). The control and survival results obtained with protons, however, occurred in the presence of markedly limited toxicities. Acute and late toxicities were evaluated prospectively in all patients. Acute toxicities were limited to mild fatigue and radiation dermatitis that was seen as mild-to-moderate erythema and required no specific medical treatment. No cases of clinical acute radiation pneumonitis were identified; this was consistent with results anticipated from prior investigations, which showed reduced pulmonary injury owing to reduced normal-tissue exposure with protons (Bush et al. 1999). No patient required steroids or anti-inflammatory therapy. No cases of acute or late esophageal or cardiac toxicity were identified.

We noted a significant difference in local disease control when comparing patients with T1 tumors treated with protons, compared to those with T2 tumors. Three years following treatment, 87% of patients with T1 disease had local tumor control compared to 49% of those with T2 tumors ($p > 0.03$). Further, we noted a significant improvement in survival in patients who received a dose of 60 CGE; the 3-year overall survival rate for an initial group that had received a total dose of 51 CGE group was 27%, versus a 55% survival rate for patients in the 60-CGE group ($p > 0.03$). Predicted mortality from concurrent disease, based on the Charlson Comorbidity Index, correlated well with observed comorbidity-specific mortality, thus helping to substantiate disease-specific survival rates in our population of lung cancer patients (Do, Bush and Slater 2010). In an effort to reduce local failures in T2 patients, and owing to the observed paucity of toxicities, the current regimen raises the total dose to 70 CGE.

A comparison of single-beam photon and proton treatment dosimetry for stage I lung cancer (Figure 9.3) demonstrates the essential reason for the difference in outcomes, especially those related to toxicity and the ability to complete treatment to the prescribed total dose. The proton dose can, in most instances, be confined entirely within the involved lung. This is substantiated by Chang and colleagues (2006) in their formal comparison between proton and photon treatment planning. They analyzed 10 patients with inoperable stage I lung cancer, comparing proton treatment plans to 3-D conformal photon plans at two total dose levels (66 and 87.5 Gy). Analysis

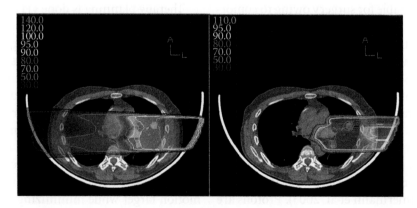

FIGURE 9.3 **(See color insert.)** Single-beam dosimetry of a lateral photon beam (left) and lateral proton beam (right). These images demonstrate how lateral proton beams stop at the surface of the mediastinum, while photons do not.

revealed an approximately 50% reduction in nontarget lung dose when protons where used. The normal lung tissue-sparing effect apparently was increased with the high-dose plans, indicating an additional benefit for dose escalation when proton beams are used. Fifteen patients with stage III lung cancer were also analyzed, comparing 3-D photon, IMRT, and proton beam plans at two total dose levels (63 and 74 Gy). Again results showed a substantial reduction in the nontarget lung dose, and again, the proton benefit seemed to be more evident with the higher-dose plans. Doses to the heart, esophagus, and spinal cord were reduced significantly. The study suggests that protons can achieve higher total doses to the target volumes, with more significant normal-tissue sparing, than 3-D conformal radiation therapy or IMRT.

Protons are also used at our institution to treat patients with locally advanced non-small-cell lung cancers. Underway at present is a phase I/II study of combined chemotherapy and high-dose, accelerated proton radiation for treating such tumors. Induction chemotherapy with paclitaxel and carboplatin is followed by weekly carboplatin, paclitaxel, and thoracic proton radiotherapy given with a high-dose, accelerated schedule to a total dose of 76 CGE in 5 weeks. The clinical target volume (CTV) includes all measurable disease as identified on thoracic CT scan, including the ipsilateral hilar and mediastinal lymph node regions. The gross target volume (GTV) includes the primary lung tumor and any clinically involved hilar and/or mediastinal lymph nodes. The dose to the CTV is 46 CGE given in 23 fractions of 2 CGE. The dose to the GTV is 30 CGE delivered in 15 2-CGE fractions; this boost is given in a twice-daily fashion during the final 15 treatments to the CTV, with daily fractions separated by an interval of at least 6 h. As in the previous protocol, the objectives of this ongoing study are to determine local control, overall and disease-free survival, toxicity, and quality of life following treatment under the combined regimen.

9.3.2 Breast Cancer

Although whole-breast radiotherapy following lumpectomy for invasive carcinoma of the breast is an established treatment, mounting evidence suggests that at least some of these patients may be well treated with partial breast irradiation (PBI). Indeed, multiple reports from phase II trials indicate that postoperative radiotherapy directed solely to the lumpectomy site may be effective in reducing local relapse rates in a subset of patients (Smith et al. 2009). Accordingly, PBI is currently being compared to standard whole-breast radiotherapy in multi-institutional randomized trials.

PBI can be administered by a variety of modalities. Among those employed have been high- and low-dose-rate brachytherapy, conformal external-beam photon radiotherapy, and intraoperative photon radiotherapy. At our institution we have used proton therapy because we believe it has several advantages over other forms of radiotherapy for PBI. Proton therapy eliminates the additional surgical procedure required for brachytherapy. In addition, it improves dose homogeneity within the target volume, which may improve cosmetic results and reduce the risk of symptomatic fat necrosis associated with brachytherapy. In comparison to 3-D conformal photon radiation therapy, the inherently superior depth dose characteristics minimize the integral dose delivered to surrounding normal tissues, particularly the lung and heart, resulting in decreased toxicity and reduced risk of radiation-induced malignancy.

We have completed two 50-subject phase II studies of proton therapy for post-lumpectomy PBI. The objectives of both studies were to evaluate the technical feasibility, acute and long-term toxicity, cosmetic outcomes, and local tumor control and survival rates associated with the procedure. Eligible patients included women with histologic evidence of invasive carcinoma (except lobular carcinoma) or (in the second study) ductal carcinoma *in situ*. Patients had primary tumors ≤3 cm in diameter. All patients with invasive carcinoma had an axillary node dissection or sentinel lymph node biopsy; a positive sentinel node indicated the need for axillary node dissection. In the first study, patients with positive axillary nodes were excluded; in the second, patients were eligible if they had no more than three nodes containing metastatic carcinoma and had no evidence of extracapsular extension. In both studies, the negative lumpectomy margins had to be at least 2 mm; there could be no evidence of distant metastasis, an extensive intraductal component, multicentric ipsilateral carcinoma or contralateral carcinoma, or prior radiation or chemotherapy for breast cancer.

Therapy planning begins with construction of an immobilization device designed to support the ipsilateral breast, compress the contralateral breast, and minimize motion due to breathing by placing the patient in the prone position (Bush et al. 2007; Figure 9.4). A planning CT scan is taken in the treatment position. On the resulting image the physician outlines the tumor bed, using clips that were implanted at lumpectomy,

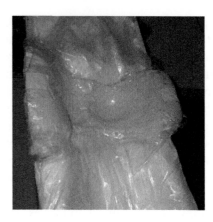

FIGURE 9.4 Immobilization device allowing the dose to the ipsilateral breast and skin to be well controlled. The patient is fitted with a treatment brassiere (Med-Tec Inc., Orange City, IA) that supports the ipsilateral breast with a rigid plastic cup while compressing the contralateral breast. Patients are then placed in a prone position in a rigid polyvinyl chloride cylinder, supported from above and below by foam bead vacuum cushions (Vac-Lok, Med-Tec Inc.).

plus relevant clinical information. A 1-cm margin is added in all directions to create the CTV, excluding the chest wall and skin. The physician then creates a multifield treatment plan, with the 90% isodose covering the CTV while the skin surface receives no more than 70%. The potential for skin toxicity is an area of concern. Accordingly, treatment planning techniques are designed to keep the 90% isodose line within the skin of the breast. Acceptable skin sparing is accomplished by using a unique immobilization system, treating multiple fields per day, using oblique beams when necessary and manually editing apertures during treatment planning when the skin dose is excessive. It appears that skin complications may be minimized by following these or similar techniques.

On each treatment day, the patient is fitted with the treatment brassiere, which is reproducibly placed on the chest using applied skin marks. The patient is positioned in her immobilization device in a prone position on the treatment couch. Orthogonal diagnostic-quality X-rays are performed before each treatment and are compared with digitally reconstructed radiographs to reproduce the body/breast position at the time of the treatment planning CT scan. Titanium clips, which are placed at the periphery of the lumpectomy cavity, are used for alignment. At least two separate fields are treated daily. The CTV receives 4.0 CGE, given daily in 10 treatment days over 2 weeks, for a total of 40 CGE.

Results have been compiled for the first study and have been submitted for publication. Results of the second study are still being accumulated. In the first study, the overall 5-year survival rate was 96%; the disease-free survival rate at 5 years was 92%. There were no recurrences within the treated area of the breast or within the treated quadrant of the ipsilateral breast. Three patients developed distant metastases. Acute toxicities were rated according to the common toxicity criteria (CTC v.2.0). These were limited to cases of radiation dermatitis: 26 patients had grade-1 skin reactions, while four patients had grade-2 sequelae. There were no cases of grade-3 or higher skin toxicity, and no other acute toxicities were observed. Late skin toxicity included three subjects with grade-1 telangiectasias. There were no cases of skin ulceration or clinical fat necrosis of the breast, nor were there instances of cardiac events, clinical radiation pneumonitis, or rib fractures. Ninety percent of patients rated their cosmetic result as good or excellent.

Control and survival outcomes with protons appear to compare favorably to photon-based modalities, and toxicity and cosmesis outcomes appear to be, in many instances, superior. Although photon brachytherapeutic methods can reduce radiation doses to nontargeted tissues compared to external-beam techniques, they require invasive procedures; sequelae can include wound complications and posttreatment infections. The inherently nonuniform dose distribution within the target region leads to areas where the dose is significantly higher than that prescribed, by as much as 50% to 100% (Major et al. 2009). It is likely that the physical trauma from interstitial brachytherapy, along with areas of excessive dose, has led to a measurable incidence of fat necrosis within the breast (Wazer et al. 2001). Intracavitary balloon brachytherapy, like interstitial brachytherapy, has been reported to run a risk of wound complications and infections, as well as seroma formation and fat necrosis (Nelson et al. 2009). The technique is suitable only for patients with certain lumpectomy cavity geometries and treatment areas that are sufficiently distant from the skin surface (Chen and Vicini 2007).

Three-dimensional conformal or IMRT photon techniques can lead to superior dose uniformity within the targeted area when compared to brachytherapy, but this outcome is obtained at the expense of increasing volumes of nontargeted tissues to low and moderate doses. At least one recent publication has questioned the adequacy of the cosmetic results with these techniques; this report has correlated negative cosmetic outcome with increasing dose to nontargeted breast tissue (Jagsi et al. 2010).

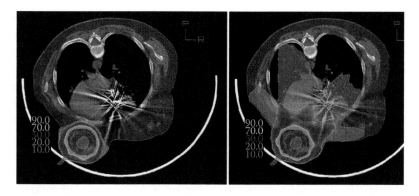

FIGURE 9.5 **(See color insert.)** Left: Typical proton dose distribution utilizing four equally weighted beams (lateral, anterior, right, and left anterior oblique). Right: Same beam arrangement using 6-MV X-rays. Although this photon treatment plan would not likely be used for treatment, the comparison demonstrates the normal tissue-sparing effects of proton beams; note that the dose distribution from the proton beams avoids the heart, lungs, and esophagus.

Our studies indicate that proton radiotherapy used for PBI produced excellent disease control within the ipsilateral breast. It was applicable to all patients enrolled in the trial without restrictions due to tumor size, location, or geometry, owing to the fact that beam-shaping methods for proton beam delivery provide a much more flexible treatment delivery process than can be given with brachytherapy. Skin toxicity is easily managed by using a multibeam technique and editing individual beam-shaping devices to control the dose delivered to the skin; doing so in the present trial led to acceptable acute and late skin toxicity outcomes. The lack of clinical fat necrosis compares favorably to other reports using brachytherapy. Proton beam radiotherapy can significantly reduce the volume of nontarget tissues exposed to irradiation when compared to photon-based IMRT treatments (Figure 9.5). The ipsilateral breast dose with proton beam appears to be significantly lower than that commonly reported with photon techniques, as was demonstrated in a recent treatment planning comparison (Moon et al. 2009).

9.3.3 Hepatocellular Carcinoma

Primary cancer of the liver has long resisted successful treatment. Surgery (hepatectomy or transplantation) is the preferred option, but most patients do not qualify owing to advanced cirrhosis and/or decompensation and strict eligibility criteria for transplantation. Several local treatments are available for patients with unresectable hepatocellular carcinoma (HCC), for those waiting liver transplantation, and as downstaging modalities for patients with tumor burdens exceeding the Milan criteria. The most commonly utilized such treatment is transarterial chemoembolization (TACE).

Historically, radiation therapy has not been part of the overall treatment paradigm for patients with unresectable HCC because the liver has a very limited ability to tolerate substantial doses of radiation.

Because proton beam radiotherapy has unique properties that allow delivery of high doses to targets within the body while significantly reducing doses to surrounding healthy tissues, we are conducting an ongoing prospective phase II trial of protons for HCC. A preliminary report was published in 2004 (Bush et al. 2004a). Since that time we have continued to enroll subjects in this trial to further evaluate the efficacy and toxicity of proton radiation therapy for patients with HCC. Eligible subjects have preexisting cirrhosis, as diagnosed by tissue biopsy or with abdominal imaging (CT scan or MRI) performed within 3 months of enrollment and showing characteristics of HCC. An independent radiologist verifies and confirms the diagnosis of HCC. Patients are eligible regardless of tumor size, transplant candidacy, or alpha fetoprotein (AFP) level. It is required, however, that the 50% isodose line encompass no more than one-third of the nontumor hepatic tissue. Patients with extrahepatic spread, >3 lesions, absence of cirrhosis, and tense ascites are excluded. The diagnosis of cirrhosis is determined by the study hepatologist with clinical and radiologic examinations. Patients are followed in the hepatology and radiation medicine clinics to assess overall survival, progression-free survival, patterns of recurrence, and treatment-related toxicity.

Treatment planning begins with construction of a full-body immobilization device that is used to create the planning CT and for all treatment sessions. Subjects undergo CT simulation in the immobilization device; images are acquired with IV contrast enhancement during the arterial phase, and acquisition takes place only when the subject is actively breath holding at end-inspiration. The GTV is contoured on each CT image and includes all areas of known disease as identified on the planning images and/or seen on diagnostic CT or MRI. A CTV is created to extend 1 cm beyond the GTV in all dimensions, unless edited if the volume extends beyond the liver parenchyma so that it will not include organs known not to be involved by the tumor (i.e., abdominal wall, kidney, bowel). All organs and normal tissue regions not targeted for treatment are contoured for dose-volume-histogram calculations. The entire liver is contoured, taking care to omit the gall bladder and portal vessels. Bowel lying near the target volume and the right kidney also is contoured.

A computerized 3-D treatment plan is generated. The target region is the CTV with the additional margin to account for daily setup variations and energy-specific beam penumbra. Beam angles typically used include right lateral, posterior, and posterior oblique beams; anterior beams are used only in selected patients, as they will not be within the immobilization device.

All doses are prescribed to a point at or near the center of the CTV target region. The total dose delivered is 63 CGE, given in 15 fractions over 3 weeks. The uniformity of dose across the target region is designed to not vary by more than 10% of the prescribed dose. At least two beams are delivered each day to reduce the dose at the beam entry site. The volume of liver receiving more than 25 Gy is limited to 33% or less whenever possible. If any portion of the bowel falls within the 90% isodose volume, field reductions are done. Treatments are planned so that at least 50% of the right kidney receives no more than 20 Gy; the left kidney is spared. Each week, all patients are clinically evaluated by the treatment radiation oncologist to assess treatment tolerance and to monitor for acute toxicities.

Seventy-six subjects were enrolled between April 1998 and October 2006. Fifty-four percent of the patients were outside the Milan criteria, 24% were Child class C, and 16% had a MELD score >15. Sixty-five percent of patients were within the San Francisco criteria. Although 86% had a solitary lesion, in 48% the tumor size exceeded 5 cm in diameter. Patients in the study represented a cohort with relatively advanced-stage HCC, and a significant number had advanced decompensated liver disease. The median progression-free survival time for the entire group of patients was 36 months, with a 95% confidence interval of 30 to 42 months. Survival was significantly better for patients

within the Milan criteria, with a 3-year progression-free survival rate of 60%, as opposed to 22% for those outside the criteria. Eighteen subjects underwent liver transplantation following treatment. The 3-year survival rate for transplanted patients following proton therapy was 70%, compared to 10% for those not transplanted. Among patients who had elevated AFP levels prior to treatment, 62% demonstrated a decrease.

Local treatment failure—defined as radiographic signs that the size of the treated tumor was increasing or if AFP increased following treatment without any radiographic disease progression outside of the primary treatment area—occurred in 20% of patients. Only three patients experienced local treatment failure without also having other new lesions develop in other parts of the liver; 36% of subjects experienced recurrent tumor in other nontreated portions of the liver.

All explanted livers were subject to pathologic analysis. Six of eighteen (33%) showed no evidence of residual gross or microscopic hepatocellular carcinoma. Thirty-nine percent showed microscopic residual HCC, while 28% showed gross residual HCC.

Acute toxicity during proton therapy was minimal and included mild fatigue and skin reactions consisting of erythema (grade 1). No acute toxicities required treatment to be interrupted or discontinued. Five patients experienced grade 2 gastrointestinal side effects following treatment; all cases were managed medically. Radiation-induced liver disease (RILD) was evaluated by clinical and laboratory evaluation; overall, no statistically significant change in aspartate aminotransferase, alanine aminotransferase, alkaline phosphatase, bilirubin, albumin, and prothrombin time was observed. The treatment paradigm was determined to be safe and efficacious (Bush et al. 2011).

Although treatment of HCC remains challenging, proton therapy offered good outcomes for many of the patients we treated. The primary reason lies in its ability to spare normal tissue and to permit the use of few treatment portals. Few comparisons are possible with photons, inasmuch as photon radiation historically has been little used to treat HCC because of the greater volume of liver tissue that must receive at least some dose of radiation when photon beams are used (Figure 9.6). Proton therapy for localized HCC has also been reported from proton treatment centers in Japan: outcomes from those studies suggest that proton therapy can be a safe and effective treatment for patients with unresectable HCC (Chiba et al. 2005; Hata et al. 2006). However, photon-based therapy is being increasingly utilized as treatment for patients with unresectable HCC, thanks to technologies such as

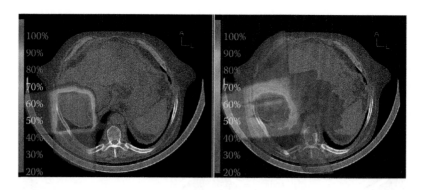

FIGURE 9.6 **(See color insert.)** Treatment planning comparison of a two-field proton plan (left) and a five-field IMRT plan (right) for a large hepatocellular carcinoma (outlined in red).

3-D treatment planning and IMRT, which have allowed higher doses to be given to targets within the liver. An excellent review of such treatment indicates that 2- to 3-year overall survival rates from photon regimens, most of them in combination with TACE, are in the range of 20% to 35% (Hawkins and Dawson 2006).

The role of radiation therapy in HCC is changing; proton therapy is one manifestation of that change. It is still uncertain, however, how localized radiotherapy treatments compare to standard treatments. TACE is the most commonly used treatment for patients with unresectable HCC. Due to competing comorbid conditions such as cirrhosis that exist in this group of patients, it is likely that the relative effectiveness of a new mode of therapy can only be demonstrated through a well-designed randomized clinical trial. Accordingly, we have designed and initiated such a trial, which compares proton radiotherapy to TACE for patients with newly diagnosed HCC.

9.3.4 Adenocarcinoma of the Prostate

As noted above, our institution has a long and evolutionary history with proton therapy for prostate cancer. The treatment protocols we use now are the result of almost two decades of investigation. Treatment of prostate cancer with protons has migrated from perineal ports delivered with fixed beams to multifield approaches and increasing doses. Initial randomized trials in conjunction with MGH revealed increase biochemical disease-free survival in early prostate cancer. This was seen in patients receiving the high-dose arm as compared to the conventional dose (Zietman et al. 2010). No increased late toxicity was seen in the high-dose arm of the trial. Based on these favorable outcomes, trials now include hypofractionation versus standard fractionation for low-risk prostate cancer,

dose escalation utilizing hypofractionation in patients at intermediate risk, and combined therapy of hormones, photons, and protons for high-risk prostate cancer. Patients are categorized based on stage, PSA levels, and pathology. The latter is being compared with our standard protocol, which calls for a high total dose, delivered with protons alone for most patients with stage T1–T2C disease, Gleason scores ranging from 2 to 6, and PSA values of 10 or less.

Pretreatment evaluation includes history and physical examination, including digital rectal examination; baseline urinary status evaluation, including urinalysis; evaluation of pelvic lymph nodes, pretreatment prostate specific antigen (PSA) determination; and, optionally, endorectal MRI/magnetic resonance spectroscopy (MRS), recommended in patients who are clinically stage T2B–T2C.

Each patient is immobilized in a full-body pod, while positioned supine. A water balloon is placed into the rectum at the time of immobilization; thereafter, a treatment planning CT is performed. Each patient is encouraged to fill his bladder by drinking 8 to 16 ounces of fluid prior to the CT scan. These immobilizations are repeated prior to each daily treatment, and localization is achieved by acquiring orthogonal radiographs immediately prior to each daily treatment; each daily patient alignment is performed with reference to bony landmarks. Fiducials have been placed into the prostate and motion measured relative to bony landmarks as part of clinical trials, but to date, using our external and internal immobilization, no significant prostate motion has been seen outside of our PTV margins.

In treatment planning the physician defines the GTV as all known disease being encompassed in the prostate gland, as indicated by the planning CT scan, additional imaging studies (ultrasound, MRI), and

Chapter 9

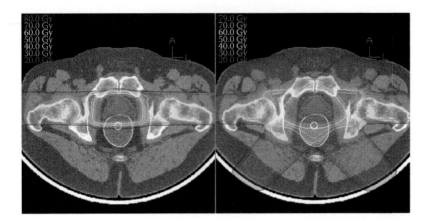

FIGURE 9.7   **(See color insert.)** Treatment planning comparison of a two-field proton plan (left) and a six-field IMRT plan (right) for carcinoma of the prostate (outlined in red).

clinical information. If evidence exists that disease extends beyond the prostate, the patient is not eligible for the protocol. The CTV is defined as the GTV plus areas considered to be at risk for containing microscopic tumor, such as the seminal vesicles in intermediate and high-risk patients. For low-risk patients the CTV is the same as the GTV. The physician then defines a planning target volume (PTV), which provides a margin around the CTV to compensate for patient or organ motion, and variations in daily setup position (i.e., the CTV + 5 mm in all directions).

Treatment is administered to the PTV via multiple conformal proton beams, with one or all fields treated daily, depending on dose delivered per fraction. For standard fractionation the total delivered dose is 81 CGE in 45 fractions. A lateral beam arrangement is employed for most patients. The bladder, rectum, and femoral heads are critical structures, to which the prescribed dose is limited. Each patient is seen weekly during treatment by his attending radiation oncologist.

Results obtained at LLUMC with the various proton protocols for prostate cancer have, over the years, been consistent: grade 3 or 4 complications are rare; rates of grade 2 sequelae rarely have exceeded 25% and are usually much lower. Control and long-term survival rates depend on several factors, chief among them posttreatment PSA nadir and stage at presentation, but these rates have consistently been at 70% or higher, generally the latter, again depending primarily on initial presentation and posttreatment PSA (Slater et al. 2004; Slater, Rossi and Yonemoto 1999; Rossi et al. 2004; Slater 2006). The ability to spare normal tissue (Figure 9.7) underlies the ability to devise dose-escalation protocols with proton beams; we have done this several times over the years at Loma Linda, either by raising the total dose delivered to the tumor or by increasing the effective biologic dose through measures such as hypofractionation. As noted above, long-term follow-up of men treated with a higher total dose to the prostate demonstrated superior long-term cancer control compared to those receiving conventional doses, without increases in grade 3 or higher late morbidity (Zietman et al. 2010).

9.4   Future Treatment Delivery Enhancements

Three-dimensional conformal therapy in protons has allowed us to increase from just a few tumor sites to over 50 in the span of less than two decades, despite still having a very limited number of facilities worldwide. New developments, including large-field scanning, will allow the number of clinical applications of proton therapy to increase significantly faster than has been the case in the preceding two to three decades. Imaging advances, whether at the diagnostic front or the treatment verification side, and software advances such as electronically produced intensity modulation, will continue to allow us to advance and perfect capabilities, much as has been done in the past 30 years.

As we have seen in the past, these changes over the next few decades will enhance what we can do. As has long been true in the photon world, however, change will not eliminate what has been learned and developed over this time. The modality is young, and has a long way to go.

References

Austin-Seymour, M., Munzenrider, J., Goitein, M. et al. 1989. Fractionated proton radiation therapy of chordoma and low-grade chondrosarcoma of the base of the skull. *J. Neurosurg.* 70(1): 13–17.

Bush, D. A., Dunbar, R. D. and Bonnet, R. et al. 1999. Pulmonary injury from proton and conventional radiotherapy as revealed by CT. *Am. J. Roentgenol.* 172: 735–739.

Bush, D. A., Hillebrand, D. J., Slater, J. M. and Slater, J. D. 2004a. High-dose proton beam radiotherapy of hepatocellular carcinoma; preliminary results of phase II trial. *Gastroenterology* 127(Suppl. 1): S189–S193.

Bush, D. A., Slater, J. D. and Shin, B. B. et al. 2004b. Hypofractionated proton beam therapy for stage I lung cancer. *Chest* 126: 1198–1203.

Bush, D. A., Slater, J. D. and Garberoglio, C. et al. 2007. A technique of partial breast irradiation utilizing proton beam radiotherapy: Comparison with conformal x-ray therapy. *Cancer J.* 13: 114–118.

Bush, D. A., Kayali, Z., Grove, R. and Slater, J. D. 2011. The safety and efficacy of high-dose proton beam radiotherapy for hepatocellular carcinoma: A phase 2 prospective trial. *Cancer* 117(13): 3053–3059. (Epub ahead of print; doi:10.1002/cncr.25809.)

Chang, J. Y., Zhang, X. and Wang, X. et al. 2006. Significant reduction of normal tissue dose by proton radiotherapy compared with three-dimensional conformal or intensity-modulated radiation therapy in stage I or stage III non-small-cell lung cancer. *Int. J. Radiat. Oncol. Biol. Phys.* 65: 1087–1096.

Chen, P. Y. and Vicini, F. A. 2007. Partial breast irradiation. Patient selection, guidelines for treatment, and current results. *Front. Radiat. Ther. Oncol.* 40: 253–271.

Chiba, T., Toyuuye, K. and Matsuzaki, Y. et al. 2005. Proton beam therapy for hepatocellular carcinoma: A retrospective review of 162 patients. *Clin. Cancer Res.* 11(10): 3799–3805.

Do, S. Y., Bush, D. A. and Slater, J. D. 2010. Comorbidity-adjusted survival in early stage lung cancer patients treated with hypofractionated proton therapy. *J. Oncol.* 2010: 251208. (Epub December 1, 2010.)

Goitein, M. and Abrams, M. 1983. Multi-dimensional treatment planning: I. Delineation of anatomy. *Int. J. Radiat. Oncol. Biol. Phys.* 9: 777–787.

Goitein, M., Abrams, M. and Rowell, D. et al. 1983. Multi-dimensional treatment planning—Part II: Beam's eye-view, back projection, and projection through CT sections. *Int. J. Radiat. Oncol. Biol. Phys.* 9: 789–797.

Gragoudas, E. S., Goitein, M. and Seddon, J. et al. 1984. Preliminary results of proton beam irradiation of macular and paramacular melanomas. *Br. J. Ophthalmol.* 68(7): 479–485.

Hata, M., Tokuuye, K. and Sugahara, S. et al. 2006. Proton beam therapy for hepatocellular carcinoma with limited treatment options. *Cancer* 107(3): 591–598.

Hawkins, M. A. and Dawson, L. A. 2006. Radiation therapy for hepatocellular carcinoma: From palliation to cure. *Cancer* 106: 1653–1663.

Jagsi, R., Ben-David, M. A. and Moran, J. M. et al. 2010. Unacceptable cosmesis in a protocol investigating intensity-modulated radiotherapy with active breathing control for accelerated partial-breast irradiation. *Int. J. Radiat. Oncol. Biol. Phys.* 76: 71–78.

Kjellberg, R. N. 1986. Stereotactic Bragg peak proton beam radiosurgery for cerebral arteriovenous malformations. *Ann. Clin. Res.* 18(Suppl. 47): 17–19.

Kjellberg, R. N., Shintani, A., Frantz, A. G. and Kliman, B. 1968. Proton beam therapy in acromegaly. *N. Engl. J. Med.* 278: 689–695.

Major, T., Frohlich, G., Lovey, K. et al. 2009. Dosimetric experience with accelerated partial breast irradiation using image-guided interstitial brachytherapy. *Radiother. Oncol.* 90: 48–55.

Marucci, L., Ancukiewicz, M., Lane, A. M. et al. 2011. Uveal melanoma recurrence after fractionated proton beam therapy: Comparison of survival in patients treated with reirradiation or with enucleation. *Int. J. Radiat. Oncol. Biol. Phys.* 79(3): 842–846.

Moon, S. H., Shin, K. H. and Kim, T. H. et al. 2009. Dosimetric comparison of four different external beam partial breast irradiation techniques: Three-dimensional conformal radiotherapy, intensity-modulated radiotherapy, helical tomotherapy, and proton beam therapy. *Radiother. Oncol.* 90: 66–73.

Munzenrider, J. E., Gragoudas, E. S. and Seddon, J. M. et al. 1988. Conservative treatment of uveal melanoma: probability of eye retention after proton treatment. *Int. J. Radiat. Oncol. Biol. Phys.* 15(3): 553–558.

Neilsen, I. R., Slater, J. M. and Shreyer, D. W. 1980. CT scanner assumes key role in computer-based radiotherapy planning system. In *Medinfo 80: Proceedings of the 3rd World Conference on Medical Informatics.* D. A. B. Lindberg and S. Kaihara (eds.). Amsterdam, New York, and Oxford: North Holland Publishing, pp. 25–28.

Nelson, J. C., Beitsch, P. D. and Vicini, F. A. et al. 2009. Four-year clinical update from the American Society Surgeons MammoSite brachytherapy trial. *Am. J. Surg.* 198: 83–91.

PTCOG Web site, summary of current ion beam therapy facilities. (Available at http://ptcog.web.psi.ch/patient_statistics.html; accessed 15 September 2010. The figure cited in the text is a reasonable minimum extrapolation of data given on the Web site for patients treated as of the end of 2009 (67,092).)

Rossi, C. J. Jr., Slater, J. D. and Yonemoto, L. T. et al. 2004. Influence of patient age on biochemical freedom from disease in patients undergoing conformal proton radiotherapy of organ-confined prostate cancer. *Urology* 64: 729–732.

Rowell, N. P. and Williams, C. J. 2001. Radical radiotherapy for stage I/II non-small-cell lung cancer in patients not sufficiently fit for or declining surgery (medically inoperable): A systematic review. *Thorax* 56: 628–638.

Shipley, W. U., Prout, G. R. Jr. and Coachman, N. M. et al. 1988. Radiation therapy for localized prostate carcinoma: Experience at the Massachusetts General Hospital (1973–1981). *NCI Monogr.* 7: 67–73.

Slater, J. D. 2006. Clinical applications of proton radiation treatment at Loma Linda University: Review of a 15-year experience. *Technol. Cancer Res. Treat.* 5: 81–89.

Slater, J. M., Archambeau, J. O. and Miller, D. W. et al. 1992. The proton treatment center at Loma Linda University Medical Center: Rationale for and description of its development. *Int. J. Radiat. Oncol. Biol. Phys.* 22: 383–389.

Slater, J. M., Miller, D.W. and Archambeau, J. O. 1988. Development of a hospital-based proton beam treatment center. *Int. J. Radiat. Oncol. Biol. Phys.* 14: 761–775.

Slater, J. M., Neilsen,. I. R. and Chu, W. T. et al. 1974. Radiotherapy treatment planning using ultrasound-sonic graph pen-computer system. *Cancer* 34: 96–99.

Slater, J. D., Rossi, C. J. Jr. and Yonemoto, L. T. 1999. Conformal proton therapy for early-stage prostate cancer. *Urology* 53: 978–984.

Slater, J. D., Rossi, C. J. Jr. and Yonemoto, L. T. et al. 2004. Proton therapy for prostate cancer: The initial Loma Linda University experience. *Int. J. Radiat. Oncol. Biol. Phys.* 59: 348–352.

Smith, B. D., Arthur, D. W., Buchholz, T. A. et al. 2009. Accelerated partial breast irradiation consensus statement from the American Society for Radiation Oncology (ASTRO). *Int. J. Radiat. Oncol. Biol. Phys.* 74: 987–1001.

Suit, H., Goitein, M. and Munzenrider, J. et al. 1982. Evaluation of the clinical applicability of proton beams in definitive fractionated radiation therapy. *Int. J. Radiat. Oncol. Biol. Phys.* 8: 2199–2205.

Tobias, C. A., Roberts, J. E. and Lawrence, J. H. et al. 1956. Irradiation hypophysectomy and related studies using 340 MeV protons and 190 MeV deuterons. *Proceedings of the International Conference on the Peaceful Uses of Atomic Energy.* United Nations Publications. 10: 95–96.

Wazer, D. E., Lowther, D. and Boyle, T. et al. 2001. Clinically evident fat necrosis in women treated with high-dose-rate brachytherapy alone for early-stage breast cancer. *Int. J. Radiat. Oncol. Biol. Phys.* 50: 107–111.

Zietman, A. L., Bae, K. and Slater, J. D. et al. 2010. Randomized trial comparing conventional-dose with high-dose conformal radiation therapy in early-stage adenocarcinoma of the prostate: Long-term results from Proton Radiation Oncology Group/American College of Radiology 95–09. *J. Clin. Oncol.* 28: 1106–1111.

Zimmermann, F. R., Bamberg, M. and Molls, M. et al. 2003. Radiation therapy alone in early stage non-small-cell lung cancer. *Semin. Surg. Oncol.* 21: 91–97.

Zietman, A. L., DeSilvio, M. L. and Slater, J. D. et al. 2008. Comparison of conventional-dose versus high-dose conformal radiation therapy in clinically localized adenocarcinoma of the prostate: A randomized controlled trial. *J. Am. Med. Assoc.* 294: 1233–1239 (erratum: *J. Am. Med. Assoc.* 299, 899–900).

10. Clinical Experience with Particle Radiotherapy

Alexandra D. Jensen, M. W. Münter, Anna V. Nikoghosyan, and Jürgen Debus

In 1946, barely 27 years after the proton had been discovered by Ernest Rutherford, physicist Robert Wilson was the first to propose treatment with protons and heavy ions for malignant tumors in medicine. Berkeley Radiation Laboratory pioneered particle treatment with the first patients receiving treatment in 1954. More than 2500 patients with various indications were treated with both protons and heavy ions at this institution until closure of its clinical project in 1992. The physical properties of particle beams allowing sharp dose gradients and hence relative dose escalation with reduction of dose to normal organs seems especially appealing in otherwise relatively radioresistant tumors. Early data from the Lawrence Berkeley Labs seemed to clinically support this physical advantage (Castro et al. 1994, 1997; Linstadt et al. 1988; Saunders et al. 1985).

Meanwhile, 35 particle therapy centers worldwide, including two centers in Germany, are in clinical operation (Table 10.1) and 22 new centers are planned or about to start clinical treatment.

While radiotherapy with protons and ions has various advantages regarding physical dose distribution and in the case of heavy charged particles also relative biological effectiveness, official approval of this treatment modality has only been obtained for a few indications such as chordoma, chondrosarcoma, and malignant salivary gland tumors. Accessibility of particle therapy is still limited; therefore, this chapter reviews available clinical evidence of the selected indications based on a Medline search of original contributions from active particle centers.

10.1 Review of Clinical Evidence

10.1.1 Uveal Melanoma

Proton therapy for uveal melanoma has been well established over the years, achieving local controls of over 90% at 5 years. Hence, this technique developed into an equal method besides surgical enucleation and brachytherapy with ^{106}Ru-applicators for this disease wherever

Proton and Carbon Ion Therapy Edited by C.-M. Charlie Ma and Tony Lomax © 2013 Taylor & Francis Group, LLC. ISBN: 978-1-4398-1607-3.

available. Eye preservation rates range between 75% and 92% (Dendale et al. 2006; Gragoudas et al. 2000; Egger et al. 2003; Desjardins et al. 2003; Castro et al. 1997; Damato et al. 2005; Höcht et al. 2004; Mosci et al. 2009; Caujolle et al. 2010; Tsuji et al. 2007) using doses of around 60 GyE in 4 fractions. A recent update of the Nicoise data confirmed these results with a 10-year local control rate of 92.1% (Tsuji et al. 2007). Gragoudas et al. (2000) tested 50 versus 60 GyE in a prospective phase III trial and could not find a significant difference in local control but significantly higher field of view with

Chapter 10

Table 10.1 Particle Centers in Operation as of 26.06.2011 from the PTCOG Web Site

Facility, Location	Country	Particle	Maximum Clinical Energy (MeV)	Beam Direction	Start of Treatment	Total Patients Treated	Date of Total
ITEP, Moscow	Russia	p	250	1 horizontal	1969	4246	Dec-10
St. Petersburg	Russia	p	1000	1 horizontal	1975	1362	Dec-10
PSI, Villigen	Switzerland	p	72	1 horizontal	1984	5458	Oct-10
Dubna	Russia	p	200[a]	horizontal	1999	720	Dec-10
Uppsala	Sweden	p	200	1 horizontal	1989	1000	Dec-10
Clatterbridge	England	p	62	1 horizontal	1989	2021	Dec-10
Loma Linda	CA, USA	p	250	3 gantry, 1 horizontal	1990	15000	Jan-11
Nice	France	p	65	1 horizontal	1991	4209	Dec-10
Orsay	France	p[b]	230	1 gantry, 1 horizontal	1991	5216	Dec-10
iThemba Labs	South Africa	p	200	1 horizontal	1993	511	Dec-09
MPRI(2)	IN, USA	p	200	2 gantry, 1 horizontal	2004	1145	Dec-10
UCSF	CA, USA	p	60	1 horizontal	1994	1285	Dec-10
HIMAC, Chiba	Japan	C-ion	800/u	horizontal, vertical	1994	5497	Aug-10
TRIUMF, Vancouver	Canada	p	72	1 horizontal	1995	152	Dec-10
PSI, Villigen	Switzerland	p[c]	250[d]	1 gantry, 1 horizontal	1996	772	Dec-10
G.S.I. Darmstadt	Germany	C-ion[c]	430/u	1 horizontal	1997	440	Nov-09
HZB (HMI), Berlin	Germany	p	72	1 horizontal	1998	1660	Dec-10
NCC, Kashiwa	Japan	p	235	2 gantry	1998	772	Dec-10
HIBMC, Hyogo	Japan	p	230	1 gantry	2001	2382	Nov-09
HIBMC, Hyogo	Japan	C-ion	320	horizontal, vertical	2002	638	Nov-09
PMRC(2), Tsukuba	Japan	p	250	gantry	2001	1849	Dec-10
NPTC, MGH Boston	USA	p	235	2 gantry, 1 horizontal	2001	4967	Dec-10
INFN-LNS, Catania	Italy	p	60	1 horizontal	2002	174	Mar-09
Shizuoka	Japan	p	235	gantry, horizontal	2003	986	Dec-10
WERC, Tsuruga	Japan	p	200	1 horizontal, vertical	2002	62	Dec-09
WPTC, Zibo	China	p	230	2 gantry, 1 horizontal	2004	1078	Dec-10
MD Anderson Cancer Center, Houston, TX	USA	p[e]	250	3 gantry, 1 horizontal	2006	2700	Apr-11
UFPTI, Jacksonville, FL	USA	p	230	3 gantry, 1 horizontal	2006	2679	Dec-10
NCC, IIsan	South Korea	p	230	2 gantry, 1 horizontal	2007	648	Dec-10

(continued)

Table 10.1   Particle Centers in Operation as of 26.06.2011 from the PTCOG Web Site (Continued)

Facility, Location	Country	Particle	Maximum Clinical Energy (MeV)	Beam Direction	Start of Treatment	Total Patients Treated	Date of Total
RPTC, Munich	Germany	p[c]	250	4 gantry, 1 horizontal	2009	446	Dec-10
ProCure PTC, Oklahoma City, OK	USA	p	230	1 gantry, 1 horizontal, 2 horizontal/60°	2009	21	Dec-09
HIT, Heidelberg	Germany	p[c]	250	2 horizontal	2009	Treatment started	May-11
HIT, Heidelberg	Germany	C-ion[c]	430/u	2 horizontal	2009	395	May-11
UPenn, Philadelphia	USA	p	230	4 gantry, 1 horizontal	2010	150	Dec-10
GHMC, Gunma	Japan	C-ion	400/u	3 horizontal, vertical	2010	Treatment started	Mar-10
IMPCAS, Lanzhou	China	C-ion	400/u	1 horizontal	2006	126	Dec-10
CDH Proton Center, Warrenville, IL	USA	p	230	1 gantry, 1 horizontal, 2 horizontal/60°	2010	Treatment started	Oct-10
IFJ PAN, Krakow	Poland	p	60	1 horizontal	2011	9	Apr-11

[a] Degraded beam.

[b] New cyclotron with fixed beam operational since July 2010; the gantry operational since October 2010.

[c] With beam scanning.

[d] Degraded beam from 1996 to 2006; dedicated 250 MeV cyclotron since 2007; OPTIS2 patients included (since November 2010).

[e] With spread beam and beam scanning (MD Anderson, since 2008).

the lower doses; hence, this dose has since been established. Even for larger uveal melanomas, high rates of eye and vision preservation could be shown by the use of carbon ion therapy in a Japanese trial (Marucci et al. 2011). In the event of local relapse, Marucci and colleagues were able to show that reirradiation can be a good option; no significant difference in overall survival could be detected in a retrospective comparison of enucleation and reirradiation for recurrent uveal melanoma (Marucci et al. 2011). With recurrent tumors slightly smaller in the reirradiation group, these trended even toward higher overall survival. An overview of the various studies can be found in Table 10.2.

10.1.2   Base of Skull Tumors: Chordoma and Chondrosarcoma

Chordoma and chondrosarcoma of the skull base have traditionally been a challenge for radiation oncologists. Treatment is primarily surgical; however, complete resection of tumors in the skull base very often is not achievable. Moreover, these tumors are considered comparatively radioresistant and are mostly located directly adjacent to critical and radiosensitive structures. Techniques using conventional radiation have therefore shown only disappointing local control rates between 17% and 50% for chordoma of the skull base. Among these, the best results could be achieved by the introduction of fractionated stereotactic radiotherapy (Romero et al. 1993). Therefore, these tumors are an ideal target for particle therapy considering the superior dose distribution and in the case of C12 therapy, increased RBE. Figures 10.1 through 10.3 show an exemplary carbon ion dose distribution in the case of a 63-year-old female with extensive chordoma of the skull base.

Various working groups were able to achieve local control rates between 46% and 74% at 5 years for chordoma and between 78% to 98% for chondrosarcoma applying particle radiation. However, because

Chapter 10

Table 10.2 Overview of Particle Therapy for Uveal Melanoma

	Journal	Patients	RT	Local Control at 5a	Eye Preservation Rate at 5a
Char et al.	*Ophthalmology* 1993	184	He vs. [125]I	He: improved local control	Data not given
Castro et al.	*IJROBP* 1997	347	He	96%	81%
Courdi et al.	*IJROBP* 1999	538	Protons	89%	88%
Gragoudas et al.	*Arch Ophthalmol.* 2000	188	Protons (70 GyE)	97%	96%
Fuss et al.	*IJROBP* 2001	78	Protons	90.5%	75.2%
Desjardins et al.	*J. Fr. Ophthalmol.* 2003	1272	[125]I (A) vs. protons (B)	A: 96.2%; B: 96%	A: 92.8%; B: 88.8%
Hocht et al.	*Strahlenther Onkol.* 2004	245	Protons	95.5%/3a	87.5%/3a
Damato et al.	*IJROBP* 2005	88	Protons	96.7%/4a	90.6%/4a
Dendale et al.	*IJROBP* 2006	1406	Protons	96%	92.3%
Tsuji et al.	*IJROBP* 2007	57	[12]C	97.4%/3a	91.1%/3a
Mosci et al.	*Eur. J. Ophthalmol.* 2009	368	Protons	91.6%/at evaluation	88.4%/at evaluation
Caujolle et al.	*IJROBP* 2010	886	Protons	93.9%/5a; 92.1%/10a	91.1%/5a; 87.3%/10a
Marucci et al.	*IJROBP* 2011	73	Protons for re-RT (A) vs. Sx (B) at local relapse	Not given	5a-OS: A: 36%; B: 63%

chordoma and chondrosarcoma tumors are rare, very often pooled data is published. An overview of the various studies and results is displayed in Tables 10.3 (chordoma) and 10.4 (chondrosarcoma).

Especially heavy ions seem advantageous for these tumors, which is supported by early data from the Lawrence Berkeley Laboratory (LBL) using helium and neon ions, showing local control rates of 78% for chondrosarcoma and 63% for chordoma (Castro et al. 1994). Terahara et al. (1999) reported slightly lower local controls by a combination of photon and proton radiation (median dose 68.9 GyE) (59% at 5a; 44% at 10a) in 115 patients; however, treatment was mostly carried out in the pre-3-D era (1978 and 1993). Another analysis

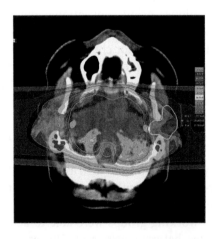

FIGURE 10.1 **(See color insert.)** Sixty-three-year-old female with extensive chordoma of the skull-base; axial dose distribution, prescribed dose: 66 GyE carbon ions in 3 GyE per fraction.

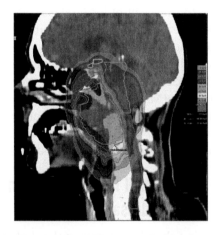

FIGURE 10.2 **(See color insert.)** Sixty-three-year-old female with extensive chordoma of the skull-base; sagittal dose distribution, prescribed dose: 66 GyE carbon ions in 3 GyE per fraction.

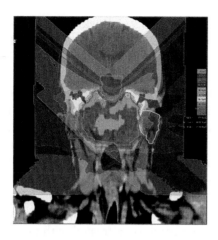

FIGURE 10.3 (See color insert.) Sixty-three-year-old female with extensive chordoma of the skull-base; coronal dose distribution, prescribed dose: 66 GyE carbon ions in 3 GyE per fraction.

by the colleagues from the Paul Scherrer Institute (PSI Villigen) in 18 patients with chordoma and 11 patients with chondrosarcoma (total doses applied: 74 and 68 GyE protons) reported 3-year local controls of 87.5% and 100%, respectively (Schulz-Ertner et al. 2007b). Recently published long-term results showed a 5-year local control of 81% for chordoma and 94% for chondrosarcoma of the skull base (spiro 1986). Similar results were seen by the application of C12 heavy ions: 54 patients with chondrosarcoma (median dose 60 GyE) achieved local controls of 96.2% and 89.8% at 3 and 4 years Schulz-Ertner et al. 2007a). This treatment was accompanied by very mild toxicity (≥3: 1.9%). Within the same time, 96 patients with macroscopical remnant chordoma were also treated at the same institution (median dose 60 GyE). Here again, local controls of 80.6% and 70%

Table 10.3 Overview of Particle Therapy for Chordoma of the Skull Base

Author	Journal	Patients	Radiotherapy	Local Control
Romero et al.	*Radiother. Oncol.* 1993	18	conv. RT	17%/5a
Debus et al.	*IJROBP* 2000	45	FSRT	50%/5a
Munzenrider and Liebsch	*Strahlenther Onkol.* 1999	229	p+/photons	73%/5a
Castro et al.	*IJROBP* 1994	53	He	63%/5a
Noel et al.	*IJROBP* 2001	34	p+/photons	83.1%/3a
Weber et al.	*IJROBP* 2005	11	p+	87.5%/3a
Schulz-Ertner et al.	*IJROBP* 2007	96	C12	70%/5a
Rutz et al.	*IJROBP* 2008	8	p+	100%/3a (children)
Habrand et al.	*IJROBP* 2008	26	p+	77%/5a (children)
Combs et al.	*Cancer* 2009	7	C12	85.7%/median f/u: 49 mo (children)
Ares et al.	*IJROBP* 2009	42	p+	81%/5a

Table 10.4 Overview of Particle Therapy for Chondrosarcoma of the Skull Base

Author	Journal	Patients	Radiotherapy	Local Control
Munzenrider and Liebsch	*Strahlenther Onkol.* 1999	290	p+/photons	98%/5a
Castro et al.	*IJROBP* 1994	27	He	78%/5a
Noel et al.	*IJROBP* 2001	11	p+/photons	90%/3a
Weber et al.	*IJROBP* 2005	18	p+	100%/3a
Schulz-Ertner et al.	*IJROBP* 2007	54	C12	89.8%/4a
Rutz et al.	*IJROBP* 2008	8	p+	100%/3a (children)
Habrand et al.	*IJROBP* 2008	3	p+	100%/5a (children)
Combs et al.	*Cancer* 2009	10	C12	100%/4a (children)
Ares et al.	*IJROBP* 2009	22	p+	94%/5a

Chapter 10

at 3 and 5 years were achieved (Terahara et al. 1999). Also in this series only two cases of three late toxicities were found (4.1%; opticus-neuropathy, necrosis of fat implant). A retrospective comparison of control rates with results from other working groups suggested a clear dose-dependency of local control, further underlining the advantage of particle therapy in the treatment of these tumors. Children and young adults especially seem to profit from particle therapy: in addition to excellent control rates for these indications, accompanying toxicity is minimal without any reported severe late effects reported so far (Habrand et al. 2008; Schneider et al. 2008; Combs et al. 2009; Ares et al. 2009).

10.1.3 Cancer of the Head and Neck: Malignant Salivary Gland and Paranasal Sinus Tumors

Malignant salivary gland tumors are a rare and heterogenous group of tumors accounting for about 3% to 5% of head and neck cancers. High-grade tumors such as mucoepidermiod carcinoma (35%) and adenoid cystic carcinoma (25%) are the most common histologic subtypes (Chen et al. 2007b), characterized by a rather slow pattern of growth, perineural spread, and high potential of hematogenous metastases. So far, standard therapy of high-grade salivary gland carcinoma consists of complete surgical resection and adjuvant radiation in a high-risk situation (R+ or close margin, perineural spread, neural infiltration, large tumors (T3/4), or nodal metastases (Chen et al. 2007b; Gurney et al. 2005; Mendenhall et al. 2004). To achieve local control, radiation doses of >60 Gy are recommended (Chen et al. 2008; Garden et al. 1994; Terhaard et al. 2004).

Regarding radiation therapy, local control was also significantly improved by the application of high-precision techniques, dose-escalation and high-LET radiation (Terhaard 2004; Schulz-Ertner et al. 2005; Douglas et al. 2003; Mizoe et al. 2004; Huber et al. 2001; Pommier et al. 2006). Intensity-modulated radiation therapy (IMRT) as well as fractionated stereotactic RT could already improve local control as compared to conventional RT techniques achieving 3-year progression-free survival (PFS) rates of 38% (Münter et al. 2006).

So far, the highest local control rates at 75% to 100% (Douglas et al. 2003; Huber et al. 2001) were achieved by neutron radiation albeit at the cost of significant late toxicity.

Dose escalation (Mizoe et al. 2004) though only in a small group of patients suggests treatment with heavy

ions at 70.2 GyE (3 × 3.9 GyE/wk) or 64 GyE (4 × 4 GyE/wk). Despite high doses per fraction, no CTC III late toxicities and very few III acute reactions occurred. Local control at 5 years was still 100%; however, various histologic subtypes of salivary gland tumors were included in the trial. Pommier et al. (2006) treated 23 patients with adenoid cystic carcinoma with protons at 75.9 GyE (median) in various fractionation schemes. Overall local control at 5 years was 93%; however, the authors noted several 3 as well as one 5 late toxicity (temporal lobe necrosis). Douglas et al. (2003) published a retrospective analysis of 159 patients with adenoid cystic carcinoma treated with neutrons (total dose: 19.2 Gy), but only reported somewhat disappointing local control rates of 57% at 5 years accompanied with significant late toxicity (14% > CTC III). Results of the combined IMRT-C12 treatment therefore compare favorably (Schulz-Ertner et al. 2005) with only very mild treatment-related side effects (III late toxicity in only one patient). Fifty-four Gy IMRT (2Gy/Fx) plus 18 GyE C12-boost (3 GyE/Fx) yielded local control rates of about 78% at 4 years (Schulz-Ertner et al. 2005). A recent update of all patients treated with this regimen at the GSI Darmstadt even reported a 5-year control rate of 82% (Münter et al. 2010). An exemplary dose carbon ion dose distribution (24 GyE boost) and treatment outcome for a 69-year-old patient with adenoid cystic carcinoma is depicted in Figures 10.4 through 10.7.

In view of the outcome and low toxicity profile, photon IMRT plus C12-boost has been accepted as the standard treatment and method of choice for adenoid cystic carcinoma in Germany. Table 10.5 gives an overview of treatment results in larger patient series for

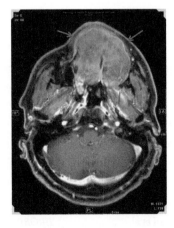

FIGURE 10.4 (See color insert.) Sixty-nine-year-old patient with large adenoid cystic carcinoma, pretherapeutic contrast-enhanced MRI.

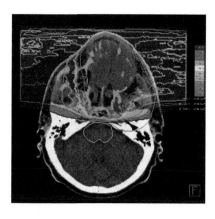

FIGURE 10.5 (See color insert.) Sixty-nine-year-old patient with adenoid cystic carcinoma; axial carbon ion boost dose distribution (24 GyE C12 + 50 Gy IMRT).

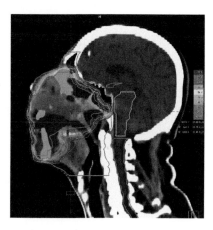

FIGURE 10.6 (See color insert.) Sixty-nine-year-old patient with adenoid cystic carcinoma; sagittal carbon ion boost dose distribution (24 GyE C12 + 50 Gy IMRT).

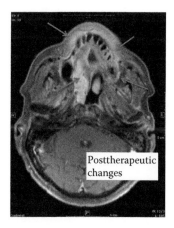

Posttherapeutic changes

FIGURE 10.7 Sixty-nine-year-old patient with complete remission of a large adenoid cystic carcinoma 6 weeks post-completion of treatment (50 Gy IMRT + 24 Gy C12), posttherapeutic contrast-enhanced MRI.

malignant salivary gland tumors as well as paranasal sinus cancers.

Sinonasal malignancies include malignant tumors of various histologies in the nasal cavity and paranasal sinuses. Squamous cell carcinomas account for the majority of these tumors; however, adenoidcystic carcinoma, esthesioneuroblastoma, and mucosal melanoma are also found. Due to limited accessibility of these sites and late occurrence of symptoms, patients are mostly diagnosed with advanced disease. Local relapse rates following surgical treatment are consequently high therefore adjuvant radiotherapy is often recommended. In conventional treatment techniques, sufficient dose application in radiation therapy has been limited by dose to surrounding organs at risk and subsequent early and late toxicity leading to loss of vision in approximately one third of patients. With the advent of more sophisticated radiation treatment techniques, toxicities were effectively reduced while local control remained more or less stable (Chen et al. 2007a). First and predominant site of treatment failure in paranasal sinus and nasal cavity cancer remains in-field, stressing the need for further dose escalation in RT (Chen et al. 2008; Hoppe et al. 2008). Increased biological efficiency (RBE) and physical properties of dose distributions argue strongly in favor of particle therapy for this indication. Acute toxicity and initial treatment response of various extensive paranasal sinus malignancies treated with IMRT and carbon ion boost are promising in a recent series Jensen et al. 2011a) while grade III acute toxicities were roughly 24%, objective response rate at the first follow-up was already 50% (CR and PR) in this cohort. Late toxicity of patients treated with protons for advanced malignant sinonasal tumors was acceptable: 36% of patients developed radiation-induced late effects (grades 1–3), which included only 2.8% grade 3 reactions (Weber et al. 2006). No difference between proton and carbon ion therapy has yet been demonstrated with regard to potential late effects at the optic structures (Demizu et al. 2009). Initial treatment results have shown high rates of 86% local control at 2 years posttreatment (Truong et al. 2009).

For mucosal melanoma of the paranasal sinus, particle therapy seems to be superior to standard techniques, recent reports of three groups using either carbon ions alone (Yanagi et al. 2009), in combination with chemotherapy (Jingu et al. 2011), or protons (Zenda et al. 2011) at hypofractionated doses between 52.8 and 64 GyE have reported remarkable control rates up to 84.1% (Yanagi et al. 2009) (Table 10.5).

Chapter 10

Table 10.5 Overview of Particle Therapy for Malignant Head and Neck Cancers Including Malignant Salivary Gland and Paranasal Sinus Cancers

Author	Journal	Patients	Indication	Radiotherapy	Local Control at 5a	OS	Late Toxicity (°3/4)
Chen et al.	*IJROBP* 2007	207	ACC	Surgery, no RT	86%	83%	dna
Chen et al.	*IJROBP* 2008	140	ACC	Sx (A) vs. 64 Gy photons (B)	A: 80%; B: 92%	85%	not given
Douglas et al.	*IJROBP* 2000	159	ACC	Neutron, 19.2 Gy; A: R2; B: R1-2	A: 57%; B: 100%	A: 75%; B: 90%	9.4%
Mendenhall et al.	*Head Neck* 2004	101	ACC	A: photons (50-72.4 Gy) vs B: RT+Sx	A: 56%; B: 94%	A: 75%; B: 90%	12.9%
Garden et al.	*Cancer* 1994	160	ACC	Photons/60 Gy	96%	81%	33%
Gurney et al.	*Laryngoscope* 2005	33	ACC	Photons/60 Gy	94%		not given
Münter et al.	*Radiat. Oncol.* 2006	25	large ACC	Photon IMRT/66Gy	38%	72%/3a	0%
Huber et al.	*Radiother. Oncol.* 2001	75	ACC	A: neutrons/ 16 Gy; B: photons/ 64 Gy; C: mixed/8 Gy neutrons+ photons 32 Gy	A: 75%; B: 32%; C: 32%		A: 19%; B: 4%; C: 10%
Mizoe et al.	*IJROBP* 2004	36	various, including ACC	C12/48.6 – 52.8 Gy	50%		0%
Schulz-Ertner et al.	*Cancer* 2005	29	ACC	C12+IMRT/ 72 GyE	77.5%/4a	75.8%/4a	3.4%
		34	ACC	IMRT/66Gy	24.6%/4a	77.9%/4a	5.9%
Pommier et al.	*Arch. Otolaryngol. Head Neck Surg.* 2006	23	ACC	Protons/75.9 GyE	93%		17%
Münter et al.	*J. Clin. Oncol.* 2010	57	ACC	C12+IMRT/72 GyE	82%	78%	0%
Weber et al.	*Radiother. Oncol.* 2006	36	various	p+/69.6 GyE, RT bid	DFS: 73.1%	80.8%	36%
Yanagi et al.	*IJROBP* 2009	72	mucosal melanoma	C12/52.8–64 Gy in 16 Fx	84.1%	27.0%	0%

(*continued*)

Table 10.5   Overview of Particle Therapy for Malignant Head and Neck Cancers Including Malignant Salivary Gland and Paranasal Sinus Cancers (Continued)

Author	Journal	Patients	Indication	Radiotherapy	Local Control at 5a	OS	Late Toxicity (°3/4)	
Truong et al.	*Head Neck* 2009	20	various	p+/76 GyE	86%/2a	53%/2a		combined platin-based chemoradiation in 85% of patients
Zenda et al.	*IJROBP* 2010	14	mucosal melanoma	p+/60GyE in 15 Fx	PFS: 43.7%/2a	58%/3a	14.3%	
Jingu et al.	*Radiother. Oncol.* 2011	37	mucosal melanoma	C12/57.6 Gy in 16 Fx	81.1%/3a	65.3%/3a		combined chemoradiation with DTIC, ACNO, and vincristin

Hence, further exploration of this technique for mucosal melanoma in this anatomical site seems warranted.

Reirradiation in the paranasal sinus is hampered by close proximity of critical organs to the area of relapse. Application of high doses is therefore rarely possible without exceeding normal tissue tolerance. While confidence has improved re-treating squamous cell carcinoma of the head and neck, reirradiation in the paranasal sinus represents a therapeutic challenge. The application of scanned particle beams has so far shown only mild side effects and respectable response rates exceeding 50% at first follow-up (Jensen et al. 2011b); however, there is no data on potential late effects and long-term tumor control as of now.

10.1.4   Non-Small Cell Lung Cancer (NSCLC)

Dose dependency of local control in lung tumors was recognized a long time ago (Perez et al. 1987). International guidelines recommend doses of 60–70 Gy for fractionated (photon) RT; however, very often in clinical routine patients already have a pretherapeutic impaired pulmonary function. These doses are rarely achievable without compromising target volumes in extensive tumors. Dose escalation using photon radiation is therefore only possible for small tumors: maximum reported doses per (single-) fraction are 37 GyE (T1/T2), resulting in clinical outcomes comparable to surgical interventions. Improvement of treatment results for lung tumors historically came with the introduction of more conformal treatment techniques in the 1990s and application of IMRT for lung tumors in the beginning of the new millennium.

There is extensive data and experience in the treatment of early-stage lung cancer (T1/T2) with protons (Bush et al. 1999, 2004b) as well as C12 ions (Nihei et al. 2006; Miyamoto et al. 2007a, b; Koto et al. 2004). Various fractionation regimen have been evaluated.

Results for treatment of T1/T2 tumors at the facility at Chiba with doses between 79.2 GyE (8.8 Gy/Fx) and 95.4 GyE (5.3 GyE/Fx) published local control rates of 79% at 5 years with the only significant factor on multivariate analysis being total dose delivered (Koto et al. 2004; Miyamoto et al. 2007b; Nishimura et al. 2003; Kadono et al. 2002). Most of the local treatment failures occurred within the high-dose area, suggesting still further dose escalation to be necessary for improvement of treatment outcome.

Treatment results from Loma Linda on 68 patients (Bush et al. 2007b) suggested dose-dependency also for the overall survival (51 GyE: 27%; 60 GyE: 55%).

Local control following particle irradiation exceeds results achieved by fractionated photon RT. Total doses of >86.4 GyE (18 Fx) or 72 GyE (9 Fx) yield local controls of over 90% at 5 years (Miyamoto et al. 2003), which apparently also translate into a favorable overall survival.

Although almost all patients showed pulmonary changes in their follow-up CT, only 9.8% (8/81) of patients had RTOG 2/3 acute pulmonary reactions that completely resolved, hence pneumonitis rates compare favorably to those commonly reported for photon RT (>15% at 12 mo/17–20% at 24 mo). A phase-I dose-escalation reported one case of CTC 3 acute pneumonitis leading to their standard dose of 94 GyE (Nihei

Chapter 10

et al. 2006). Carbon ion therapy has also been shown to be feasible in patients aged 80 years and above: local control and overall survival at a consistently low toxicity profile were shown to be high despite advanced age (Sugane et al. 2009).

Correct dose delivery in particle RT however, is dependent on breathing motion and hence considerable range uncertainties. Therefore, complicated patient positioning including application of modern gating techniques necessitate investigation of hypofractionated treatment regimes. Another trial from Chiba showed promising results after hypofractionated C12 RT (total dose 72 GyE at 8 GyE/Fx) with local control at 94.7% (5a) and cause-specific survival 50% (5a; T1: 55.2% and T2: 42.9%) (Miyamoto et al. 2007b).

Doses per fraction were further increased (Miyamoto et al. 2007a) (52.8/60 GyE in 4 Fx), maintaining excellent

local control without increased toxicity. Reduction of pulmonary function parameters was marginal (FeV1 −8%, FVC −7%) compared to 20% to 50% in standard photon techniques. These results were confirmed for protons by another working group (Hata et al. 2007a).

The working group at Tsukuba, Japan, evaluated proton-therapy in 51 patients with NSCLC of all stages. Patients with stage II/III disease also received elective nodal photon irradiation and a boost to positive nodes up to 76 GyE (3 GyE/Fx). Stages I/II showed survival rates of 55% and 23% at 2 and 5 years; expectably, patients with stage III/IV disease did less well with an OS of 62% and 0%, respectively. Again, no RTOG°2/3 late toxicities were seen, though as described previously, almost all patients (92%) showed radiogenic pulmonary changes in their CT (Shioyama et al. 2003).

Table 10.6   Overview of Particle Therapy for Non-Small Cell Lung Cancer

	Journal	Patients	Stage		Local Control	OS	Late Toxicity°3/4
Koto et al.	*Radiother. Oncol.* 2004		T1	¹²C		T1: 64.4%/5a	
Nishimura et al.	*IJROBP* 2003		T2	95,4 GyE a 5,3 GyE		T2: 22%/5a	°3: 14.8%, no°4/°5
Kadono et al.	*Chest* 2002	81		79,2 GyE a 8,8 GyE	79%/5a		
Miyamto et al.	*Radiother. Oncol.* 2003						
Nihei et al.	*IJROBP* 2006	37	T1/T2	p+/70 GyE a 3,5 GyE bis 98 GyE a 4,9 GyE	80%/2a	82%/2a	>°2: 16.2%
Miyamoto et al.	*IJROBP* 2007	50	T1/T2	¹²C/72 GyE a 8 GyE	94,7%/5a	T1: 89.4%/5a; T2: 55.1%/5a	°3: 2%
Miyamoto et al.	*J. Thorac. Oncol.* 2007	80	T1/T2	¹²C/T1: 52.8 GyE a 13.2 GyE; T2: 60 GyE a 15 GyE	T1: 97%/5a; T2: 80%/5a	T1: 62%/5a; T2: 25%/5a	No°3 or higher
Hata et al.	*IJROBP* 2007	21	T1/T2	p+/50 GyE to 60 GyE in 10 Fx	T1: 100%/2aT2: 90%/2a	T1: 100%/2aT2: 47%/2a	No°3 or higher
Sugane et al.	*Lung Cancer* 2009	28	T1/T2	¹²C	95.8%/5a	30.7%/5a	
Shioyama et al.	*IJROBP* 2003	51	I-IV	p+ or p+/ photons/ 70–78GyE		I/II: 55%/2a; 23%/5a; III/IV: 62%/2a, 0%/5a	No°2/°3 or higher
Bush et al.	*Chest* 2004	68	T1/T2	p+/51–60 GyE in 10 Fx	T1: 87%/3a; T2: 49%/3a	60GyE: 55%/3a; 51 GyE: 27%/3a; overall: 74%/3a	No°2/°3 or higher

Table 10.6 gives a summary of the trials as described. While this data is indicative of potential benefits of particle therapy in NSCLC, treatment-related toxicity (especially pneumonitis rates) of all reported regimens has been low. As local control has been proven to be dose-dependent and high doses are needed for long-term tumor control, carbon ion therapy may add benefit by increased RBE. This concept hence warrants further exploration, especially in patients unfit for or unwilling to undergo complex surgical procedures and subsequent morbidity.

10.1.5   Esophageal Carcinoma

Radiooncological treatment of esophageal carcinoma usually consists of platinum-containing chemoradiotherapy. Based on the idea of normal tissue sparing, the Tsukuba working group developed a combined photon/proton treatment regimen. Radiation treatment consisted of photon radiation (48 Gy) and a proton boost (median 31.7 GyE). Forty patients received a median total dose of 76 Gy and six patients received protons only. The trial recruited patients of all tumor stages, however, 50% of the patients had T1 tumors, and 85% had no evidence of nodal disease. In this series, local control at 5 years for T1 tumors was 83% and 29% for T2-4 tumors (Sugahara et al. 2005). This difference proved statistically significant, further underlining results seen in established (photon) regimen. However, the authors further noted a clear dose dependency in contrast to the results of the Intergroup Trial INT 0123 (Minsky et al. 2002). Cases of local relapse mostly occurred at the craniocaudal field edges or out of field, suggesting changes in the applied target volume concept that have since been implicated. Moreover, almost half of the patients (48%) developed therapy-related esophageal ulcerations, which completely resolved in only one third of these patients. The high complication rate was assumed to be due to the high doses per fraction (median 3 GyE [2.5–3.7 GyE]). Current study protocols were adapted accordingly. A recent update of 51 patients treated with a median dose of 79 GyE protons for locally advanced esophageal cancer showed an objective response rate as assessed by endoscopy of 92.0% and a local control of 38% at 5 years (Mizumoto et al. 2010).

Treatment of esophageal carcinoma with particle therapy is therefore possible without major complications.

In another analysis, the authors have even intensified local therapy and given treatment in a hyperfractionated concomitant boost concept to a median total dose of 78 GyE. Treatment was tolerated well; only one patient developed a grade 3 esophageal late effect. Despite approximately 53% T3/T4 tumors in this cohort, local control and overall survival are impressive with 42.8% and 84.4%, respectively. Objective response rate was 100% (CR: 89%, PR: 11% at 4 months posttreatment) (Mizumoto et al. 2011).

One important message from this experience needs to be considered: while the commonly accepted dose in esophageal cancer has been derived from grade V toxicities in the Minsk trial (Sugahara et al. 2005) occurring in the experimental arm but below 56 Gy, these reports show higher doses to the esophagus are possible with acceptable toxicity, yet particle therapy for esophageal cancer needs further exploration.

10.1.6   Hepatocellular Carcinoma

There is extensive literature on the treatment of hepatocellular carcinoma (HCC) with particle therapy (Table 10.7). HCC is a very radioresistant tumor while surrounding normal liver tissue is very radiosensitive, hence conventional RT techniques achieving sufficient doses levels are only feasible for small tumors.

Various trials using proton therapy in various fractionation schemes achieved impressive local control rates of 75% to 96% at 2a and OS rates between 55% to 66% at 2a (refer to Table 10.6 for further details). Therapy-related side effects were generally mild: the trial by Bush et al. (2004a) reported only three cases of GI bleeding; in all of these cases, the respective bowel part was directly adjacent to the tumor. Other trials found no late toxicity >II at all (Mizumoto et al. 2007). Even in the elderly population, good clinical results could be maintained without increased toxicity (Hata et al. 2007b).

Dose and fractionation vary depending on tumor site and size. However, even in close proximity to the porta hepatis or digestive organs, treatment can be carried out without major side effects (Fukumitsu et al. 2009; Sugahara et al. 2009, 2010; Imada et al. 2010). Treatment outcome significantly correlated with tumor volume (Nakayama et al. 2009) and site (Imada et al. 2010). However, even in patients with portal vein tumor thrombosis or tumors larger than 10 cm, response rates of around 89% (Sugahara et al. 2009) and local control rates of 87% at 2 years (Sugahara et al. 2010) are achievable.

No significant change in liver function was found after C12 therapy (Kato et al. 2004); liver tolerance

Chapter 10

Table 10.7   Overview of Particle Therapy for Hepatocellular Carcinoma

Author	Journal	Patients	Radiotherapy	Tumor Diameter	Local Control	OS
Bush et al.	*Gastroenterology* 2004	34	p+/63 GyE/15 Fx	5.7 cm	75%/2a	55%/2a
Kato et al.	*IJROBP* 2004	24	C12/49.5–79.5 GyE a 3.3–5.3 GyE/Fx		81%/5a	25/5a
Kawashima et al.	*J. Clin. Oncol.* 2005	30	p+/70 GyE/20 Fx	4.5 cm	PFS: 96%/2a	66%/2a
Chiba et al.	*Clin. Cancer Res.* 2005	162	p+/72 GyE/16 Fx		90%/5a	23.5%/5a
Hata et al.	*Strahlenther Onkol.* 2007	21	p+/24 GyE/Fx		Not given, objective response rate 100%	
Mizumoto et al.	*Jpn. J. Clin. Oncol.* 2007	3	p+/72.6 GyE/22 Fx		86%/3a	45.1%/3a
Fukumitsu et al.	*IJROBP* 2009	51	p+/66 GyE/10 Fx	>2 cm from porta hepatis or digestive organs	87.8%/5a	38.7%/5a
Nakayama et al.	*Cancer* 2009	318	p+/77Gy in 10 Fx (tumors within 2 cm of digestive organs); 72.6 GyE in 22 Fx (within 2 cm of porta hepatis); 66.0 GyE in 10 Fx (peripheral tumors)		Not given	64.7%/3a; 44.6%/5a
Sugahara et al.	*Strahlenther Onkol.* 2009	35	p+/72.6 GyE/22 Fx	2.5–13.0 cm associated with portal vein thrombosis	Response rate: CR: 89%; local PFS 46%/2a	48%/2a
Sugahara et al.	*IJROBP* 2010	22	p+/72.6 GyE/22 Fx	>10 cm	PFS: 24%/2a	36%/2a
Imada et al.	*Radiother. Oncol.* 2010	64	C12/52.8 GyE/4Fx	A: within 2 cm of porta hepatis; B: >2cm of porta hepatis	A: 87.8%/5a; B: 95.7%/5a	A: 22.2%/5a; B: 34.8%/5a

was evaluated retrospectively for patients with liver cirrhosis and after proton therapy (Ohara et al. 1997). Similar to surgical treatment concepts, posttherapeutic hypertrophic changes were found in normal liver tissue after treatment. Hypertrophy varied between 19% and 51% and correlated significantly to the irradiated liver volume and initial functioning liver volume.

Even reirradiation after prior proton RT (median dose 72 GyE) was tolerated reasonably well. In 27 patients another course of radiation (median interval 24.5 months) with a median dose of 66 GyE was

applied. However, in view of liver reserves, patients undergoing reirradiation for HCC should not suffer from cirrhotic changes > Child-Pugh A (Hashimoto et al. 2006).

Probably the largest analysis of particle therapy in 162 patients with this disease was performed by Chiba and coworkers (Chiba et al. 2005). Control rates supported the excellent results (please see Table 10.7), while acute and late side-effects were negligible and without influence on the patients' prognosis.

Compared to photon therapy results, particle therapy yields superior therapeutic outcomes, with only the most modern techniques (i.e., stereotactic radio-surgery) getting anywhere near this range (Robertson et al. 1993; Blomgren et al. 1995), though the amount of data concerning RT for HCC is still more abundant for particle techniques. Future trials need to explore the role of particle therapy in this disease. Like in NSCLC though, carbon ion therapy may be an equally good treatment option for patients unfit for surgical procedures.

10.1.7   Pancreatic Carcinoma

While the role of radiotherapy in the treatment of pancreatic carcinoma still needs to be determined, it is also one of the rather radioresistant tumors; hence, application of particle therapy with the advantage of higher RBE seems obvious. Recent Japanese trials are still in follow-up and therefore not yet completely published. All in all, three phase I/II trials for carbon ion therapy have been initiated in Chiba so far.

One dose-escalation (44.8–48 GyE in 16 Fx) trial is evaluating neoadjuvant C12-therapy for operable pancreatic carcinomas. So far, side effects have generally been mild, with no acute toxicities >3 reported. Only two grade 3 toxicities were found; however, these were postoperative portal vein stenoses that cannot safely be attributed to radiation therapy only. Local control at 12 and 24 months was an impressive 100% and overall survival was 62%. Resectability after heavy ion therapy proved to be 68%; actuarial OS in this group was much higher (90.1%) (Yamada et al. 2005).

Another protocol also evaluated hypofraction-ated carbon ion therapy in a neoadjuvant setting. Preoperative PT consists of 8 Fx in 2 weeks. However, the trial is still recruiting, hence initial results are pending.

A recently published phase II trial evaluating hypofractionated neoadjuvant chemoradiation using capecitabine and proton beam therapy in resectable pancreatic carcinoma was able to establish a 5 × 5 GyE regimen in this setting. Overall survival was respect-able with 75% at 1 year; all noted failures proved to be distant failures. With four grade 3 treatment-related toxicities (pain in one patient, stent obstruction in three patients), treatment-related side effects were acceptable (Hong et al. 2011).

The group from Chiba investigated the applica-tion of C12 therapy (38.4–52.8 GyE/12 Fx dose esca-lation) in locally advanced, inoperable pancreatic carcinoma. Recruitment of patients was completed with 47 patients in 2007. Seven grade 3 acute reac-tions (anorexia 6 pts, cholangitis 1 pt) and only one 3 late reaction were reported. Follow-up of this trial is still short, hence there is only preliminary data: OS at 1 a is 43%, local control at doses > 45.6 GyE is 95% (Yamada et al. 2008).

Early data from Berkeley comparing He and pho-ton irradiation in 49 patients reported superior local control rates for He therapy though increased local control did not translate into increase survival in this cohort (Linstadt et al. 1988). Though the protocol did apply combined chemoradiation, treatment regimen consisted of a split course radiation treatment, which would not be state of the art today. Together with the fact that treatment planning was not performed three-dimensionally (He and photons), one needs to be care-ful applying these results today.

Based on these results, a final assessment as to par-ticle therapy for pancreatic carcinoma is not possible. However, the very systematic approach to the question by the Japanese investigators could clearly demon-strate the feasibility of this treatment for operable and locally advanced disease. Clinical results seem so far to be superior to comparable photon trials with a very favorable toxicity profile.

10.1.8   Carcinoma of the Uterine Cervix

Radiation therapy as chemoradiation has long been established for the treatment of cervical carcinoma. Standard treatment includes external beam RT and/or brachytherapy. Further dose escalation by the applica-tion of particle therapy was investigated in the follow-ing reports with regard to the effect on local control rates in this disease.

Kato and coworkers published two phase I-II tri-als (44 and 94 pts) on treatment with carbon ions (Shioyama et al. 2003; Miyamoto et al. 2007a; Sugahara et al. 2005). All patients were diagnosed with locally advanced carcinoma of the cervix (median diameter 6.5 cm). Treatment consisted of pelvic irradiation (16 Fx) followed by a boost to the primary tumor in 8 Fx. Total doses varied between 52 and 72 GyE (Kato et al. 2006; Matsushita et al. 2006), yielding 5-year local control rates of 45% (Matsushita et al. 2006) and 79% (Kato et al. 2006). There were no serious acute treatment-related toxicities >2; however, two patients needed surgical interventions at GI late complications in patients receiving doses >60 GyE to the intestine, and therefore doses have been limited since to under 60 GyE.

Proton therapy in this setting has been evaluated by Kagei et al. (2003): patients receiving pelvic irradiation (photons and 44.8 GyE) plus 24 or 28 GyE proton boost to the tumor. Local control at 5 years and OS at 10 years in this trial were 100% and 89% (IIB) and 61% and 40% (IIIB/IVA).

Clinical influence of oxygenation for photon therapy could be clearly demonstrated by Dunst et al. (2003). However, many preclinical investigations were able to exclude therapy-relevant influence of tumor hypoxia in high-LET radiation. Clinical data published by Kato et al. (2006) in advanced cervical carcinoma could now prove this thesis clinically.

Though we lack major trials to firmly establish particle therapy in this indication, initial results are very promising for the particle therapy approach. However, independence of tumor oxygenation in carbon ion therapy may be one of the major advantages in a selected patient group warranting further exploration.

10.1.9 Extracranial Sarcoma/chordoma and Chondrosarcoma

Sarcomas and chordomas again are marked by their relative radioresistance, thus presenting ideal targets for particles. However, for these comparatively rare tumors there is scarce prospective evidence. Retrospective data though, indicates high control rates achieved by a combination of photon and proton irradiation especially in the primary situation as opposed to patients being treated for local relapse. With a median follow-up of 8.8 years, 12/14 patients were still locally controlled. A dose > 73 GyE led to continuing local control (Park et al. 2006). Postoperative proton therapy (median 72 GyE) for extracranial chordoma in 26 patients (15 pelvic tumors) resulted in an actuarial OS of 84% at 3 years and PFS of 77% (Rutz et al. 2007). A combination of protons/IMRT and electron IORT for retroperitoneal sarcoma of various histologies was able to achieve a 90% local control at a median follow-up of 33 months in the primary setting (Yoon et al. 2010).

DeLaney and coworkers have achieved control rates of 78% at 5 years with a preoperative regimen of high-dose photon/proton RT in a phase II trial for spinal sarcomas (DeLaney et al. 2009). Thirty patients with unresectable sacral chordoma were treated with carbon ions at 52.8–73.6 GyE. Overall survival and local control at 5 years were reported with 52% and 96%, respectively (Imai et al. 2004). Pooled data of

prospective phase I/II and II trials applying a median dose of 70.4 GyE carbon ions in 16 fractions resulted even in a local control of 89% at 5 years (Imai et al. 2010). Retrospective comparison of surgical and purely radiotherapeutic series for sacral chordoma even suggests a higher local control rate with an impressive 100% at 5 years as opposed to 62.5% at 5 years for the surgical series (Nishida et al. 2011). While these tumors can gain considerable sizes, treatment and normal tissue sparing can pose a challenge. The case of a carbon ion boost for a 69-year-old patient with inoperable sacral chordoma is shown in Figures 10.8 through 10.12.

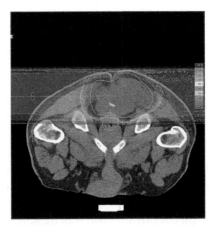

FIGURE 10.8 (See color insert.) Sixty-nine-year-old patient with sacral chordoma; axial carbon ion boost dose distribution (24 GyE C12 + 50 Gy IMRT).

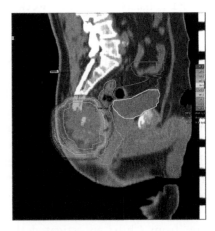

FIGURE 10.9 (See color insert.) Sixty-nine-year-old patient with sacral chordoma; sagittal carbon ion boost dose distribution (24 GyE C12 + 50 Gy IMRT).

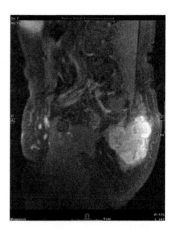

FIGURE 10.10 Sixty-nine-year-old patient with sacral chordoma; pretherapeutic contrast-enhanced MRI.

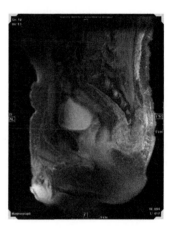

FIGURE 10.11 Sixty-nine-year-old patient with sacral chordoma; contrast-enhanced MRI 6 months posttreatment already showing treatment response.

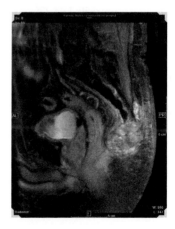

FIGURE 10.12 Sixty-nine-year-old patient with sacral chordoma and contrast-enhanced MRI 12 months posttreatment: Further tumor shrinkage.

10.1.10 Prostate Cancer

There is a variety of clinical data regarding particle therapy of prostate carcinoma, among them two phase III trials (Zietman et al. 2005, 2008, 2010; Shipley et al. 1995). Zietman et al. (2005, 200, 2010) treated 393 patients with prostate carcinoma (PSA <5 ng/ml) with either photons alone or combined photon/proton therapy, and the trial by Shipley et al. (1995) had a similar design but recruited 202 patients with stages T3–4 N0–2. Both of these trials could demonstrate superior freedom from biochemical relapse/local control in the proton group, albeit the proton groups in both trials were treated to higher total doses (see Table 10.7 for details). Other researchers (Yonemoto et al. 1997; Rossi et al. 2004; Slater et al. 2004) published data on 1255 patients also treated with a combined approach using 45 Gy photons to the pelvis followed by 30 GyE proton boost to the prostate. This regimen also yielded very good results with a freedom from biochemical relapse rate of 73% or even 90% in patients with initial PSA <90%. Another trial (Mayahara et al. 2007) employed protons alone; however, follow-up is not yet mature, hence only toxicity data is currently available. Though no phase III trials are available for treatment with carbon ions, there is a considerable amount of data for heavy ion therapy in the treatment of prostate carcinoma. Doses employed are between 54 and 72 GyE in various fractionations (Akakura et al. 2004; Ishikawa et al. 2008; Tsuji et al. 2005), yielding similar results. However, 54 GyE C12 was found to be insufficient to obtain lasting local control whereas doses of 72 GyE seemed to cause higher acute toxicity rates in a dose escalation trial (Akakura et al. 2004). Treatment-related toxicity was generally very mild in all the trials (Table 10.8), grade 3 late toxicities were very rare (1% @ 5/10 year in the Loma Linda series); hence treatment outcome as well as toxicity profiles are very much in favor of particle therapy. Biochemical relapse-free survival is dose-dependent; the necessity for dose escalation at tolerable side effects has hence led to the routine use of IMRT for primary prostate cancer at many institutions. Further stepwise dose escalation will largely be limited by accompanying toxicity. Prostate cancer is a common disease, so clinical trials will need to determine significance of observed benefits in terms of relapse-free survival and toxicity in controlled clinical settings.

Chapter 10

Table 10.8   Overview of Particle Therapy for Prostate Cancer

	Journal	Patients	Stage/PSA	Dose	Freedom from Biochemical Failure	Toxicity	
Zietman et al.	*J. Am. Med. Assoc.* 2005/2008	393	<5 ng/ml	A: 70.2 Gy photons vs. B: 50.4 Gy photons + 28.8 GyE protons	A: 78.8%; B: 91.3% (5a)	>°3: 2% (A); 1% (B)	
Zietman et al.	*J. Clin. Oncol.* 2010	393			A: 67.6%; B: 83.3% (10a)	2% (A or B)	
Shipley et al.	*IJROBP* 1995	202	T3/4 N0-2	50.4 Gy photons plus 16.8 Gy photons vs. 25.2 GyE protons	Local control a8a (G3): 19% (photons) vs. 84% (photons/protons)	Photon/proton: acute rectal toxicity, urethral stenosis (both ns)	
Mayahara et al.	*IJROBP* 2007	287		protons 74 GyE	81%/5a	No > °1 acute rectal toxicity; °2 acute GU toxicity: 39%; °3 acute GU toxicity: 4%	
Ishikawa et al.	*IJROBP* 2008	175		C12 66 GyE/20 Fx	89.5% at median f/u of 48 mo	Increased GU late toxicity with ADT > 24 mo	
Shimazaki et al.	*Anticancer Res.* 2010	254		C12 66 GyE/20 Fx	Low-risk: 76%; intermediate risk: 91%; high-risk: 76%		No consistent antihormonal therapy
Nihei et al.	*Jpn. J. Clin. Oncol.* 2005	30	T1-3 N0	50 Gy photons + 26 GyE protons	80% at median f/u of 30 mo	Acute°1/2 GU toxicity 80%; GI toxicity 57%; late °1/2 GU toxicity 17%; GI toxicity 37%; no °3 acute/late toxicity	
Yonemoto et al.	*IJROBP* 1997	106		45 Gy pelvic RT (photons) + 30 GyE protons	73%; 90% with PSA < 40 ng/ml; local control at 2a 97%; age-independent	°1/2 late toxicity 12%, °3/4 GI/GU late toxicity 1% @ 5 + 10a	
Rossi et al.	*Urology* 2004						
Slater et al.	*IJROBP* 2004	1255					

(*continued*)

Table 10.8   Overview of Particle Therapy for Prostate Cancer (Continued)

	Journal	Patients	Stage/PSA	Dose	Freedom from Biochemical Failure	Toxicity
Akakura et al.	*Prostate* 2004	96	locally advanced	54–72 GyE C12	No local control @ 54 GyE	66 GyE: 1°3 acute toxicity; 72 GyE: 5°3 acute toxicity
Shimazaki et al.	*Jpn. J. Clin. Oncol.* 2006	37		60–66 GyE C12	85% @5a; low-risk: 96%	
Tsuji et al.	*IJROBP* 2005	201		C12; mostly 66 GyE/20 Fx	83.2% @ 5a	No > °3 GU/GI toxicity

10.2   Summary and Outlook

In summary, irradiation with protons and heavy ions does indeed seem to have advantages such as a larger series demonstrating benefits and establishing particle therapy for uveal melanoma, chordoma/chondrosarcoma of the skull base, and adenoid cystic carcinoma/paranasal sinus cancers. There is also data showing promising results in single-fraction treatment of early-stage NSCLC, HCC and prostate carcinoma. New data is emerging in the study of pancreatic carcinoma. However, phase III trials comparing photon and particle therapy or even proton and heavy ions are largely pending. First, this is due to the fact that particle therapy is still complex and not easily available. Second, a lot of these tumors have a comparatively low incidence so systematic randomized controlled trials are not possible even in the photon world.

So the question remains, do we need more trials, among them randomized controlled trials, to prove the clinical benefit of particle therapy?

This discussion is repetitively revived whenever new treatment techniques become available, first with the introduction of linear accelerators and again with the more common availability of IMRT. While there is definitely good reason to apply more sophisticated new techniques, (i.e., particle therapy), it still needs to be proven whether the advantages in dose distribution translate into measurable clinical benefits such as improvement in overall survival or quality of life. Physicians are

required to make evidence-based decisions in almost any field of medicine. New techniques also come with certain challenges: motion management and radiobiological properties of particle beams as well as fractionation patterns are still issues in this comparatively new technique. Additionally, more widespread introduction of particle therapy still brings more financial pressure onto the respective country's health system.

Though there is considerable evidence, albeit mostly level 2b, published data is very heterogenous even within the same indications. For many tumors, there is evidence both for proton and carbon ion therapy. We do need to determine which quality of radiation to choose for the individual patient if both are available. Chemoradiation has been established in the treatment of various tumors throughout the years, and this has yet to be evaluated concomitantly with proton/heavy ion therapy. Very little is known about the clinical interactions between systemic therapy and particle irradiation in terms of efficacy and toxicity. New indications can be opened up to treatment with particle beams, hence the need to have phase I/II trials for these entities. However, we would strongly emphasize the scientific aspiration to produce meaningful results for patients and physicians for the future. Transnational efforts may be necessary to achieve this goal and therefore various projects to this effect are welcome and currently under preparation.

References

Akakura, K., Tsujii, H., Morita, S., Tsuji, H., Yagashita, T., Isaka, S. et al. 2004. Phase I/II clinical trials of carbon ion therapy for prostate cancer. *Prostate* 58(3): 252–258.

Ares, C., Hug, E. B., Lomax, A. J., Bolsi, A., Timmermann, B., Rutz, H. P. et al. 2009. Effectiveness and safety of spot scanning proton radiation therapy for chordomas and chondrosarcomas of the skull base: First long-term report. *Int. J. Radiat. Oncol. Biol. Phys.* 75: 1111–1118.

Blomgren, H., Lax, I., Naslund, I. and Svanstrom, R. 1995. Stereotactic high dose fraction radiation therapy of extracranial tumors using an accelerator. Clinical experience of the first thirty-one patients. *Acta Oncol.* 34(6): 861–870.

Bush, D. A., Slater, J. D., Bonnet, R., Cheek, G. A., Dunbar, R. D., Moyers, M. et al. 1999. Proton-beam radiotherapy for early-stage lung cancer. *Chest* 116: 1313–1319.

Bush, D. A., Hillebrand, D. J., Slater, J. M. and Slater, J. D. 2004a. High-dose proton beam radiotherapy of hepatocellular carcinoma: preliminary results of a phase II trial. *Gastroenterology* 127(5 Suppl 1): S189–S193.

Bush, D. A., Slater, J. D., Shin, B. B., Cheek, G. C., Miller D. W. and Slater, J. M. 2004b. Hypofractionated proton beam radiotherapy for stage I lung cancer. *Chest* 126: 1198–1203.

Castro, J. R., Char, D. H., Petti, P. L., Daftari, I. K., Quivey, J. M., Singh, R. P. et al. 1997. 15 years experience with helium ion radiotherapy for uveal melanoma. *Int. J. Radiat. Oncol. Biol. Phys.* 39(5): 989–996.

Castro, J. R., Linstadt, D. E., Bahary, J. P., Petti, P. L., Daftari, I., Collier, J. M. et al. 1994. Experience in charged particle irradiation of tumors of the skull base: 1977–1992. *Int. J. Radiat. Oncol. Biol. Phys.* 29(4): 647–655.

Caujolle, J. P., Mammar, H., Chamorey, E., Pinon, F., Herault, J. and Gastaud, P. 2010. Proton beam radiotherapy for uveal melanomas at NICE teaching hospital: 16 years' experience. *Int. J. Radiat. Oncol. Biol. Phys.* 78: 98–103.

Char, D. H., Quivey, J. M., Castro, J. R., Kroll, S. and Phillips, T. 1993. Helium ions versus iodine 125 brachytherapy in the management of uveal melanoma. A prospective, randomized, dynamically balanced trial. *Ophthalmology* 100(10): 1547–1554.

Chen, A. M., Daly, M. E., Bucci, M. K., Xia, P., Akazawa, C., Quivey, J. M., Weinberg, V. et al. 2007a. Carcinomas of the paranasal sinuses and nasal cavity treated with radiotherapy at a single institution over five decades: Are we making improvement? *Int. J. Radiat. Oncol. Biol. Phys.* 69: 141–147.

Chen, A. M, Daly, M. E., El-Sayed, I., Garcia, J., Lee, N. Y., Bucci, M. K. and Kaplan, M. K. 2008. Patterns of failure after combined-modality approaches incorporating radiotherapy for sinonasal undifferentiated carcinoma of the head and neck. *Int. J. Radiat. Oncol. Biol. Phys.* 70: 338–343.

Chen, A. M., Granchi, P. J., Garcia, J., Bucci, M. K., Fu, K. K. et al. 2007b. Local-regional recurrence after surgery without postoperative irradiation for carciomas of the major salivary glands: implications for adjuvant therapy. *Int. J. Radiat. Oncol. Biol. Phys.* 67(4): 982–987.

Chiba, T., Tokuuye, K., Matsuzaki, Y., Sugahara, S., Chugannji, Y., Kagei, K. et al. 2005. Proton beam therapy for hepatocellular carcinoma: A retrospective review of 162 patients. *Clin. Cancer Res.* 11(10): 3799–3805.

Combs, S. E., Nikoghosyan, A., Jaekel, O., Karger, C. P., Haberer, T., Münter, M. W. et al. 2009. Carbon ion radiotherapy for pediatric patients and young adults treated for tumors of the skull base. *Cancer* 115: 1348–1355.

Courdi, A., Caujolle, J. P., Grange, J. D., Diallo-Rosier, L., Sahel, J., Bacin, F. et al. 1999. Results of proton therapy of uveal melanomas treated in Nice. *Int. J. Radiat. Oncol. Biol. Phys.* 45(1): 5–11.

Damato, B., Kacperek, A., Chopra, M., Campbell, I. R. and Errington, R. D. 2005. Proton beam radiotherapy of iris melanoma. *Int. J. Radiat. Oncol. Biol. Phys.* 63(1): 109–115.

Debus, J., Schulz-Ertner, D., Schad, L., Essig, M., Rhein, B., Thillmann, C. O. et al. 2000. Stereotactic fractionated radiotherapy for chordomas and chondrosarcomas of the skull base. *Int. J. Radiat. Oncol. Biol. Phys.* 47(3): 591–596.

DeLaney, T. F., Liebsch, N. J., Pedlow, F. X., Adams, J., Dean, S., Yeap, B. Y. et al. 2009. Phase II study of high-dose photon/proton radiotherapy in the management of spine sarcomas. *Int. J. Radiat. Oncol. Biol. Phys.* 74: 732–739.

Demizu, Y., Murakami, M., Miyawaki, D., Niwa, Y., Akagi, T., Sasaki, R. et al. 2009. Analysis of vision loss caused by radiation-induced optic neuropathy after particle therapy for head-and-neck and skull-base tumors adjacent to optic nerves. *Int. J. Radiat. Oncol. Biol. Phys.* 75: 1487–1492

Dendale, R., Lumbroso-Le Rouic, L., Noel, G., Feuvret, L., Levy, C., Delacroix, S. et al. 2006. Proton beam radiotherapy for uveal melanoma: Results of Curie Institut-Orsay proton therapy center (ICPO). *Int. J. Radiat. Oncol. Biol. Phys.* 65(3): 780–787.

Desjardins, L., Lumbroso, L., Levy, C., Mazal, A., Delacroix, S., Rosenwald, J. C. et al. 2003. [Treatment of uveal melanoma with iodine 125 plaques or proton beam therapy: Indications and comparison of local recurrence rates]. *J. Fr. Ophthalmol.* 26(3): 269–276 (in French).

Douglas, J. G., Koh, W. J., Austin-Seymour, M. and Laramore, G. E. 2003. Treatment of salivary gland neoplasms with fast neutron radiotherapy. *Arch Otolaryngol. Head Neck Surg.* 129(9): 944–948.

Dunst, J., Kuhnt, T., Strauss, H. G., Krause, U., Pelz, T., Koelbl, H. et al. 2003. Anemia in cervical cancers: Impact on survival, patterns of relapse, and association with hypoxia and angiogenesis. *Int. J. Radiat. Oncol. Biol. Phys.* 56(3): 778–787.

Egger, E., Zografos, L., Schalenbourg, A., Beati, D., Böhringer, T., Chamot, L. et al. 2003. Eye retention after proton beam radiotherapy for uveal melanoma. *Int. J. Radiat. Oncol. Biol. Phys.* 55(4): 867–880.

Fukumitsu, N., Sugahara, S., Nakayama, H., Fukuda, K., Mizumoto, M., Abei, M, et al. 2009. A prospective study of hypofractionated proton beam therapy for patients with hepatocellular carcinoma. *Int. J. Radiat. Oncol. Biol. Phys.* 74: 831–836.

Fuss, M., Loredo, L. N., Blacharski, P. A., Grove, R. I. and Slater, J. D. Proton radiation therapy for medium and large choroidal melanoma: preservation of the eye and its functionality. 2001. *Int. J. Radiat. Oncol. Biol. Phys.* 49(4): 1053–1059.

Garden, A. S., Weber, R. S., Ang, K. K., Morrison, W. H., Matre, J. and Peters, L. J. 1994. Postoperative radiation therapy for malignant tumors of minor salivary glands. Outcome and patterns of failure. *Cancer* 73(10): 2563–2569.

Gragoudas, E. S., Lane, A. M., Regan, S., Li, W., Judge, H. E., Munzenrider, J. E. et al. 2000. A randomized controlled trial of varying radiation doses in the treatment of choroidal melanoma. *Arch. Ophthalmol.* 118(6): 773–778.

Gurney, T. A., Eisele, D. W., Weinberg, V., Shin, E. and Lee, N. 2005. Adenoid cystic carcinoma of the major salivary glands treated with surgery and radiation. *Laryngoscope* 115(7): 1278–1282.

Habrand, J. L., Schneider, R., Alapetite, C., Feuvret, L., Petras, S., Datchary, J. et al. 2008. Proton therapy in pediatric skull base and cervical canal low-grade bone malignancies. *Int. J. Radiat. Oncol. Biol. Phys.* 71: 672–675.

Hashimoto, T., Tokuuye, K., Fukumitsu, N., Igaki, H., Hata, M., Kagei, K. et al. 2006. Repeated proton beam therapy for hepatocellular carcinoma. *Int. J. Radiat. Oncol. Biol. Phys.* 65(1): 196–202.

Hata, M., Tokuuye, K., Kagei, K., Sugahara, S., Nakayama, H., Fukumitsu, N. et al. 2007. Hypofractionated high-dose proton beam therapy for stage I non-small-cell lung cancer: Preliminary results of a phase I/II clinical study. *Int. J. Radiat. Oncol. Biol. Phys.* 68(3): 786–793.

Hata, M., Tokuuye, K., Sugahara, S., Tohno, E., Fukumitsu, N., Hashimoto, T. et al. 2007. Proton irradiation in a single fraction for hepatocellular carcinoma patients with uncontrollable ascites. Technical considerations and results. *Strahlenther Onkol.* 183(8): 411–416.

Höcht, S., Bechrakis, N. E., Nausner, M., Kreusel, K. M., Kluge, H., Heese, J. et al. 2004. Proton therapy of uveal melanomas in Berlin. 5 years of experience at the Hahn-Meitner Institute. *Strahlenther Onkol.* 180(7): 419–424.

Hong, T. S., Ryan, D. P., Blaszkowsky, L. S., Mamon, H. J., Kwak, E. L., Mino-Kenudson, M. et al. 2011. Phase I study of preoperative short-course chemoradiation with proton beam therapy and capecitabine for resectable pancreatic ductal adenocarcinoma of the head. *Int. J. Radiat. Oncol. Biol. Phys.* 79: 151–157.

Hoppe, B. S., Nelson, C. J., Gomez, D. R., Stegman, L. D., Wu, A. J., Wolden, S. L., Pfister, D. G. et al. 2008. Unresectable carcinoma of the paranasal sinuses: Outcome and toxicities. *Int. J. Radiat. Oncol. Biol. Phys.* 72: 763–769.

Huber, P. E., Debus, J., Latz, D., Zierhut, D., Bischof, M., Wannenmacher, M. et al. 2001. Radiotherapy for advanced adenoid cystic carcinoma: Neutrons, photons or mixed beam? *Radiother. Oncol.* 59(2): 161–167.

Imada, H., Kato, H., Yasuda, S., Yanagi, T., Kishimoto, R., Kandatsu, S. et al. 2010. Comparison of efficacy and toxicity of short-course carbon ion radiotherapy for heaptocellular carcinoma depending on their proximity to the porta hepatis. *Radiother. Oncol.* 96: 231–235.

Imai, R., Kamada, T., Tsuji, H., Yanagi, T., Baba, M., Miyyamoto, T. et al. 2004. Carbon ion radiotherapy for unresectable sacral chordomas. *Clin. Cancer Res.* 10(17): 5741–5746.

Imai, R., Kamada, T., Tsuji, H., Sugawara, S., Serizawa, I., Tsujii, H. et al. 2010. Effect of carbon ion radiotherapy for sacral chordoma: Results of phase I-II and phase II clinical trials. *Int. J. Radiat. Oncol. Biol. Phys.* 77: 1470–1476.

Ishikawa, H., Tsuji, H., Kamada, T., Hirasawa, N., Yanagi, T. Mizoe, J.E. et al. 2008. Adverse effects of androgen deprivation therapy on persistent genitourinary complications after carbon ion radiotherapy for prostate cancer. *Int. J. Radiat. Oncol. Biol. Phys.* 72(1): 78–84.

Jensen, A. D., Nikoghosyan, A. V., Ecker, S., Ellerbrock, M., Debus, J. and Münter, M. W. 2011a. Carbon ion therapy for advanced sinonasal malignancies: Feasibility and acute toxicity. *Radiat. Oncol.* 6: 30.

Jensen, A. D., Nikoghosyan, A. V., Ellerbrock, M., Ecker, S., Debus, J. and Münter, M. W. 2011b. Re-irradiation with scanned particle beams in recurrent tumours of the head and neck: Acute toxicity and feasibility. *Radiother. Oncol.* 101(3): 383–387.

Jingu, K., Kishimoto, R., Mizoe, J., Hasegawa, A., Bessho, H., Tsujii, H. et al. 2011. Malignant mucosal melanoma treated with carbon ion radiotherapy with concurrent chemotherapy: Prognostic value of pretreatment apparent diffusion coefficient (ADC). *Radiother. Oncol.* 98: 68–73.

Kadono, K., Homma, T., Kamahra, K., Nakayama, M., Satoh, H., Sekizawa, K. et al. 2002. Effect of heavy-ion radiotherapy on pulmonary function in stage I non-small cell lung cancer patients. *Chest* 122(6): 1925–1932.

Kagei, K., Tokuuye, K., Okumura, T., Ohara, K., Shioyama, Y., Sugahara, S. et al. 2003. Long-term results of proton beam therapy for carcinoma of the uterine cervix. *Int. J. Radiat. Oncol. Biol. Phys.* 55(5): 1265–1271.

Kato, H., Tsujii, H., Miyamoto, T., Mizoe, J. E., Kamada, T., Tsuji, H. et al. 2004. Results of the first prospective study of carbon ion radiotherapy for hepatocellular carcinoma with liver cirrhosis. *Int. J. Radiat. Oncol. Biol. Phys.* 59(5): 1468–1476.

Kato, S., Ohno, T., Tsujii, H., Nakano, T., Mizoe, J. E., Kamada, T. et al. 2006. Dose escalation study of carbon ion radiotherapy for locally advanced carcinoma of the uterine cervix. *Int. J. Radiat. Oncol. Biol. Phys.* 65(2): 388–397.

Kawashima, M., Furuse, J., Nishio, T., Konishi, M., Ishii, H., Kinoshita, T. et al. 2005. Phase II study of radiotherapy employing proton beam for hepatocellular carcinoma. *J. Clin. Oncol.* 23(9): 1839–1846.

Koto, M., Miyamoto, T., Yamamoto, N., Nishimura, H., Yamada, S., Tsujii, H. et al. 2004. Local control and recurrence of stage I non-small cell lung cancer after carbon ion radiotherapy. *Radiother. Oncol.* 71(2): 147–156.

Linstadt, D., Quivey, J. M., Castro, J. R., Andejeski, Y., Phillips, T. L., Hannigan, J., et al. 1988. Comparison of helium-ion radiation therapy and split-course megavoltage irradiation for unresectable adenocarcinoma of the pancreas. Final report of a Northern California Oncology Group randomized prospective clinical trial. *Radiology* 168(1): 261–264.

Marucci, L., Ancukiewicz, M., Lane, A. M., Collier, J. M., Gragiudas, E. and Munzenrider, J. E. 2011. Uveal melanoma recurrence after fractionated proton beam therapy: comparison of survival in patients treated with reirradiation or with enucleation. *Int. J. Radiat. Oncol. Biol. Phys.* 79: 842–846.

Matsushita, K., Ochiai, T., Shimada, H., Kato, S., Ohno, T., Nikaido, T. et al. 2006. The effects of carbon ion irradiation revealed by excised perforated intestines as a late morbidity for uterine cancer treatment. *Surg. Today* 36(8): 692–700.

Mayahara, H., Murakami, M., Kagawa, K., Kawaguchi, A., Oda, Y., Miyawaki, D. et al. 2007. Acute morbidity of proton therapy for prostate cancer: The Hyogo Ion Beam Medical Center experience. *Int. J. Radiat. Oncol. Biol. Phys.* 69(2): 434–443.

Mendenhall, W. M., Morris, C. G., Amdur, R. J., Werning, J. W., Hinerman, R. W. and Villaret, D. B. 2004. Radiotherapy alone or combined with surgery for adenoid cystic carcinoma of the head and neck. *Head Neck* 26(2): 154–162.

Minsky, B. D., Pajak, T. F., Ginsberg, R. J., Pisansky, T. M., Martenson, J., Komaki, R. et al. 2002. INT 0123 (Radiation Therapy Oncology Group 94-05) phase III trial of combined-modality therapy for esophageal cancer: High-dose versus standard-dose radiation therapy. *J. Clin. Oncol.* 20(5): 1167–1174.

Miyamoto, T., Yamamoto, N., Nishimura, H., Koto, M., Tsujii, H., Mizoe, J. E. et al. 2003. Carbon ion radiotherapy for stage I non-small cell lung cancer. *Radiother. Oncol.* 66(2): 127–140.

Miyamoto, T., Baba, M., Sugane, T., Nakajima, M., Yashiro, T., Kagei, K. et al. 2007a. Carbon ion radiotherapy for stage I non-small cell lung cancer using a regimen of four fractions during 1 week. *J. Thorac. Oncol.* 2(10): 916–926.

Miyamoto, T., Baba, M., Yamamoto, N., Koto, M., Sugawara, T., Yashiro, T. et al. 2007b. Curative treatment of stage I non-small-cell lung cancer with carbon ion beams using a hypofractionated regimen. *Int. J. Radiat. Oncol. Biol. Phys.* 67(3): 750–758.

Mizoe, J. E., Tsujii, H., Kamada, T., Matuoka, Y., Tsuji, H., Osaka, Y. et al. 2004. Dose escalation study of carbon ion radiotherapy for locally advanced head-and-neck cancer. *Int. J. Radiat. Oncol. Biol. Phys.* 60(2): 358–364.

Chapter 10

Mizumoto, M., Tokuuye, K., Sugahara, S., Hata, M., Fukumitsu, N., Hashimoto, T. et al. 2007. Proton beam therapy for hepatocellular carcinoma with inferior vena cava tumor thrombus: Report of three cases. *Jpn. J. Clin. Oncol.* 37(6): 459–462.

Mizumoto, M., Sugahara, S., Nakayama, H., Hashii, H., Nakahara, A., Terashima, H. et al. 2010. Clinical results of proton-beam therapy for locoregionally advanced esophageal cancer. *Strahlenther Onkol.* 186: 482–488.

Mizumoto, M., Sugahara, S., Okumura, T., Hashimoto, T., Oshiro, Y., Fukumitsu, N. et al. 2011. Hyperfractionated concomitant boost proton beam therapy for esophageal carcinoma. *Int. J. Radiat. Oncol. Biol. Phys.* 81(4): e601–e606.

Mosci, C., Mosci, S., Barla, A., Squarcia, S., Chauvel, P. and Iborra, N. 2009. Proton beam radiotherapy of uveal melanoma: Italian patients treated in Nice, France. *Eur. J. Ophthalmol.* 19: 654–660.

Münter, M. W., Schulz-Ertner, D., Hof, H., Nikoghosyan, A., Jensen, A., Nill, A. et al. 2006. *Radiat. Oncol.* 1:17.

Münter, M., Umathum, V., Jensen, A. D., Nikoghosyan, A., Hof, H., Jaekel, O. et al. 2010. Combination of intensity modulated radiation therapy (IMRT) and a heavy ion (C12) boost for sub-total resected or inoperable adenoid cystic carcinomas (ACCs) of the head and neck region. *J. Clin. Oncol.* 28: e16010.

Munzenrider, J. E. and Liebsch, N. J. 1999. Proton therapy for tumors of the skull base. *Strahlenther Onkol.* 175 Suppl 2: 57–63.

Nakayama, H., Sugahara, S., Tokita, M., Fukuda, K., Mizumoto, M., Abei, M. et al. 2009. Proton beam therapy for hepatocellular carcinoma. The Univeristy of Tsukuba Experience. *Cancer* 115: 5499–5506.

Nihei, K., Ogino, T., Ishikura, S., Kawashima, M., Nishimura, H., Arahira, S. et al. 2005. Phase II feasibility study of high-dose radiotherapy for prostate cancer using proton boost therapy: First clinical trial of proton beam therapy for prostate cancer in Japan. *Jpn. J. Clin. Oncol.* 35(12): 745–752.

Nihei, K., Ogino, T., Ishikura, S. and Nishimura, H. 2006. High-dose proton beam therapy for Stage I non-small-cell lung cancer. *Int. J. Radiat. Oncol. Biol. Phys.* 65(1): 107–111.

Nishida, Y., Kamada, T., Imai, R., Tsukushi, S., Yamada, Y., Sigiura, H. et al. 2011. Clinical outcome of sacral chordoma with carbon ion radiotherapy compared with surgery. *Int. J. Radiat. Oncol. Biol. Phys.* 79: 110–116.

Nishimura, H., Miyamoto, T., Yamamoto, N., Koto, M., Sugimura, K. and Tsujii, H. 2003. Radiographic pulmonary and pleural changes after carbon ion irradiation. *Int. J. Radiat. Oncol. Biol. Phys.* 55(4): 861–866.

Noel, G., Habrand, J. L., Mammar, H., Pontvert, D., Haie-Meder, C., Hasboun, D. et al. 2001. Combination of photon and proton radiation therapy for chordomas and chondrosarcomas of the skull base: The Centre de Protontherapie D'Orsay experience. *Int. J. Radiat. Oncol. Biol. Phys.* 51(2): 392–398.

Ohara, K., Okumura, T., Tsuji, H., Chiba, T., Min, M., Tatsuzaki, H. et al. 1997. Radiation tolerance of cirrhotic livers in relation to the preserved functional capacity: Analysis of patients with hepatocellular carcinoma treated by focused proton beam radiotherapy. *Int. J. Radiat. Oncol. Biol. Phys.* 38(2): 367–372.

Park, L., DeLaney, T. F., Liebsch, N. J., Hornicek, F. J., Goldberg, S., Mankin, H. et al. 2006. Sacral chordomas: Impact of high-dose proton/photon-beam radiation therapy combined with or without surgery for primary versus recurrent tumor. *Int. J. Radiat. Oncol. Biol. Phys.* 65(5): 1514–1521.

Perez, C. A., Pajak, T. F., Rubin, P., Simpson, J. R., Mohiuddin, M., Brady, L. W. et al. 1987. Long-term observations of the patterns of failure in patients with unresectable non-oat cell carcinoma of the lung treated with definitive radiotherapy. Report by the Radiation Therapy Oncology Group. *Cancer* 59(11): 1874–1881.

Pommier, P., Liebsch, N. J., Deschler, D. G., Lin, D. T., McIntyre, J. F., Barker, F. G. et al. 2006. Proton beam radiation therapy for skull base adenoid cystic carcinoma. *Arch. Otolaryngol. Head Neck Surg.* 132(11): 1242–1249.

Robertson, J. M., Lawrence, T. S., Dworzanin, L. M., Andrews, J. C., Walker, S., Kessler, M. L. et al. 1993. Treatment of primary hepatobiliary cancers with conformal radiation therapy and regional chemotherapy. *J. Clin. Oncol.* 11(7): 1286–1293.

Romero, J., Cardenes, H., la Torre, A., Valcarcel, F., Magallon, R., Regueiro, C. et al. 1993. Chordoma: Results of radiation therapy in eighteen patients. *Radiother. Oncol.* 29(1): 27–32.

Rossi, C. J. Jr., Slater, J. D., Yonemoto, L. T., Jabola, B. R., Bush, D. A., Levy, R. P. et al. 2004. Influence of patient age on biochemical freedom from disease in patients undergoing conformal proton radiotherapy of organ-confined prostate cancer. *Urology* 64(4): 729–732.

Rutz, H. P., Weber, D. C., Sugahara, S., Timmermann, B., Lomax, A. J., Bolsi, A. et al. 2007. Extracranial chordoma: Outcome in patients treated with function-preserving surgery followed by spot-scanning proton beam irradiation. *Int. J. Radiat. Oncol. Biol. Phys.* 67(2): 512–520.

Rutz, H. P., Weber, D. C., Goitein, G., Ares, C., Bolsi, A., Lomax, A. J. et al. 2008. Postoperative spot-scanning proton radiation therapy for chordoma and chondrosarcoma in children and young adolescents: Initial experience at Paul Scherrer Institute. *Int. J. Radiat. Oncol. Biol. Phys.* 71: 220–225.

Saunders, W., Castro, J. R., Chen, G. T., Collier, J. M., Zink, S. R., Pitluck, S. et al. 1985. Helium-ion radiation therapy at the Lawrence Berkeley Laboratory: recent results of a Northern California Oncology Group Clinical Trial. *Radiat. Res. Suppl.* 8: S227–S234.

Schulz-Ertner, D., Nikoghosyan, A., Didinger, B., Munter, M., Jakel, O., Karger, C. P. et al. 2005. Therapy strategies for locally advanced adenoid cystic carcinomas using modern radiation therapy techniques. *Cancer* 104(2): 338–344.

Schulz-Ertner, D., Karger, C. P., Feuerhake, A., Nikoghosyan, A., Combs, S. E., Jakel, O. et al. 2007a. Effectiveness of carbon ion radiotherapy in the treatment of skull-base chordomas. *Int. J. Radiat. Oncol. Biol. Phys.* 68(2): 449–457.

Schulz-Ertner, D., Nikoghosyan, A. et al. 2007b. Carbon ion radiotherapy of skull base chondrosarcomas. *Int. J. Radiat. Oncol. Biol. Phys.* 67(1): 171–177.

Shimazaki, J., Akakura, K., Suzuki, H., Ichikawa, T., Tsuji, H., Ishikawa, H. et al. 2006. Monotherapy with carbon ion radiation localized prostate cancer. *Jpn. J. Clin. Oncol.* 36(5): 290–294.

Shimazaki, J., Tsujii, H., Ishikawa, H., Okada, T., Akakura, K., Suzuki, H., Harada, M., Tsujii, H. 2010. Carbon ion radiotherapy for treatment of prostate cancer and subsequent outcomes after biochemical failure. *Anticancer Res.* 30: 5105–5111.

Shioyama, Y., Tokuuye, K., Okumura, T., Kagei, K., Sugahara, S., Ohara, K. et al. 2003. Clinical evaluation of proton radiotherapy for non-small-cell lung cancer. *Int. J. Radiat. Oncol. Biol. Phys.* 56(1): 7–13.

Shipley, W. U., Verhey, L. J., Munzenrider, J. E., Suit, H. D., Urie, M. M., McManus, P. L. et al. 1995. Advanced prostate cancer: The results of a randomized comparative trial of high dose irradiation boosting with conformal protons compared with conventional dose irradiation using photons alone. *Int. J. Radiat. Oncol. Biol. Phys.* 32(1): 3–12.

Slater, J. D., Rossi, C. J. Jr., Yonemoto, L. T., Bush, D. A., Jabola, B. R., Levy, R. P. et al. 2004. Proton therapy for prostate cancer: The initial Loma Linda University experience. *Int. J. Radiat. Oncol. Biol. Phys.* 59(2): 348–352.

Spiro, R. H. 1986. Salivary neoplasms: Overview of a 35-year experience with 2,807 patients. *Head Neck Surg.* 8(3): 177–184.

Sugahara, S., Tokuuye, K., Okumura, T., Nnakahara, A., Saida, Y., Kagei, K. et al. 2005. Clinical results of proton beam therapy for cancer of the esophagus. *Int. J. Radiat. Oncol. Biol. Phys.* 61(1): 76–84.

Sugahara, S., Nakayama, H., Fukuda, K., Mizumoto, M., Tokita, M., Abei, M. et al. 2009. Proton-beam therapy for hepatocellular carcinoma associated with portal vein tumor thrombosis. *Strahlenther Onkol.* 185: 782–788.

Sugahara, S., Oshiro, Y., Nakayama, H., Fukuda, K., Mizumoto, M., Abei, M. et al. 2010. Proton beam therapy for large hepatocellular carcinoma. *Int. J. Radiat. Oncol. Biol. Phys.* 76: 460–466.

Sugane, T., Baba, M., Imai, R., Nakajima, M., Yamamoto, N., Miyamoto, T. et al. 2009. Carbon ion therapy for elderly patients 80 years and older with stage I non-small-cell lung cancer. *Lung Cancer* 64: 45–50.

Terahara, A., Niemierko, A., Goitein, M., Finkelstein, D., Hug, E., Liebsch, N. et al. 1999. Analysis of the relationship between tumor dose inhomogeneity and local control in patients with skull base chordoma. *Int. J. Radiat. Oncol. Biol. Phys.* 45(2): 351–358.

Terhaard, C. H. J., Lubsen, H., Van der Tweel, I., Hilgers, F. J. M., Eijkenboom, W. M. H., Marres, H. A. M. et al. 2004. Salivary gland carcinoma: Independent prognostic factors for locoregional control, distant metastases, and overall survival: Results of the Dutch head and neck oncology cooperative group. *Head Neck* 26(8): 681–92; Discussion 692–369.

Truong, M. T., Kamat, U. R., Liebsch, N. J., Curry, W. T., Lin, D. T., Barker, F. G. et al. 2009. Proton radiation therapy for primary sphenoid sinus malignancies: Treatment outcome and prognostic factors. *Head Neck* 31: 1297–1308.

Tsuji, H., Yanagi, T., Ishikawa, H., Kamada, T., Mizoe, J. E., Kanai, T. et al. 2005. Hypofractionated radiotherapy with carbon ion beams for prostate cancer. *Int. J. Radiat. Oncol. Biol. Phys.* 63(4): 1153–1160.

Tsuji, H., Ishikawa, H., Yanagi, T., Hirasawa, N., Kamada, T., Mizoe, J. E. et al. 2007. Carbon-ion radiotherapy for locally advanced or unfavorably located choroidal melanoma: A Phase I/II dose-escalation study. *Int. J. Radiat. Oncol. Biol. Phys.* 67(3): 857–862.

Weber, D. C., Ritz, H. P., Pedroni, E. S., Bolsi, A., Timmermann, B., Verwey, J. et al. 2005. Results of spot-scanning proton radiation therapy for chordoma and chondrosarcoma of the skull base: The Paul Scherrer Institut experience. *Int. J. Radiat. Oncol. Biol. Phys.* 63(2): 401–409.

Weber, D. C., Chan, A. W., Lessell, S., McIntyre, J. F., Goldberg, S. I., Bussiere, M. R. et al. 2006. Visual outcome of accelerated fractionated radiation for advanced sinonasal malignancies employing photons/protons. *Radiother. Oncol.* 81: 243–249.

Yamada, S., Kato, H., Yamaguchi, K., Kitabayashi, H., Tsujii, H., Saisyo, H. et al. 2005. Carbon ion therapy for patients with localized, resectable adenocarcinoma of the pancreas. *Proc. ASCO GI* Abstr. 130.

Yamada, S., Hara, R. et al. Carbon Ion radiotherapy for pancreas cancer. Proc. NIRS-MD Anderson Symposium on Clinical Issues for particle therapy 2008.

Yanagi, T., Mizoe, J., Hasegawa, A., Takagi, R., Bessho, H., Onda, T., et al. 2009. Mucosal malignant melanoma of the head and neck treated by carbon ion radiotherapy. *Int. J. Radiat. Oncol. Biol. Phys.* 74: 15–20.

Yonemoto, L. T., Slater, J. D., Rossi, C. J. Jr., Antoine, J. E., Loredo, L., Archambeau, J. O. et al. 1997. Combined proton and photon conformal radiation therapy for locally advanced carcinoma of the prostate: Preliminary results of a phase I/II study. *Int. J. Radiat. Oncol. Biol. Phys.* 37(1): 21–29.

Yoon, S. S., Chen, Y. L., Kirsch, D. G., Maduekwe, U. N., Rosenberg, A. E., Nielsen, G. P. et al. 2010. Proton-beam, intensity-modulated, and/or intraoperative electron radiation therapy combined with aggressive anterior surgical resection for retroperitoneal sarcomas. *Ann. Surg. Oncol.* 17: 1515–1529.

Zenda, S., Kawashima, M., Nishio, T., Kohno, R., Nihei, K., Onozawa, M. et al. 2011. Proton beam therapy as a nonsurgical approach to mucosal melanoma of the head and neck: A pilot study. *Int. J. Radiat. Oncol. Biol. Phys.* 81(1):135–139.

Zietman, A.L., DeSilvio, M. L., Slater, J. D., Rossi, C.J. Jr., Miller, D. W., Adams, J. A. et al. 2005. Comparison of conventional-dose versus high-dose conformal radiation therapy in clinically localized adenocarcinoma of the prostate: A randomized controlled trial. *J. Am. Med.* Assoc. 294(10): 1233–1239.

Zietman, A. L. 2008. Correction: Inaccurate analysis and results in a study of radiation therapy in adenocarcinoma of the prostate. *J. Am. Med. Assoc.* 299(8): 898–899.

Zietman, A. L., Bae, K., Slater, J. D., Shipley, W. U., Efasthiou, J. A., Coen, J. J. et al. 2010. Randomized trial comparing conventional-dose with high-dose conformal radiation therapy in early-stage adenocarcinoma of the prostate: long-term results from proton radiation oncology group/American College of Radiology 95–09. *J. Clin. Oncol.* 28: 1106–1111.

Chapter 10

11. Future Prospects for Particle Therapy

C.-M. Charlie Ma

As discussed in detail in previous chapters of this book, many factors have contributed to greatly increase interest in particle therapy in the United States, Europe, Asia, and other parts of the world (Smith 2006; Schulz et al. 2007; Moyers et al. 2007; Maughan et al. 2008). In fact, there has been such an explosion of interest in particle therapy during the past 5 years one could argue that the current enthusiasm is overheated. There is an indication that some institutions seeking particle therapy facilities are doing so without critical evaluation of the clinical role of particle therapy and the relatively large financial, clinical, and personnel commitments required for developing and operating these facilities. Moreover, there is an acute shortage of trained particle therapy staff and few training programs. Therefore, institutions that are building new particle therapy facilities may have a difficult time recruiting trained, proficient staff for clinical operations. In this chapter, we will briefly discuss the potential technological developments in particle therapy and their therapeutic implications or clinical issues that need to be considered when we move forward with this exceptional treatment modality for cancer management.

11.1 Technological Advances in Particle Therapy

11.1.1 Single-Room Particle Therapy Systems

There is a general consensus that smaller particle therapy facilities are better suited for community hospitals and large private practices. Commercial manufacturers have been developing the technologies necessary for "single-room solutions" for proton therapy systems. Single-room proton therapy systems may provide a lower cost option for implementing proton therapy. All or most components (accelerator, beam transport, energy selection, and beam delivery) can be mounted on a rotating gantry or very near the gantry, thus reducing the size of the complete proton therapy system enough that it can be installed in its entirety (or nearly so) in a single treatment room. There will be no competition with other treatment rooms for proton beams because

Proton and Carbon Ion Therapy Edited by C.-M. Charlie Ma and Tony Lomax © 2013 Taylor & Francis Group, LLC. ISBN: 978-1-4398-1607-3.

Chapter 11

each room has its own accelerator. If multiple rooms are used, the entire proton treatment capability is not lost if one of the accelerators goes down and patients can be transferred to other rooms for the treatment.

There are potential disadvantages associated with single-room particle therapy systems. For example, the cost of these systems may not necessarily be cheaper per treatment room as compared to large systems that use a single accelerator to feed several treatment rooms. Additional rooms may have to be installed if demand for particle therapy increases in a location with a single-room system. If several treatment rooms are eventually needed, the cost effectiveness of the single-room concept may be lost; for example, an additional accelerator will be needed for each treatment room. Maintaining multiple accelerators will also increase maintenance costs. Furthermore, the pulse structure of some accelerators (e.g., synchrocyclotrons) chosen for the proposed single-room systems may make it difficult to use spot scanning techniques to deliver IMPT treatments. IMPT could be delivered using a multileaf collimator (MLC); however, this method may have problems such as poor efficiency in the use of particle beams and increased neutron production. Finally, these single-room systems are still in their developmental stage, and thus their operability, reliability, and maintainability are not known.

11.1.2 Advances in Accelerator Technology

Particle accelerators are designed to control electric and magnetic fields in such a way as to efficiently accelerate charged particles. A linear accelerator (linac) accelerates particles in a linear path, and therefore its length is proportional to the electric field and the gain in particle energy. Conventional linacs cannot provide sufficient electric field strengths to accelerate heavy charged particles such as protons or light ions with therapeutic energies for radiation oncology applications. For a given electric field strength, alternative solutions are cyclotrons and synchrotrons that repeatedly steer charged particles across the same electric field to achieve desired energies. A cyclotron provides a constant magnetic field to steer charged particles (different energies corresponding to different radii), and its size and weight are proportional to the strength of the magnetic field and therefore to the size and weight of the magnets used. A synchrotron provides a variable magnetic field to steer charged particles of different energies through the same trajectory, and it is generally larger and heavier than cyclotrons.

For therapeutic applications, protons must be accelerated to 230–270 MeV (corresponding to 33–43 cm range in water). The Francis H. Burr Proton Therapy Facility at Massachusetts General Hospital (MGH) uses a room-temperature cyclotron containing 3 Tesla magnets, which is 4m in diameter and its iron core and copper coils weigh more than 220 tons. Conventional cyclotrons are still too large and heavy for gantry-mounted proton therapy systems but they have been used for large proton therapy facilities to provide beams for multiple treatment rooms. Synchrotrons have been used for proton therapy applications and they are the only sources used for ion therapy facilities. Conventional accelerators have been optimized in terms of performance, stability, and cost, and are being improved with new technical capabilities for beam scanning, gating, and IMPT dose delivery.

Several developments already underway show considerable potential to provide a new generation of more compact, more efficient, and less expensive accelerators for proton therapy. Some developments will improve upon current technologies, while others will provide entirely new types of accelerators, some of which are highlighted here. Chapter 2 provides an excellent overview of new accelerator technologies.

11.1.2.1 Superconducting Cyclotrons and Synchrocyclotrons

Particle accelerators employing room-temperature magnets are generally too large and too heavy to be placed on a treatment gantry. They are, therefore, kept in separate locations and used to feed multiple treatment rooms to maximize the use of such an expensive accelerator. Novel designs of superconductor accelerators are being developed that are compact enough to be used for single-room proton therapy systems. The ACCEL Comet cyclotron weighs only 80 tons, and has been used in compact proton therapy designs for both single-room and multiple-room facilities. The Mevion (previously Still River) superconductor cyclotron weighs about 17 tons, and is placed directly on the proton treatment gantry. The gantry only rotates 180 degrees while the 360-degree beam incidence is facilitated by a robotic couch. A portable c-arm cone-beam computed tomography (CT) is used for accurate target localization prior to a treatment. The proton system only provides broad, passive scattered beams and a secondary collimator with a separate, more precise rotation system is used to deliver shaped proton fields at accurately defined incident angles. Such compact designs allow for single-room proton therapy systems with a reduced shielding and construction cost. The Mevion proton therapy system requires a 50 ft × 50 ft × 40 ft room and the entire facility, including the proton therapy system and the room, will cost more than $20 million.

11.1.2.2 FFAG Accelerators

A compromised accelerator design has combined the best aspects of both cyclotrons and synchrotrons; a fixed-field alternating-gradient (FFAG) accelerator (Craddock 2004) consisting of a ring of magnets (like a synchrotron) that provide a large, constant magnetic field to allow for different trajectory radii (like a cyclotron). FFAG accelerators combine many of the positive features of cyclotrons and synchrotrons with fixed magnetic fields as in cyclotrons and pulsed acceleration as in synchrotrons. FFAG accelerators can be cycled faster than synchrotrons, limited only by the rate of the RF modulation. Higher duty factors will permit higher average beam currents and high repetition rates for spot scanning. Fixed fields require simpler and cheaper power supplies and are more easily operated than synchrotrons. With respect to fixed field cyclotrons, nonscaling FFAG accelerators allow strong focusing, and hence, smaller aperture requirements, which leads to low beam losses and better control over the beam. FFAG accelerators (like synchrotrons) have a magnetic ring that allows beam extraction at variable energies rather than just a single energy, as in a cyclotron. Their superconducting magnets and compact size make FFAG accelerators attractive for proton therapy applications.

The first proton FFAG was built in 1999 at KEK in Japan (Aiba et al. 2000). This machine was the proof of principle (POP) for proton FFAG accelerators. After this success, a 150-MeV FFAG accelerator was developed at KEK. To date, several FFAG accelerators have been built in the energy range 150–250 MeV and others have been proposed. The nonscaling FFAG accelerator was also developed in 1999 (Johnstone et al. 1999), The magnet design of this accelerator provides a variation of orbit length with energy, which can be arranged to greatly compress the range of the orbit radii and thus the magnet aperture, while maintaining linear magnetic field dependence. In addition, the small apertures and linear fields allow simplification and cost reduction compared with scaling FFAG accelerators. A particle therapy system was proposed using a nonscaling FFAG accelerator and gantry composed of nonscaling FFAG cells, which would accelerate carbon ions up

to 400 MeV/μ and protons up to 250 MeV (Keil et al. 2007).

11.1.2.3 Laser Acceleration

Laser acceleration was first suggested for electrons. (Tajima and Dawson 1979) Rapid progress in laser-electron acceleration followed the development of chirped pulse amplification and high fluence solid-state laser materials such as Ti:sapphire. Recently, charged particle acceleration using laser-induced plasmas has illuminated the search for a compact, cost-effective proton or ion source for radiotherapy (Ma and Maughan 2006; Ma et al. 2001; Bulanov and Khoroshkov 2002; Fourkal et al. 2002). The laser acceleration of ions provides a several-orders-of-magnitude-larger acceleration gradient than conventional acceleration, of the order of 1 TeV/m. Several options exist in terms of target configurations and acceleration mechanisms. Energetic proton and ion beams with high beam quality have been produced in the last few years from thick metallic foils (e.g., few micrometers thick aluminum) irradiated by ultraintense short laser pulses. The results from most previous experiments are based on the Target Normal Sheath Acceleration model (TNSA). Because these targets are relatively thick, the laser pulse is mostly reflected and the conversion efficiency of laser pulse energy to ion kinetic energy is normally less than 1%. The dependence of maximum ion energy on laser intensity is less than a linear function. The maximum proton energy based on the TNSA mechanism has somewhat improved since its first discovery: from 58 MeV in the year 2000 to, more recently, 78-MeV cutoff energy for the exponential spectrum, with 6×10^{13} particles (Snavely et al. 2000). The possibility to accelerate quasi-monoenergetic ion bunches has already been demonstrated within the TNSA regime by restricting the ion source to a small volume, where the sheath field is homogenous. However, a very high laser intensity of $>10^{22}$ W/cm^2 is required to accelerate protons to 200 MeV and above.

Because of the advantage in accelerating limited mass by laser pressure, high-energy ion acceleration experiments using thin (submicrometer to nanometer) targets and ultrahigh contrast (UHC) short-pulse lasers have garnered more interest. A new mechanism for laser-driven ion acceleration was proposed, where particles gain energy directly from the radiation pressure acceleration or phase stable acceleration (RPA/PSA) (Schwoerer et al. 2006). By choosing the laser intensity, target thickness, and density such that the radiation pressure equals the restoring force established

Chapter 11

by the charge separation field, ions can be bunched in a phase-stable way and efficiently accelerated to a higher energy. In recent years, experiments with quasi-monoenergetic peaks for C^{6+} at ~30 MeV were observed at MPQ/MBI (Steinke et al. 2010) and C^{6+} at >500 MeV (exponential) and 100 MeV (quasi-monoenergetic) at LANL (Henig et al. 2009). Furthermore, at LANL, quasi-monoenergetic protons at ~40 MeV were generated from nanometer-thin diamond-like carbon foils. Interpretation of these experiments in terms of RPA is, however, not conclusive. Theoretical studies show that the required medical proton/carbon beams can be generated from hydrogen/carbon foil (of submicron thickness) with a laser intensity of ~10^{21} to 10^{22} W/cm^2 with sufficient ion abundance and a quasi-monoenergetic (peaked) energy distribution.

A step beyond the conventional TNSA mechanism is the so-called breakout afterburner (BOA) mechanism, which was discovered theoretically in 2006. The main difference between TNSA and BOA (or RPA) is the decoupling of the ion acceleration from the driving laser field due to the thickness of the target. In contrast, for the RPA and BOA mechanism, the electrons that are accelerating the ions are still interacting with the laser field. To use the maximum number of available electrons, the target is required to be dense enough so that the laser beam is not initially penetrating the target but is coupled with the electrons. At some point the target will become relativistically transparent to the laser light, so that the light can directly interact with electrons, accelerating the ions at the rear surface together. So the BOA mechanism starts as normal TNSA, but then during the rising edge of the laser pulse the intensity couples with the already moving electron-ion front at the rear side of the target. Numerical simulations predict ion energies of hundreds of megaelectron volts for existing laser parameters and up to the gigaelectron volt range for currently planned laser systems. Recently, acceleration of protons up to energies of 120 MeV existed using the TRIDENT laser at LANL (LANL 2011).

A significant difference from TNSA is that in a mixture of target atoms, all of the accelerated ions propagate at the same particle velocity, governed by the slowest (i.e., the heaviest) species present. Thus, for high-energy proton acceleration, a pure hydrogen target is the ideal choice. For each laser pulse duration and intensity as well as for each target composition, one can determine an optimum target thickness based on the above mentioned physics. Another method for laser-proton acceleration is to use a hydrogen gas jet with

density just above the critical density, which is $10^{19}/cm^3$ for a CO_2 laser. This method has the characteristic feature of creating very narrow energy spreads and practically monoenergetic beams. Recent experiments using this method has produced >20 MeV monoenergetic protons with a narrow energy spread of about 1% and low emittance (Haberberger et al. 2011).

Starting from 1998, researchers at LLNL and Stanford University investigated the feasibility of using laser-accelerated protons for radiation therapy. As a result, two research initiatives were put forward focusing on target configuration to achieve therapeutic proton energies (Tajima 1999) and on particle selection and beam collimation to solve the problem of broad energy and angular distributions of laser-accelerated protons, (Ma 2000), respectively. Research continued at both Japanese Atomic Energy Agency (JAEA) and Fox Chase Cancer Center (FCCC) along these directions as the principal investigators moved to new institutions. With the support of the U.S. Department of Health and Human Services, an experimental facility dedicated to laser-proton acceleration for cancer treatment was established at FCCC and the research continued under this initiative (Ma 2000; Fourkal et al. 2003; Luo et al. 2005; Ma et al. 2006; Fan et al. 2007; Veltchev et al. 2007; Luo et al. 2008). The system being developed is based on the design of Ma (Ma 2000) with the following three key elements: (1) a compact laser-proton source to produce high-energy protons, (2) a compact particle selection and beam collimating device for accurate beam delivery, and (3) a treatment optimization algorithm to achieve conformal dose distributions using laser-accelerated proton beams. The system was reviewed in detail by Ma et al. (Ma et al. 2006).

11.1.2.4 Dielectric Wall Accelerators

Another accelerator design has recently been proposed based on a new class of insulators called high-gradient insulators (HGIs), which feature significantly improved voltage holding ability over conventional insulators. This new type of induction linac is the so-called dielectric wall accelerator (DWA). A conventional induction linac has an accelerating field only in the gap of the accelerating cells, which represents only a small fraction of the length. By replacing the conducting beam pipe by an insulating wall, accelerating fields can be applied uniformly over the entire length of the accelerator, yielding a much higher gradient (around 100 MeV/m) and allowing the design of much more compact linear accelerators. The pulsed acceleration field is developed by a series of so-called asymmetric Blumlein structures incorporated into the insulator. The DWA technique is being incorporated in new compact proton therapy systems. The LLNL prototype DWA is designed for a maximum of 250-MeV proton energy with an overall length of less than 3 m (Caporaso et al. 2008; Chen and Paul 2007). This DWA will produce a 100-mA beam current in a short pulse of 1 ns with a 50-Hz pulse repetition rate and variable energy and focus. LLNL is collaborating with Tomotherapy, Inc. (Madison, WI), which has recently been acquired by Accuray Inc., to develop a single-room proton therapy system based on the DWA technique (Mackie 2007). The LLNL compact accelerator was designed to be mounted directly on a gantry. The gantry will rotate 200 degrees while the 360-degree beam incidence will be provided by proper couch rotation. The vault dimensions are 20 ft × 20 ft × 14 ft. This DWA-based proton therapy system is estimated to cost >\$20 million for a single-room facility.

11.1.3 Pencil Beam (Spot) Scanning Techniques

Until recently, particle therapy patients have mainly been treated with passive scattering techniques that have been improved continuously for better efficiency and versatility. In parallel to photon therapy advances, efforts have been made to develop MLC for passive scattering applications. So far, MLCs used for particle therapy have been less than optimal (e.g., they have small field sizes, and for some, their size does not permit positioning the MLC close to the patient surface, resulting in large air gaps that degrade the lateral penumbra of the treatment beam). MLCs have not been used to deliver IMPT as those used in photon IMRT. Instead, pencil beam (spot) scanning techniques have been developed for IMPT treatments, which provide a substantial dosimetric improvement over passive-scattering techniques.

Pencil-scanning techniques were first tested at the PSI in Switzerland, where one set of scanning magnets allowed for 1-D scanning, and couch shifts were used together to deliver pencil beam BPs to a target volume (Pedroni et al. 1995). To realize 3-D scanning, the magnets and the energy selection system were first used to scan a section of the target volume and the couch was then shifted to scan another section. This scanning technique has been used to deliver both SFUD and IMPT treatments. More recent pencil-scanning systems have been investigated to use two sets of scanning magnets to scan the beam in a plane perpendicular to the nozzle axis to allow for 2-D scanning. The target volume is divided into energy layers (highest to lowest energy), and each layer is sequentially scanned with pencil beams. There are two ways to scan the beam spot throughout the target volume: (1) dynamic (or continuous) scanning in a raster or line pattern, where either the scan speed or beam intensity, or both, are varied to produce the intensity distribution prescribed by the treatment plan; and (2) discrete spot-scanning, where the beam is cut off between spots and the dose at each spot is varied to achieve the prescribed dose pattern.

A clinical challenge for the implementation of pencil-beam scanning techniques is to reduce or eliminate treatment uncertainties caused by range errors or target motions. Errors due to respiratory motion are significant for several treatment situations, particularly for the treatment of lung tumors, which may be reduced in two primary ways: (1) beam gating, where the beam is turned off when the target has moved out of the target position for which the treatment plan was calculated, and (2) repainting spots, repainting layers, or repainting the entire treatment volume (i.e., repainting means to scan a particular spot position, layer, or volume two or more times). The latter will require a considerable amount of new technology to be developed for beam delivery, primarily in the area of very rapid scanning and energy changes, especially when compared with beam gating or target tracking (Lu et al. 2007; Furukawa et al. 2007; Bert et al. 2007). More issues related to target motion will be discussed in the next section.

11.2 Clinical Application Issues of Particle Therapy

11.2.1 Therapeutic Effectiveness

For proton therapy, there has been an ongoing debate in the radiation oncology community on the appropriateness of prospective, randomized clinical trials on its superiority over photon therapy. One argument is that the dose distributions of protons will always be superior to those for photons when both modalities are optimized and one cannot justify randomized clinical trials of protons versus photons because there would not be equipoise between the two arms of the trial (Goitein and Cox 2008; Suit et al. 2008; Zietman et al. 2005). On the other hand, some argue that many proton facilities have been using simple treatment techniques which have produced dose distributions that are sometimes inferior to state-of-the-art photon IMRT dose distributions. A consensus is the need for additional clinical data; it is not feasible to draw statistically meaningful comparisons between photon and proton treatment techniques without data acquired through carefully controlled clinical studies in an expanded range of disease sites.

Following the proton therapy dose escalation studies on prostate cancer at MGH and LLUMC (Zietman et al. 2005), the National Cancer Institute has approved funding for a P01 grant application submitted by MGH and M. D. Anderson Cancer Center (MCACC) to conduct clinical studies on proton therapy with the objective of applying advanced proton radiation planning and delivery techniques to improve outcomes for patients with non-small cell lung cancer (proton dose escalation and proton vs. photon randomized trials), liver tumors, pediatric medulloblastoma and rhabdomyosarcoma, spine/skull base sarcomas, and paranasal sinus malignancies. The University of Florida and University of Pennsylvania facilities are also expected to participate in multi-institutional clinical studies.

For carbon ion therapy, a number of clinical trials have been carried out for different treatment sites using the passive beam delivery (Castro 1995). For pencil-beam scanning systems, clinical trials have also been performed but the experience is limited to a few disease sites where organ motion is not important (Schulz-Ertner and Tsujii 2007; Schulz-Ertner et al. 2005). As the depth modulation for active techniques varies over the tumor cross section, the RBE has to be calculated locally at each point in the treatment field rather than by using fixed RBE profiles. As compared to passive techniques, this may introduce an additional source of uncertainty. Further clinical studies are being planned at newly established hospital-based facilities with pencil-beam delivery systems on other disease sites including other malignant brain tumors, prostate carcinomas, pancreatic tumors, and if motion compensation techniques for scanned beams are better understood, also lung tumors. For some treatment sites, the use of ion beams only may be problematic due to the involved range uncertainties. Combined treatments of photon IMRT with an ion beam boost may be of significant impact. New dose escalation studies aim to maximize local control at acceptable normal tissue toxicity rates. For hypofractionated treatment schemes, caution must be exercised in evaluating the high biologically effective doses to the normal tissues as the fractionation effect of heavy ions in the plateau region is known to be similar as for photons.

A potential advantage of ion beam therapy is the treatment of radioresistant tumors (e.g., hypoxic tumors), which are known to be highly radioresistant against low-LET radiation. The increase of the required dose for hypoxic tumors is described by the so-called oxygen enhancement ratio (OER), which has a value of 1 for well oxygenated tumors and values of up to 3 for hypoxic tumors for photons and protons. So far, only very few data is available on the comparison of the effectiveness of heavy ions for hypoxic and normoxic tumors (Nakano et al. 2006). As the treatment of hypoxic tumors is still a major challenge in radiotherapy, future studies with heavy ions have to focus on these tumor types.

A long-lasting question has been what is the best ion type for particle therapy. The cost-effectiveness of an ion type is only one aspect of the clinical evaluation. The synergetic effect of heavy ion therapy with additional chemotherapy, hormone, or targeted therapies has to be investigated, which is especially important in cases where a significant improvement in local control can be achieved with ion beams, but the overall survival is still limited because of distant metastases (e.g., for adenoidcystic carcinoma) (Schulz-Ertner et al. 2005). In addition to the therapeutic advantage, the possibility of secondary malignancy due to neutrons produced by fragmentations is an important issue, especially for the treatment of pediatric patients. Generally, more neutrons are produced in a carbon beam as compared to a proton beam. For pencil-beam scanning techniques, the biologically equivalent neutron dose is estimated to be 0.4% in the carbon ion beam (Gunzert-Marx et al. 2004). Even more neutrons are produced in a passive beam delivery system compared to the scanned beam system. Therefore, the pencil-beam scanning technique should be used for pediatric patients whenever possible.

11.2.2 Treatment Planning

For passive-scattering techniques, range uncertainties and changes in scattering conditions caused by heterogeneities are partially offset by expanding the margins around the clinical target volume. In IMPT, the dose distribution is sensitive to the uncertainties in each individual beam of the potentially several thousand BPs used in the treatment plan, and uncertainties in the spot placements can result in unacceptable hot and cold regions in the target volume. One approach to improving IMPT treatment planning is to include errors in the optimization calculations in such a way as to ensure that the treatment plan predicts actual dose distributions more reliably. Work in this area is underway and these approaches have the potential to improve the accuracy and robustness of IMPT treatment plans (Pflugfelder et al. 2008).

The prototype treatment planning system for a scanned carbon ion beam developed for the GSI facility already incorporates all features needed for IMPT. To achieve a homogenous biological effect even for a single field, it is necessary to optimize and control the number of ions at each scan spot. This intrinsically implies intensity (or fluence) modulation of the ion beam at each beam spot. Since the depth dose modulation in that case is varying throughout the radiation field, a detailed biological modeling of the RBE had to be included in the treatment planning system.

An important problem associated with IMPT optimization is its memory/CPU time requirement since the particle numbers of all beam spots of all fields have to be optimized at the same time rather than separately. Due to the large number of beam spots per field (typically 50,000 for a skull base tumor), the required computer memory and computing time are considerable (>1 GB and >1 h on a 1.2-GHz Power4-CPU, respectively). IMPT has already been clinically applied at GSI with the existing scanning technology and without additional efforts for quality assurance. It has been shown that setup errors do not lead to worse treatment plans for IMPT, and therefore, robustness to setup-related range uncertainties will be included as an additional constraint in the IMPT dose optimization process. Since the number of fields applied in particle therapy is generally smaller than in photon IMRT, the selection of proper beam angles becomes more important. As the range uncertainty may vary significantly for different beam angles, automatic optimization algorithms for the beam angles should be designed to reduce these uncertainties to a minimum (Jäkel et al. 2008).

Conventional proton treatment planning has been using pencil beam algorithms for passive scattering techniques. These algorithms can have appreciable errors in some situations where tissue interfaces or large heterogeneities exist in the beam path or when metallic implants are present adjacent to or in the target volume. Monte Carlo calculations have the potential to improve the accuracy of dose calculations in those situations. Monte Carlo treatment planning

has been developed for proton therapy applications (Paganetti et al. 2005, 2008; Ilic et al. 2005). However, Monte Carlo calculations are time-intensive and have not been implemented in commercial systems for routine application, although they have been used in basic and clinical research (e.g., for benchmarking pencil beam calculations or comparisons of proton therapy treatment plans using conventional proton accelerators and laser-proton accelerators) (Luo et al. 2008).

Fast Monte Carlo calculations utilizing advanced computational techniques are being developed at research centers aiming at routine clinical applications for particle therapy (Li et al. 2005; Veltchev et al. 2010). Monte Carlo calculations have also been used to improve the efficiency of proton beam commissioning by eliminating a large number of measurements, especially in the case of spot beam scanning where approximately 100 energies need to be commissioned (Newhauser et al. 2005, 2007). Monte Carlo calculations are more efficient than conventional dose calculations for IMPT treatment planning since the dose calculation uncertainty in the target volume is determined by the total number of particles used in the plan calculation, not by the number of gantry angles, beamlets, or BPs.

11.2.3 Motion Management

Image guidance has long been used for patient setup and target localization in particle therapy and the state-of-the-art imaging devices are being used in modern particle therapy facilities (Haberer et al. 2004; Amaldi and Kraft 2007). These devices can be used to measure interfractional motion by planar or cone-beam imaging. If no significant deformations are present, translational setup corrections can be applied to the treatment couch. In the case of a robotic couch, rotational corrections around three axes can also be obtained. If deformations are significant, an adaptive approach would be necessary that includes a real-time modification of contours and reoptimization of the treatment plan, which have not been routinely available in either photon IMRT or particle therapy. As an alternative, different treatment plans for different representative geometries may be prepared. In the actual treatment situation, the most suitable plan can then be selected to deliver the best treatment.

Intrafractional target movements are caused by respiratory motion, although motion of heart or prostate (induced, e.g., by peristaltic motion and variation of organ filling) may also be important. In contrast to photon RT, motion-related changes of the required particle range may lead to large margins in the beam direction, and hence, to additional high doses in normal tissues (Minohara et al. 2003). For scanned beams, target coverage may be disturbed by the interplay effect between beam scanning and target movements, which may lead to hot or cold spots, and hence cannot be considered by increased margins. A rescanning technique can be used to average the effect of irradiations at different breathing phases (Furukawa et al. 2007).

More advanced techniques have been developed to compensate respiratory motion by gating on a stable breathing phase or by tracking of the tumor motion (Minohara et al. 2000; Bert et al. 2007). For passive beam delivery systems, gating is already clinically applied (Minohara et al. 2000). For scanned beams, irradiation time may increase significantly due to the interplay between respiratory motion and duty cycle of the synchrotron. This delivery time may be reduced to a certain extent by gating and also the extraction of the beam from the synchrotron. Remaining interplay effects due to residual motion of the target may be compensated by rescanning (Furukawa et al. 2007). Tumor tracking has only been achieved with pencil-beam scanning systems, in which the change of the lateral tumor position is compensated by the scanner magnets in real time and the changes of radiological depth are corrected by a computer-controlled range shifter consisting of opposing PMMA wedges (Bert and Rietzel 2007; Grozinger et al. 2006). Depending on the actual breathing phase, the beam delivery system switches to one out of several treatment plans, each of them optimized on a different phase of a 4-D CT. The necessity to calculate the range on a 4-D CT may involve additional uncertainties as 4-D CT protocols differ considerably from conventional CT protocols.

An important issue for tumor tracking is the real-time determination of the actual tumor position. There is a potential mismatch between the external surrogate signals and the tumor position during treatment planning CT and the actual treatment. Real-time fluoroscopic images may be required to ensure that the tumor position is as expected. This requires fast image analysis, and in contrast to photon therapy, implantation of radio-opaque fiducials as landmarks may have a larger impact on the range of the particles and thus the dose distribution (depending on the size and location of the marker). In view of these problems, it is likely that only gating will be clinically implemented in the near future as this technique solely requires reproducibility of a single breathing phase rather than that of the complete breathing cycle (Jäkel et al. 2008).

11.2.4 Dosimetry Issues

11.2.4.1 Dose Measurement

The Code of Practice for dosimetry (TRS-398) has been used for particle therapy dose calibration and clinical measurements with an uncertainty of 3% (Andreo et al. 2000). The basic quantities, like stopping powers for various particles and fragmentation spectra and *w* values, and chamber-specific correction factors have to be determined more precisely to arrive at the same level of accuracy as for photon therapy. Currently, absorbed dose to water is determined using ionization chambers calibrated in a ^{60}Co beam. In the near future, water calorimetry may be developed as a primary standard for particle beams in the same way as already done for megavoltage X-rays today in some countries.

With the introduction of dynamic beam-scanning systems, more efficient multichannel dosimetry systems need to be developed. There are already a number of detector systems developed for particle physics applications that are currently investigated in view of their use in particle therapy dosimetry. For example, gas electron multiplier (GEM) chambers (Sauli 1997) can be used as large area tracking detectors with high temporal and spatial resolution (1 ms and 40 μm, respectively). In combination with a scintillator for residual energy measurements, they were already used for time-resolved radiography or as beam profile monitors (Chen et al. 2006). Silicon pixel detectors with pixel sizes around 50 μm can be used to measure single particle events with little to no background (Zhang et al. 2008; Jakubek et al. 2005). The detector signal is proportional to the energy loss in the detector and may be used to discriminate particles and thus allow for a direct dose determination. Extremely thin (less than 50 μm) polycrystalline diamond detectors with millimeter resolution can yield a 100% efficiency for particle counting even at high count rates, which makes them interesting for monitoring of the beam position, intensity, and profile during application while minimizing the interaction with the beam (Rebisz et al. 2007).

11.2.4.2 Quality Assurance for Particle Therapy

The clinical implementation of pencil-beam scanning techniques for particle therapy has generated the need for a new class of QA systems that permit rapid and accurate 2-D and 3-D data acquisition. QA systems for scanned proton beams must also be adaptable to the timing structure and energy stacking of pencil-beam scanning treatment delivery. Additionally, the dose measured at a point in the phantom can be obtained only by collecting the detector signal during the entire scan sequence. In many cases, it is also necessary to characterize a single pencil beam, which places additional requirements on the measuring system.

In 1999, Karger et al. reported a 24-ionization chamber planar array for particle therapy QA (Karger et al. 1999). In 2000, Jäkel et al. described QA procedures and methods for a treatment planning system employing scanned beams (Jäkel et al. 2000). In 2005, Pedroni et al. described several dosimetric methods for characterizing scanned beams at PSI in Switzerland including ionization chambers, films, and a charged-coupled-device (CCD) camera coupled with a scintillating screen (Pedroni et al. 2005). A linear array of 26 ionization chambers and the CCD system were also used to verify scanned dose distributions. More recently, two ionization chamber array systems have been reported: (1) a 128-chamber planar array designed to measure beam profiles in fields up to 38 cm in diameter, and (2) an array of 122 small-volume multilayer ionization chambers used for depth-dose measurements in clinical proton beams (Nichiporov et al. 2007). As more proton treatment facilities implement scanning techniques and the demand for appropriate dosimetry systems increases, commercial manufacturers are encouraged to develop similar commercial devices. Other dosimetric systems such as radiochromic films, alanine detectors, and polymer gels have also been used for QA measurements (Vatnitsky 1997; Nichiporov et al. 1995; Warren 2007). However, the use of these systems is limited because of saturation effects in the proton stopping region and because they are not real-time instruments. Further investigations are needed for these systems.

11.2.4.3 Positron Emission Tomography for Particle Therapy Verification

PET is a potentially useful tool for validating the dose distributions received by particle therapy patients. PET can detect annihilation γ-rays arising from the decay of small amounts of $\beta+$ emitters such as ^{11}C, ^{15}O, and ^{10}C produced by nuclear fragmentation reactions between the primary charged particle and the target nuclei in the irradiated tissue. Particle treatment delivery verification, either offline or online, can be achieved by comparing the measured $\beta+$ activity distribution with an expected pattern calculated by treatment planning or by Monte Carlo calculations of the expected $\beta+$ activity (Parodi et al. 2007, 2008; Nishio et al. 2006; Knopf et al. 2008; Fourkal et al. 2009). In particular, PET techniques offer the potential to detect and quantify range uncertainties in treatment planning calculations. The preliminary results look promising and it is expected that with further development, this technique will become an important tool for verifying particle therapy dose delivery.

11.2.5   Cost-Effectiveness

Particle therapy has significant potential to improve clinical outcomes; this potential has been borne out in a number of studies (Suit et al. 2003). Particle therapy is certainly more expensive than high-energy photon and electron beams but costs are expected to decrease over time. Reliable data on the costs as well as on the increased therapeutic effectiveness relative to conventional radiotherapy is very limited for proton and even more for carbon ion therapy. About 10 years ago, the relative cost for proton therapy versus conventional therapy currently was estimated to be around 2.5 times higher if a recovery of the capital investments is required (Goitein and Jermann 2003). By 2013, this ratio is estimated to decrease to 2.1–1.7. Without a recovery of investment costs, the ratio could be lowered to a value between 1.6 and 1.3.

In the United States, current reimbursement rates make it possible to develop and operate proton therapy facilities with a reasonable profit margin. The overall costs of proton therapy will decrease as proton therapy procedures become more efficient and patient throughput increases. Increased patient throughput will make it possible to spread the capital costs of proton therapy among a larger number of patients and the cost per patient will decrease. Accordingly, the reimbursement rates for proton therapy treatments will decrease as capital costs of hospital-based facilities will be spread among more patients as these facilities reach full-capacity operations.

A cost-effectiveness analysis for proton radiotherapy has been performed for various tumors (breast, prostate, head and neck, and childhood medulloblastoma) in Sweden, using a Markov simulation and calculating the gain in quality adjusted life years (Lundkvist et al. 2005). It was shown that proton therapy for these tumors was cost effective if appropriate risk groups were chosen. The general lack of data and large uncertainties in the analysis were questioned together with many assumptions made on the outcome of proton therapy without being based on solid clinical data (Lodge et al. 2007).

The only published cost-effectiveness analysis for carbon ion therapy was performed by the GSI group for chordoma patients (Jäkel et al. 2007). For a limited collection of 10 patients, the overall treatment costs were calculated and compared with the costs using conventional therapy. It was shown that the cost for a single fraction of carbon ion therapy is currently around 1000 € (based on a 20-fraction treatment). Interestingly, this is nearly identical to the costs per fraction for proton therapy (Goitein and Jermann 2003). It was therefore concluded that carbon ion therapy is cost effective, if local tumor control (used as indicator for therapy outcome) of more than 70% can be achieved. The main cost driver in the treatment of skull base tumors was the neurosurgical resection of recurrent tumors. The results cannot be generalized to other tumors and further investigations are needed.

11.3 Conclusion

In the last few decades, the fundamentals of particle therapy have been developed and its clinical application has shown to be feasible and effective. Particle therapy is experiencing a revolutionary transition in the last few years with remarkable advances in particle therapy treatment delivery, treatment planning, and quality assurance systems. For the next decade, further improvements are expected in particle therapy's efficiency, robustness, and accuracy. More hospital-based, state-of-the-art particle therapy facilities will be built and increasing numbers of patients will be treated. IMPT treatments will become available for routine use in an increasing number of facilities, and comprehensive dosimetry methods and QA procedures will be developed to improve the accuracy and precision of this treatment modality. With the availability of advanced IMPT techniques, it may be possible to quantify the clinical gains obtained from optimized particle therapy and to compare the clinical outcomes of high-energy photon and electron beams and proton and carbon ion beams using the best dose distributions for these modalities. For some indications, effectiveness strongly relies on successful compensation of inter- and intrafractional motion. The cost to patients will decrease, making proton therapy more financially competitive with high-energy photon and electron therapy. However, the cost effectiveness of carbon ion therapy has to be investigated further; it is probably unlikely that substantial data will be available before its therapeutic benefit has been quantified.

References

Aiba, M. et al. 2000. Development of a FFAG proton synchrotron. *Proceedings of EPAC (European Particle Accelerator Conference)*, Vienna, Austria, pp. 581–583.

Amaldi, U. and Kraft, G. 2007. European developments in radiotherapy with beams of large radiobiological effectiveness. *J. Radiat. Res. (Tokyo)* 48(Suppl A): A27–A41.

Andreo, P., Burns, D. T., Hohlfeld, K. et al. 2000. Absorbed dose determination in external beam radiotherapy: An international Code of Practice for dosimetry based on standards of absorbed dose to water. *IAEA Technical Report Series-398*. Vienna, Austria.

Bert, C. and Rietzel, E. 2007. 4D treatment planning for scanned ion beams. *Radiat. Oncol.* 2: 24.

Bert, C., Saito, N., Schmidt, A., Chaudhri, N., Schardt, D. and Rietzel, E. 2007. Target motion tracking with a scanned particle beam. *Med. Phys.* 34: 4768–4771.

Bulanov, S. V. and Khoroshkov, V. 2002. Feasibility of using laser ion accelerators in proton therapy *Plasma Phys. Rep.* 28: 453–456.

Caporaso, G. J., Mackie, T. R., Sampayan, S. et al. 2008. A compact linac for intensity modulated proton therapy based on a dielectric wall accelerator. *Phys. Med.* 24: 98–101.

Castro, J. R. 1995. Results of heavy ion radiotherapy. *Radiat. Environ. Biophys.* 34: 45–48.

Chen, G. T. Y., Mori, S., Chu, A. et al. 2006. Uncertainties in image guided radiotherapy of lung tumors. *Proceedings of NIRS-CNAO Joint Symposium on Carbon Ion Radiotherapy*, Milan Italy, pp. 192–193.

Chen, Y.-J. and Paul, A. C. 2007. *Proceedings of PAC07*. S. Hardage and C. Petit-Jean-Genaz (eds.). PAC07, Albuquerque, New Mexico, IEEE, pp. 1787–1789.

Craddock, M. K. 2004. The rebirth of FFAG. *CERN Courier* 44(6).

Fan, J., Luo, W., Fourkal, E. et al. 2007. Shielding design for a laser-accelerated proton therapy system. *Phys. Med. Biol.* 52: 3913–3930.

Fourkal, E., Fan, J. and Veltchev, I. 2009. Absolute dose reconstruction in proton therapy using PET imaging modality: feasibility study. *Phys. Med. Biol.* 54: N217–28.

Fourkal, E., Li, J. S., Xiong, W. et al. 2003. Intensity modulated radiation therapy using laser-accelerated protons: a Monte Carlo dosimetric study. *Phys. Med. Biol.* 48: 3977–4000.

Fourkal, E., Shahine, B., Ding, M. et al. 2002. Particle in cell simulation of laser-accelerated proton beams for radiation therapy. *Med. Phys.* 29: 2788–2798.

Furukawa T., Inaniwa, T., Sato, S. et al. 2007. Design study of a raster scanning system for moving target irradiation in heavy-ion radiotherapy. *Med. Phys.* 34: 1085–1097.

Goitein, M. and Cox, J. D. 2008. Should randomized clinical trials be required for proton radiotherapy? *J. Clin. Oncol.* 26: 175–176.

Goitein, M. and Jermann, M. 2003. The relative costs of proton and X-ray radiation therapy. *Clin. Oncol. (R. Coll. Radiol.)* 15: S37–S50.

Grozinger, S. O., Rietzel, E., Li, Q. et al. 2006. Simulations to design an online motion compensation system for scanned particle beams. *Phys. Med. Biol.* 51: 3517–3531.

Gunzert-Marx, K., Schardt, D. and Simon, R. S. 2004. The fast neutron component in treatment irradiations with 12C beam. *Radiother. Oncol.* 73(Suppl. 2): S92–S95.

Haberberger, D., Tochitsky, S. and Fiuza, F. 2011. Collisionless shocks in laser-produced plasma generate monoenergetic high-energy proton beams. *Nat. Phys.* DOI:10.1038/nphys2130.

Haberer, T., DeBus, J., Eickhoff, H. et al. 2004. The Heidelberg Ion Therapy Center. *Radiother. Oncol.* 73(Suppl. 2): S186–S190.

Henig, A., Steinke, S., Schnurer, M. et al 2009. Radiation-pressure acceleration of ion beams driven by circularly polarized laser pulses. *Phys. Rev. Lett.*103: 245003.

Ilic, R. D., Spasic-Jokic, V., Belicev, P. et al. 2005. The Monte Carlo SRNA-VOX code for 3D proton dose distribution in voxelized geometry using CT data. *Phys. Med. Biol.* 50: 1011–1017.

Jäkel, O., Hartmann, G. H., Karger, C. P. et al. 2000. Quality assurance for a treatment planning system in scanned ion beam therapy. *Med. Phys.* 27: 1588–600.

Jäkel, O., Karger, C. P. and DeBus, J. 2008. The future of heavy ion radiotherapy. *Med. Phys.* 35: 5653–5663.

Jäkel, O., Land, B., Combs, S. E. et al. 2007. On the cost-effectiveness of carbon ion radiation therapy for skull base chordoma. *Radiother. Oncol.* 83: 133–138.

Jakubek, J., Holy, T., Lehmann, E. et al. 2005. Spatial resolution of Medipix-2 device as neutron pixel detector. *Nucl. Instr. Meth. A* 546: 164–169.

Johnstone, C. et al. 1999. Fixed field circular accelerator designs. *Proceedings of the Particle Accelerator Conference*, New York, pp. 3068–3070.

Karger, C. P., Jakel, O. and Hartmann, G. H. 1999. A system for three-dimensional dosimetric verification of treatment plans in intensity-modulated radiotherapy with heavy ions. *Med. Phys.* 26: 2125–2132.

Keil, E., Sessler, A. M. and Trbojevic, D. 2007. Hadron cancer therapy complex using nonscaling fixed field alternating gradient accelerator and gantry design. *Phys. Rev. Spec.Top-Ac.*10: 054701-1–054701-14.

Knopf, A., Parodi, K., Paganetti et al. 2008. Quantitative assessment of the physical potential of proton beam range verification with PET/CT. *Phys. Med. Biol.* 53: 4137–4151.

LANL. September 2011. Los Alamos sets new world records in laser-ion acceleration *Physics Flash*, Los Alamos National Laboratory.

Li, J. S., Shahine, B., Fourkal, E. and Ma, C. M. 2005. A particle track-repeating algorithm for proton beam dose calculation. *Phys. Med. Biol.* 50: 1001–1010.

Lodge, M., Pijls-Johannesma, M., Stirk, L. et al. 2007. A systematic literature review of the clinical and cost-effectiveness of hadron therapy in cancer. *Radiother. Oncol.* 83: 110–122.

Lu, H. M., Brett, R., Sharp, G. et al. 2007. A respiratory-gated treatment system for proton therapy. *Med. Phys.* 34: 3273–3278.

Lundkvist, J., Ekman, M., Ericsson, S. R. et al. 2005. Proton therapy of cancer: Potential clinical advantages and cost-effectiveness. *Acta Oncol.* 44: 850–861.

Luo, W., Fourkal, E., Li, J. et al. 2005. Particle selection and beam collimation system for laser-accelerated proton beam therapy. *Med. Phys.* 32: 794–806.

Luo, W., Li, J., Fourkal, E. et al. 2008. Dosimetric advantages of IMPT over IMRT for laser-accelerated proton beams. *Phys. Med. Biol.* 53: 7151–7166.

Ma, C.-M. 2000. Compact Laser Proton Accelerator for Medicine. *SPO #22962*. Stanford University, Stanford, CA.

Ma, C.-M. et al. 2006. Development of a laser-driven proton accelerator for cancer therapy. *Laser Phys.* 16: 639–646.

Ma, C.-M., Tajima, T., Shahine, B. et al. 2001. Laser accelerated proton beams for radiation therapy. *Med. Phys.* 28: 1236.

Ma, C. M. and Maughan, R. L. 2006. Within the next decade conventional cyclotrons for proton radiotherapy will become obsolete and replaced by far less expensive machines using compact laser systems for the acceleration of the protons. *Med. Phys.* 33: 571–573.

Mackie, R. et al. 2007. A proposal for a novel compact intensity modulated proton therapy system using a dielectric wall accelerator. *Med. Phys.* 34: 2628.

Maughan, R. L., Van Den Heuvel, F. et al. 2008. Point/Counterpoint. Within the next 10-15 years protons will likely replace photons as the most common type of radiation for curative radiotherapy. *Med. Phys.* 35: 4285–4288.

Minohara, S., Endo, M., Kanai, T. et al. 2003. Estimating uncertainties of the geometrical range of particle radiotherapy during respiration. *Int. J. Radiat. Oncol. Biol. Phys.* 56: 121–125.

Minohara, S., Kanai, T., Endo, M. et al. 2000. Respiratory gated irradiation system for heavy-ion radiotherapy. *Int. J. Radiat. Oncol. Biol. Phys.* 47: 1097–1103.

Moyers, M. F., Pouliot, J. and Orton, C. G. 2007. Point/Counterpoint. Proton therapy is the best radiation treatment modality for prostate cancer. *Med. Phys.* 34: 375–378.

Nakano, T., Suzuki, Y., Ohno, T. et al. 2006. Carbon beam therapy overcomes the radiation resistance of uterine cervical cancer originating from hypoxia. *Clin. Cancer Res.* 12: 2185–2190.

Newhauser, W., Fontenot, J., Zheng, Y. et al. 2007. Monte Carlo simulations for configuring and testing an analytical proton dose-calculation algorithm. *Phys. Med. Biol.* 52: 4569–4584.

Newhauser, W., Koch, N., Hummel, S. et al. 2005. Monte Carlo simulations of a nozzle for the treatment of ocular tumours with high-energy proton beams. *Phys. Med. Biol.* 50: 5229–5249.

Nichiporov, D., Kostjuchenko, V., Puhl, J. M. et al. 1995. Investigation of applicability of alanine and radiochromic detectors to dosimetry of proton clinical beams. *Appl. Radiat. Iso.* 46: 1355–1362.

Nichiporov, D., Solberg, K., Hsi, W. et al. 2007. Multichannel detectors for profile measurements in clinical proton fields. *Med. Phys.* 34: 2683–2690.

Nishio, T., Ogino, T., Nomura, K. et al. 2006. Dose-volume delivery guided proton therapy using beam on-line PET system. *Med. Phys.* 33: 4190–4197.

Paganetti, H., Jiang, H., Parodi, K. et al. 2008. Clinical implementation of full Monte Carlo dose calculation in proton beam therapy. *Phys. Med. Biol.* 53: 4825–4853.

Paganetti, H., Jiang, H. and Trofimov, A. 2005. 4D Monte Carlo simulation of proton beam scanning: modelling of variations in time and space to study the interplay between scanning pattern and time-dependent patient geometry. *Phys. Med. Biol.* 50: 983–990.

Chapter 11

Parodi, K., Bortfeld, T. and Haberer, T. 2008. Comparison between in-beam and offline positron emission tomography imaging of proton and carbon ion therapeutic irradiation at synchrotron- and cyclotron-based facilities. *Int. J. Radiat. Oncol. Biol. Phys.* 71: 945–956.

Parodi, K., Paganetti, H., Shih, H. A. et al. 2007. Patient study of in vivo verification of beam delivery and range, using positron emission tomography and computed tomography imaging after proton therapy. *Int. J. Radiat. Oncol. Biol. Phys.* 68: 920–934.

Pedroni, E., Bacher, R., Blattmann, H. et al. 1995. The 200-MeV proton therapy project at the Paul Scherrer Institute: Conceptual design and practical realization. *Med. Phys.* 22: 37–53.

Pedroni, E., Scheib, S., Bohringer, T. et al. 2005. Experimental characterization and physical modelling of the dose distribution of scanned proton pencil beams. *Phys. Med. Biol.* 50: 541–561.

Pflugfelder, D., Wilkens, J. J. and Oelfke, U. 2008. Worst case optimization: A method to account for uncertainties in the optimization of intensity modulated proton therapy. *Phys. Med. Biol.* 53: 1689–1700.

Rebisz, M., Voss, B., Heinz, A. et al. 2007. CVD diamond dosimeters for heavy ion beams. *Diamond Relat. Mater.* 16: 1070–1073.

Sauli, F. 1997. GEM: A new concept for electron amplification in gas detectors. *Nucl. Instr. Meth. A* 386: 531–534.

Schulz-Ertner, D., Nikoghosyan, A., Didinger, B. et al. 2005. Therapy strategies for locally advanced adenoid cystic carcinomas using modern radiation therapy techniques. *Cancer* 104: 338–344.

Schulz-Ertner, D. and Tsujii, H. 2007. Particle radiation therapy using proton and heavier ion beams. *J. Clin. Oncol.* 25: 953–964.

Schulz, R. J., Smith, A. R. and Orton, C. G. 2007. Point/counterpoint. Proton therapy is too expensive for the minimal potential improvements in outcome claimed. *Med. Phys.* 34: 1135–1138.

Schwoerer, H., Pfotenhauer, S., Jackel, O. et al. 2006. Laser-plasma acceleration of quasi-monoenergetic protons from microstructured targets. *Nature* 439: 445–448.

Smith, A. 2006. Proton. *Phys. Med. Biol.* 51: R491–R504.

Snavely, R. A., Key, M. H., Hatchett, S. P. et al. 2000. Intense high-energy proton beams from Petawatt-laser irradiation of solids. *Phys. Rev. Lett.* 85: 2945–2948.

Steinke, S., Henig, A. and Schnurer, M. 2010. Efficient ion acceleration by collective laser-driven electron dynamics with ultra-thin foil targets. *Laser Part. Beams* 28: 215–221.

Suit, H., Goldberg, S., Niemierko, A. et al. 2003. Proton beams to replace photon beams in radical dose treatments. *Acta Oncol.* 42: 800–808.

Suit, H., Kooy, H., Trofimov, A. et al. 2008. Should positive phase III clinical trial data be required before proton beam therapy is more widely adopted? No. *Radiother. Oncol.* 86: 148–153.

Tajima, T. 1999. Compact laser proton accelerator beyond 100 MeV for medicine. *Lawrence Livermore National Laboratory.*

Tajima, T. and Dawson, J. M. 1979. Laser electron accelerator. *Phys. Rev. Lett.* 43: 267–270.

Vatnitsky, S. M. 1997. Radiochromic film dosimetry for clinical proton beams. *Appl. Radiat. Isot.* 48: 643–6.

Veltchev, I., Fan, J., Li, J. et al. 2010. Fast algorithm for carbon therapy dose calculation. In *Proceedings of the 16th International Conference on the Use of Computer in Radiation Therapy (ICCR)*, Amsterdam Netherlands, J.-J. Sonke (ed.). Het Nederlands Kanker Institut-Antoni van Leeuwenhoek Ziekenhuis, p. 12323.

Veltchev, I., Fourkal, E., Ma, C.-M. 2007. Laser-induced Coulomb mirror effect: Applications for proton acceleration. *Phys. Plasmas* 14: 033106.

Warren, W. B. 2007. Evaluation of ®Bang polymer gel dosimeters in proton beams. MS thesis, The University of Texas Health Science Center at Houston.

Zhang, C., Lechner, P., Lutz, G. et al. 2008. Development of DEPFET Macropixel detectors. *Nucl. Instr. Meth. A* 546: 164–169.

Zietman, A. L., DeSilvio, M. L., Slater, J. D. et al. 2005. Comparison of conventional-dose vs high-dose conformal radiation therapy in clinically localized adenocarcinoma of the prostate: A randomized controlled trial. *J. Am. Med. Assoc.* 294: 1233–1239.

Index

Printed and bound by CPI Group (UK) Ltd, Croydon, CR0 4YY

24/10/2024

01778288-0016